Handbook of Human Growth and Developmental Biology

Volume III
Developmental Biology of Organs and Systems

Part B:
Cardiovascular and Respiratory Development

Editors
**Esmail Meisami, Ph.D.
Paola S. Timiras, M.D., Ph.D.**

CRC Press
Boca Raton   Ann Arbor   Boston

Library of Congress Cataloging-in-Publication Data

Developmental biology of organs and systems / editors, Esmail Meisami,
    Paola S. Timiras.
        p.   cm. -- (Handbook of human growth and developmental biology;
    v. 3)
    Includes bibliographical references.
    Contents: Pt. A. Muscle, blood, and immunity -- pt. B.
Cardiovascular and respiratory development.
    ISBN 0-8493-3186-2 (pt. A). -- ISBN 0-8493-3187-0 (pt. B)
    1. Developmental biology. 2. Muscles--Differentiation. 3. Immune
system--Differentiation. 4. Cardiovascular system--Differentiation.
5. Respiratory organs--Differentiation. 6. Hematopoietic system-
-Differentiation. I. Meisami, Esmail. II. Timiras, Paola S.
III. Series.
    [DNLM: 1. Cardiovascular System--growth & development.
2. Hematopoiesis. 3. Human Development. 4. Immune System--growth &
development. 5. Respiratory System--growth & development.
6. Muscles--growth & development. WS 103 H236 1988 v. 3]
RJ131.H273   1988 vol. 3
[QP84]
612.6'5 s--dc20
[612.6]
DNLM/DLC
for Library of Congress   90-1522
                                                                                                                            CIP

This book represents information obtained from authentic and highly regarded sources. Reprinted material is quoted with permission, and sources are indicated. A wide variety of references are listed. Every reasonable effort has been made to give reliable data and information, but the author and the publisher cannot assume responsibility for the validity of all materials or for the consequences of their use.

All rights reserved. This book, or any parts thereof, may not be reproduced in any form without written consent from the publisher.

Direct all inquiries to CRC Press, Inc., 2000 Corporate Blvd., N.W., Boca Raton, Florida, 33431.

© 1990 by CRC Press, Inc.

International Standard Book Number 0-8493-3186-2 (Volume III, Part A)
International Standard Book Number 0-8493-3187-0 (Volume III, Part B)

Library of Congress Card Number 90-1522
Printed in the United States

This handbook is dedicated to human progeny,
with best wishes for optimal development!

PREFACE

The study of development, in particular human development, is unique and important. Not only is the unraveling of the complexities of development an exciting end in itself, but since development leads to the formation of the individual, it is a crucial means to better understand the structure and function of the adult. Indeed it would not be an exaggeration to say that the future of human society depends largely on the quality, welfare, and optimal biological development of its embryos, fetuses, neonates, infants, children, and adolescents. A proper knowledge of the biological foundation of human development and the genetic and environmental factors which regulate and/or influence its normal and abnormal course is necessary in order to optimize normal growth and development and to eradicate or minimize the harmful influences.

In the past, the study of human development has often been viewed either embryologically, stressing prenatal development, or behaviorally, focusing on postnatal psychological and social development. It is now time to consider human biological development as a continuum spanning the pre- and postnatal periods. Although the foundations of human biological development are rooted in, and share much with, animal development, there are important differences which warrant a separate and unique discipline of human growth and developmental biology. Therefore, on heuristic, altruistic, societal, scientific, and medical grounds, an urgent need exists to bring together, in one place, a comprehensive interdisciplinary treatise on human growth and development from conception to maturity, including particularly the cellular and functional aspects of developing organs and systems. It is to fulfill this need that the present handbook has been undertaken.

This handbook is intended to be useful for a wide readership, comprising professionals and researchers from diverse biological and biomedical fields, including advanced students, educators, and child development specialists. In order to facilitate its use by diverse readership, the handbook is organized along three major themes. The first theme is *neural, sensory, motor, and integrative development,* comprising the chapters in Volume I. Part A covers the basic aspects of neural development, while Part B reviews the sensory, motor, and integrative aspects of development; the genetic, hormonal, and environmental factors that regulate or influence brain development, normally and abnormally, are treated in Part C. The second theme, *endocrines, sexual development, growth, nutrition, and metabolism,* is presented in the chapters of Volume II. Part A covers the developmental aspects of endocrine glands, hormonal regulation, and reproductive and sexual development. The subjects of factors influencing growth, developmental digestion, nutrition, and metabolism are treated in Part B. The third and last theme, comprising Volume III, involves the *development biology of organs and systems.* The developmental aspects of muscle, blood, and immunity are presented in Part A and the development of cardiovascular and respiratory systems in Part B.

The authors of the individual chapters were asked to write overviews of their subject matter and not exhaustive lengthy reviews, as the latter may be found in more specialized journals, books, or series. Plus, detailed treatment may not be suitable for the readership of the handbook which is presumed to be diverse. The references provided at the end of the chapters are intended to lead the users to the appropriate original articles or review journals where the subject is discussed in more depth.

Although the authors were requested to focus on human development, this was not always possible, in which case the relevant materials from experimental animal studies are described. The inclusion of the latter materials was occasionally necessary as they elucidated mechanisms of development not known in the human or not accessible to investigation.

Although we intended the handbook to be truly comprehensive, we have not quite achieved this goal. For example, some topics, such as development of tactile senses, pain and nociception, normative growth studies, and development of the kidney, liver, and skin, are not included mainly because the original authors who undertook these parts were unable to complete their sections. As we wished to avoid further delays and were pressed for a less massive volume, we were forced to cope with these deficiencies and hope that they will be corrected in future editions.

The editors are indebted to the internationally known authors who kindly cooperated with the editors during the revisions and showed exceptional patience and forebearance over the usual delays associated with the publication of a work of this size. Special thanks are also due to the distinguished members of the Advisory Board who not only nominated prospective authors, but helped edit some of the manuscripts. For Volume III, Part A, we are particularly indebted to Profs. Mehdi Tavassoli and Rodrick Westerman, without whose invaluable contribution, the sections on blood and muscle development, respectively, would not have been possible. We are also grateful to Dr. Doherty B. Hudson for editorial assistance and to Laura Elliot, Darla Wigginton, Patricia LaForce, Sara Goering, Angela Keleher, and Angela Roberts for valuable secretarial assistance. We would like to also thank the editorial staff of CRC Press, in particular Ms. Amy Skallerup, Sandy Pearlman, Barbara Caras, Jocelyn Makepeace, and Carolyn Lea, for helping us complete this work.

E. Meisami and P. S. Timiras, Editors

THE EDITORS

Esmail (Essie) Meisami, Ph.D., was born and raised in Tehran, Iran. He attended the University of California at Berkeley where he received both the B.A. and Ph.D. degrees in Physiology. In 1971, he joined the Faculty of Science of the University of Tehran where he taught and carried out research in physiology and neuroscience and helped establish and direct the International Research Institute of Biochemistry and Biophysics and its Neurobiology Laboratory. In 1980, he returned to Berkeley where he worked until 1986 as a visiting professor and research physiologist. He is now an Associate Professor at the Department of Physiology and Biophysics and the Neuroscience Graduate Program of the University of Illinois at Urbana-Champaign. Dr. Meisami's research has been mainly concerned with the problems of brain development and the intrinsic and extrinsic factors, such as hormones and sensory stimulations, which influence it. He has pioneered the use of the olfactory system as a model for developmental neurobiology and has demonstrated the marked retarding effects of sensory deprivation on the developing olfactory bulb. In addition to authoring numerous research papers and monographs, Dr. Meisami has previously edited a book on Neural Growth and Differentiation (with Dr. M. A. B. Brazier, Raven Press, 1979). He is a member of several scientific societies and has served on the Councils of the International Brain Research Organization and International Society for Developmental Neuroscience.

Paola S. Timiras, M.D., Ph.D., is Professor of Physiology, Department of Molecular and Cell Biology, University of California, Berkeley (UCB). A native of Rome, Italy, Dr. Timiras is a graduate of the University of Grenoble, France (B.A.), University of Rome (B.A., M.D. Summa cum Laude), and the University of Montreal, Quebec, Canada (Ph.D.). Before joining the University of California (1956) she was a faculty member of the University of Montreal (1950 to 1953) and the University of Utah (1953 to 1955). She was chairman of the Department of Physiology-Anatomy at UCB from 1978 to 1984. Her interest in development and aging has led to her participation in several related societies. She has directed an innovative medical program (Health and Medical Sciences) (1973 to 1975) and a training program in developmental physiology and aging (1965 to 1976) at UCB. She has been one of the founders and president of the International Society of Developmental Neuroscience (1978 to 1981) and vice-president and president of the International Society of Psychoneuroendocrinology (1974 to 1982). Her research on the neuroendocrinology of development and aging has resulted in the publication of more than 350 papers and several books. Dr. Timiras is also a consultant for many government agencies concerned with biological research and is on the editorial board of several specialty journals. Her contributions to teaching and research have been recognized by several awards from the U.S. and abroad.

ADVISORY BOARD

Robert Balázs, M.D., Ph.D.
Netherlands Institute for Brain Research
Amsterdam, The Netherlands

Colin T. Jones, Ph.D.
Laboratory of Cellular and Developmental
 Physiology
Institute for Molecular Medicine
University of Oxford
Oxford, England

Dominick P. Purpura, M.D.
Professor and Dean
Albert Einstein College of Medicine
Bronx, New York

Mehdi Tavassoli, M.D.
Professor, Department of Medicine
University of Mississippi
Jackson, Mississippi

J. J. Van Der Werff Ten Bosch, M.D.
Professor
Department of Endocrinology, Growth
 and Reproduction
Erasmus University Medical School
Rotterdam, The Netherlands

R. A. Westerman, Ph.D.
Associate Professor
Department of Physiology
Monash University
Clayton, Victoria, Australia

CONTRIBUTORS

Alan D. Bocking, M.D.
Departments of Obstetrics, Gynecology, and Physiology
University of Western Ontario
London, Ontario, Canada

George A. Brooks, Ph.D.
Department of Physical Education
University of California
Berkeley, California

Thomas D. Fahey, Ed.D.
Department of Physical Education
California State University
Chico, California

M. Angeles Fernandez-Teran, M.D.
Department of Anatomy and Cell Biology
University of Cantabria
Santander, Spain

Claude Gaultier, M.D., Ph.D.
Laboratory of Physiology
Hospital Antoine Beclere
C. N. R. S. U. A.
Clamert, France

Richard Harding, Ph.D.
Department of Physiology
Monash University
Clayton, Victoria, Australia

Joan E. Hodgman, M.D.
Department of Pediatrics/Newborn Division, LAC/USC Medical Center
University of Southern California School of Medicine
Los Angeles, California

Toke Hoppenbrouwers, Ph.D.
Department of Pediatrics/Newborn Division, LAC/USC Medical Center
University of Southern California School of Medicine
Los Angeles, California

Jose M. Icardo, M.D.
Department of Anatomy and Cell Biology
University of Cantabria
Santander, Spain

Esmail Meisami, Ph.D.
Departments of Physiology and Biophysics
University of Illinois
Urbana, Illinois

Donna J. Messersmith, Ph.D.
Department of Embryology Research
Children's National Medical Center
Washington, D.C.

Terence E. Nicholas, Ph.D.
Department of Physiology
Flinders University of South Australia School of Medicine
Bedford Park, South Australia, Australia

Susan Ann O'Brien, M.D.
Department of Embryology Research
Children's National Medical Center
Washington, D.C.

Jose L. Ojeda, M.D.
Department of Anatomy and Cell Biology
University of Cantabria
Santander, Spain

Karel Rakusan, M.D., Ph.D.
Department of Physiology
University of Ottawa
Ottawa, Ontario, Canada

Roger N. Ruckman, M.D.
Department of Pediatrics
George Washington University School of
  Medicine
Washington, D.C.

Nicholas Sperelakis, Ph.D.
Department of Physiology and Biophysics
University of Cincinnati College of
  Medicine
Cincinnati, Ohio

Paola S. Timiras, M.D., Ph.D.
University of California at Berkeley
Berkeley, California
  and Universite de Bordeaux
Talance, France

David W. Walker, Ph.D.
Department of Physiology
Monash University
Clayton, Victoria, Australia

CRC
Handbook of Human Growth and Developmental Biology

Volume I: Neural, Sensory, Motor, and Integrative Development
Part A: Development Neurobiology
Part B: Sensory, Motor, and Integrative Development
Part C: Factors Influencing Brain Development

Volume II: Endocrines, Sexual Development, Growth, Nutrition, and Metabolism
Part A: Endocrines and Sexual Development
Part B: Growth, Nutrition, and Metabolism

Volume III: Developmental Biology of Organs and Systems
Part A: Muscle, Blood, and Immunity
Part B: Cardiovascular and Respiratory Development

TABLE OF CONTENTS

CARDIOVASCULAR DEVELOPMENT

Early Cardiac Structure and Developmental Biology 3
Jose M. Icardo, M. Angeles Fernandez-Teran, and Jose L. Ojeda

Late Heart Embryology. The Making of an Organ 25
Jose M. Icardo, M. Angeles Fernandez-Teran, and Jose L. Ojeda

Developmental Changes in Electrical Activity and Ion Channels in the Heart 51
Nicholas Sperelakis

Functional Development of the Heart: Hemodynamics 69
Roger N. Ruckman, Susan A. O'Brien, and Donna J. Messersmith

Development of Cardiovascular Function.. 85
George A. Brooks

Development of Cardiac Vasculature 101
Karel Rakusan

Malformations of the Heart and Blood Vessels....................................... 107
Roger N. Ruckman, Susan A. O'Brien, and Donna J. Messersmith

RESPIRATORY DEVELOPMENT

Fetal Lung Growth 131
R. Harding and A. D. Bocking

Development of Breathing Movements and Their Regulation Before Birth 149
David W. Walker

The Development of the Surfactant System in the Lung............................... 165
Terence E. Nicholas

Development in Lung Mechanics... 173
Claude Gaultier

Sudden Infant Death Syndrome (SIDS)... 181
Toke Hoppenbrouwers and Joan Hodgman

The Development of Respiratory Capacity During Exercise in Children and
Adolescents 209
Thomas D. Fahey

Index 221

CARDIOVASCULAR DEVELOPMENT

EARLY CARDIAC STRUCTURE AND DEVELOPMENTAL BIOLOGY

Jose M. Icardo, M. Angeles Fernandez-Teran, and Jose L. Ojeda

INTRODUCTION

Organ morphogenesis is the result of a sequence of causally related processes of ever increasing complexity, acting from the genome level up. Although the genome contains all the information necessary for the construction of organs (and organisms), there are no genes known to be specific for organ shape. Rather, it is the adequate expression of the genetic information, from the molecular to the tissue level, which ultimately results in shape.

During development, cells exhibit only a limited repertoire of behaviors. However, the ultimate shape of any organ is unique. It is the interaction between cells of similar and dissimilar developmental history, and between cells and their extracellular microenvironment, that modulates the expression of the genome. We already know a great deal about structure and we are increasing our knowledge about the different levels of organization, ranging from the molecule to the organism. However, the precise links between gene expression and shape are still to be unraveled. This is true for any organ and so it is for the heart.

When confronted with the task of studying the development of the human heart we face a number of obstacles. Human embryos are not easily available for a thorough study of development. The Carnegie and the Kyoto collections are probably the best known exceptions to this rule. The reader will be referred to them when necessary. Since experimental embryology in humans is not possible, we will inevitably have to borrow from other vertebrates, especially the chicken. Even though the ultimate shape of the heart and the developmental timing differ in man and other vertebrates, many developmental mechanisms appear to be similar, if not identical. We will try to present here the structural development of the heart, interrelating the information embryologists have obtained from the different levels of biological organization. In doing so we hope to reach a better understanding of the complex interrelationships that result in the expression of heart shape and function.

Heart morphogenesis can be subdivided into two main phases. The heart develops first as a vascular channel in close association with the rest of the vascular system. Acquisition of shape is related to the broad deformative events that result in closure of the endoderm. The heart itself is bent, deformed, and thrown into a loop. Although epithelial in origin, the heart soon develops a muscular wall, begins to beat, and acquires most of its basic functions. However, heart shape is still very primitive. In a second phase, cardiogenesis appears to depend more on mechanisms of tissue growth and tissue remodeling. It is during this period that the heart acquires a system of septa and a valve apparatus, and is transformed into a complex four-chambered organ. Thus, it seems appropriate to divide the study of heart development into two main parts. The present chapter will deal basically with developmental mechanisms, phenotype expression, and deformative events. The next chapter will deal more with complex shape transformations and the possible involvement of epigenetic factors.

EARLY CARDIOGENESIS

In the presomite human embryo some cells sort out from the splanchnic mesoderm to form angiogenic cell clusters. These clusters, located between the endodermal and mesodermal layers, extend cranially, gain a lumen, and unite to form a horseshoe-shaped vascular plexus. The lateral parts of the vascular plexus coalesce to form thin-walled endothelial tubes that will become the essential component of the endocardium. In the meantime, the

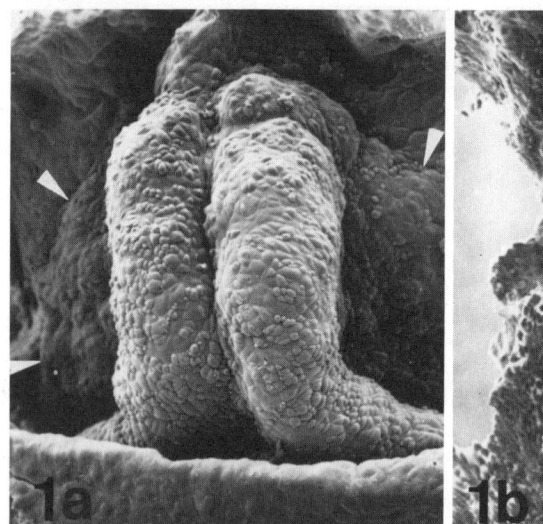

FIGURE 1. SEM micrographs depicting the heart primordia. The coelomic cavity has been opened and the heart has been exposed. (a) The paired heart anlage is on the embryonic midline. The process of fusion has already started. Arrowheads indicate the precardiac splanchnic mesoderm. (Magnification × 285.) (b) A single heart tube is formed after fusion of the paired primordium. The fusion furrow (arrow) is still visible. (Magnification × 265.)

region of the splanchnic mesoderm around the developing endothelial tubes has thickened, forming the prospective myocardium. Also, a thin layer of extracellular material has accumulated below the thickened mesoderm. This region of the splanchnic mesoderm is now called the cardiogenic plate. Thus, the earliest identifiable cardiac primordia consists essentially of a pair of troughs of splanchnic mesodermal tissue located on each side of the cardiogenic plate. The rapid cranial enlargement of the brain vesicles, and the curvature of the embryonic shield, bring the precardiac areas below the developing foregut. Closure of the endodermal foregut approaches the two cardiac primordia which meet each other and fuse (Figure 1). Thus, a single heart tube is formed in the embryonic midline. This tube is the beginning of the tubular heart.[1]

The displacement of the precardiac mesoderm toward the embryonic midline is not totally a passive process. The precardiac mesoderm appears to migrate actively upon the endoderm as a cohesive epithelial cell sheet.[2] *In vitro,* explanted precardiac mesoderms are able to migrate upon the endoderm and other epithelial cell sheets to form beating vesicular structures that resemble the early tubular heart.[3] The mechanisms involved in the migratory behavior of the precardiac cells are still under discussion. Fibronectin has been demonstrated along the basal surface of the endoderm.[4] It has been suggested[5] that the precardiac mesoderm could take directional migrating clues from the endodermal surface. Furthermore, antifibronectin antibodies and peptidic fragments of the fibronectin molecule have been shown to arrest early cardiac development.[6] However, it is still unclear whether or not these experiments demonstrate the presence of directional migrating clues: they may simply show the need for an appropriate substratum during precardiac cell migration, or they may be interfering with the expression of the myocardial phentoype (see below).

The mechanisms that regulate fusion of the cardiac primordia are unknown as yet. Different mechanisms appear to be involved in fusion of the myocardial and endocardial components. This is not to say that the two cellular components behave independently of each other. For example, cell death occurs during fusion of the primitive endocardial tubes. Cell death in the medial wall of the apposed endocardial tubes results in the formation of a

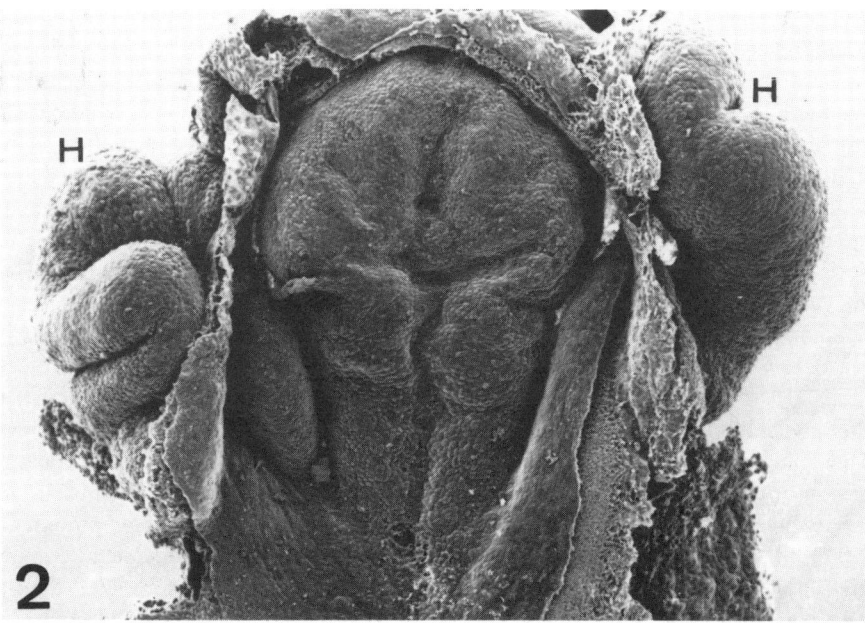

FIGURE 2. SEM micrograph of a double-hearted embryo. Fusion of the paired anlage has been prevented by cutting through the midtissues of the anterior intestinal porta. One independent heart (H) is formed on each side of the embryonic midline. (Magnification × 126.)

single tubular endocardium.[7] This process of cell death occurs in the absence of fusion,[8] suggesting that endocardial cell death is programmed. However, the presence of cell death has not been detected during fusion of the myocardial primordia. Rather, the two myocardial primordia simply appear to merge with each other, becoming a continuous cell mantle. This behavior is not very different from the behavior exhibited by two aggregates of myocardial tissue that are being pushed one against the other.[9] Myocardial cells in the zone of fusion display long processes and filopodia.[10] These cellular processes are believed to be involved in the initiation of adhesion.[11] However, although early developing myocytes have been shown to synthesize cell adhesion molecules,[12] the existence of phenomena of cell-to-cell recognition and adhesion during heart fusion remains, at best, speculative.

We should emphasize the bilateral origin of the heart. If fusion of the heart anlage is prevented,[13,14] two independent hearts, one on each side of the embryonic midline, develop (Figure 2). These hearts are mirror images of each other, are able to attain independent vascular connections, and beat at a different pace. Furthermore, they differentiate synchronously,[15] suggesting that normal heart differentiation does not depend on a midline position.

The participation of angiogenic factors in the initial development of the heart is an unexplored possibility as yet. Vessel formation, both in normal embryogenesis and in tumor development, appears to be regulated by a number of substances generically called angiogenic factors.[16,17] These factors present either mitogenic activity or chemotactic activity, or both. Further, they appear to be involved in the reorganization of vascular plexuses into definitive vessels.[18] Since the heart emerges as a part of the developing vascular system, it is tempting to speculate on the involvement of angiogenic factors in the initial steps of cardiogenesis.

DIFFERENTIATIVE EVENTS IN THE PRECARDIAC MESODERM

Differentiation involves the synthesis of specific gene products. In the heart, cytodifferentiation can be expressed by the presence of structural muscle proteins and the onset of electrical activity.

The understanding of the biochemical expression of the cardiac myocytes has recently been complicated by the fact that the same cell type can express different types of the same protein or gene product. Precardiac mesodermal cells and heart cells have been shown to express the α-, β-, and γ-isotypes of actin.[19] The synthesis of α-actin (the sarcomeric form of actin) increases drastically before fusion of the cardiac primordia, thereafter becoming the predominant type of actin synthesized.[19,20] Myosin also appears to be synthesized by the premyocardial cells.[21] However, we do not know, as is the case for actin, whether or not the premyocardial cells undergo any myosin biosynthetic transition. The increase in the synthesis of the α-form of actin correlates well with the first appearance of myofilaments[22,23] in the cytoplasm of the precardiac cells.

Spontaneous electrical activity and propagation of the excitation can be observed in the heart prior to fusion of the cardiac primordia[24] by using dyes sensitive to changes in electrical potentials. Heart beating appears to start at about the same time,[25] even though beating remains invisible to the unaided[26] observer.

Thus, there are a series of differentiation parameters (the expression of α-actin, the expression of myosin, the appearance of myofilaments, and the onset of electrical activity) that are expressed at about the same time. Troponin, another myofibrillar protein,[27] and glycogen, the main source of energy for contracting muscle, also accumulate at this time.[28] It is not known whether a single gene or a set of genes controls the expression of all these different parameters. Nor is it known whether these genes are activated at the same time or exactly when they are activated.[29] The fact that cells with heart-forming capacity can be obtained from embryos through the blastula and gastrula stages[14] suggests that the activation of the genetic information has been made much earlier in development. This is supported by experiments carried out with the thymidine analog 5-bromodeoxyuridine (BrdU). When BrdU is administered to cultures of early precardiac cells, the cells fail to differentiate into myocardial tissue.[30] If BrdU is administered at later stages, cytodifferentiation is not impaired. Therefore, the messenger RNAs responsible for the synthesis of specific macromolecules in cardiac muscle may have been formed very early in development. The BrdU studies also suggest that the entry of the precardiac cells into the differentiative cycle is not synchronous. The number of precursor heart cells insensitive to BrdU increases with age.[30]

Although the timely expression of all the differentiation parameters cited above does not appear to correspond with actual gene activation, it provides us with a starting point when trying to elucidate the mechanisms involved in cardiac differentiation. Presently, the mechanisms that trigger overt cardiac muscle differentiation are unknown. Among the possible mechanisms involved in phenotype expression, inductive interactions between tissues have been shown to play an important role.[31] In the heart, it was postulated[32,33] that contact with the endoderm induces the precardiac mesoderm to differentiate. However, when heart-forming mesoderm is cultured alone, some degree of self-differentiation occurs.[3,34] This suggests that the precardiac mesoderm has a built-in capability to differentiate. Thus, the evidence for endodermal induction is not entirely satisfactory.

Extracellular materials have also been implicated in tissue differentiation.[35] A thin layer of extracellular material is located between the mesodermal and endodermal layers at the time of cardiac primordia migration. This material is rich in glycosaminoglycans, glycoproteins, and, probably, collagen.[36] It has been suggested[37] that endodermal glycoproteins could mediate differentiative events in the precardiac mesoderm. Fibronectin has been identified as a major component of the basal lamina of the endoderm.[4] Since this glycoprotein appears to be widely associated with differentiative events,[38] it is tempting to propose a role for fibronectin in cardiac differentiation. However, the most direct evidence for the involvement of extracellular materials in cardiac differentiation comes form studies using inhibitors of collagen synthesis. When collagen synthesis is inhibited, the precardiac mesoderm fails to differentiate.[23]

Even if the precardiac extracellular matrix is directly involved in cardiac differentiation, the above studies do not imply that collagen, fibronectin, or any other single macromolecule is solely responsible for the differentiative event. The physicochemical properties of the extracellular microenvironment depend upon the high-order integration of individual molecular species. Small quantitative and/or qualitative compositional changes may drastically alter the unique properties of the matrices. The response or the precardiac mesoderm to a collagen-containing extracellular matrix may help to understand some of the discrepancies observed in culture experiments. The possibility of unspecific induction by components of the culture medium[37] should not be discarded.

THE EARLY HEART TUBE. THE FORMATION OF THE CARDIAC LOOP

The heart tube that forms on the embryonic midline presents a relatively straight tubular form (Figure 1). Contained within the pericardial cavity, the heart is connected cranially with the ventral aorta and caudally with the vitellin veins. The ventral surface of the heart is free, being bathed by the pericardial fluid. Dorsally, the heart is attached to the embryonic trunk by the dorsal mesocardium. The dorsal mesocardium represents the two folds of the precardiac mesoderm that are still unfused in the area dorsal to the endocardium.

This straight heart is not a complete heart, i.e., not all the components that can be distinguished in the mature heart are present in the primitive tube. There is experimental evidence supporting the view that the primitive heart tube represents only the future trabeculated part of the right ventricle.[2,39] Continued fusion of the heart anlage in a caudal direction brings about the merging of the future trabeculated part of the left ventricle, the atrial region, and the sinus venosus, which are thus progressively incorporated into the developing heart. Although particle-marking experiments should be interpreted cautiously, it also appears that the bulbus cordis (the truncoconal portion of the heart) is incorporated at later stages, cranial to the initial point of fusion of the paired primordia.[40] Concomitant with the merging of new heart regions the heart tube is bent, rotated to the right, and thrown into a loop (Figure 3). It is only at the end of the looping process that the different parts of the heart (bulbus cordis, ventricle, atrioventricular canal, atrium, and sinus venosus) become clearly differentiated and regional divisions can be established.[41,42] Heart looping is the first asymmetry observed in most vertebrates. As such, it has historically received a great deal of attention. However, the mechanisms involved in the regulation of looping are still under discussion.

DIFFERENTIATION AND MORPHOGENESIS

We have already discussed several aspects of the developmental biology of the precardiac areas. Although differentiation of the precardiac mesoderm may be an absolute requisite for the acquisition of heart shape, morphogenesis does not follow differentiation linearly. It has been shown clearly that cytodifferentiation (or at least a certain degree of cytodifferentiation) can be achieved in the absence of morphogenesis.[31] In fact, the degree of shape that can be obtained by culturing the precardiac areas is very poor, although the results can be improved if extracardiac areas are included in the explant.[3,34] On the other hand, differentiation does not stop with the beginning of myofibrillar assembly and the initiation of beating. Concomitant with the formation of the heart tube and heart looping, myocytes increase the number of cell contacts, organize histotypically, and a basal lamina begins to develop.[22,43] With the increase in the number of cell-to-cell contacts there is a drastic decrease in junctional resistance, better electrical coupling is achieved, and myocardial membranes start discharging rhythmically.[44,45] Both anatomic and physiologic features are assumed to maintain the phenotypic state, but the route leading from cellular activities to looping is undiscovered as yet.

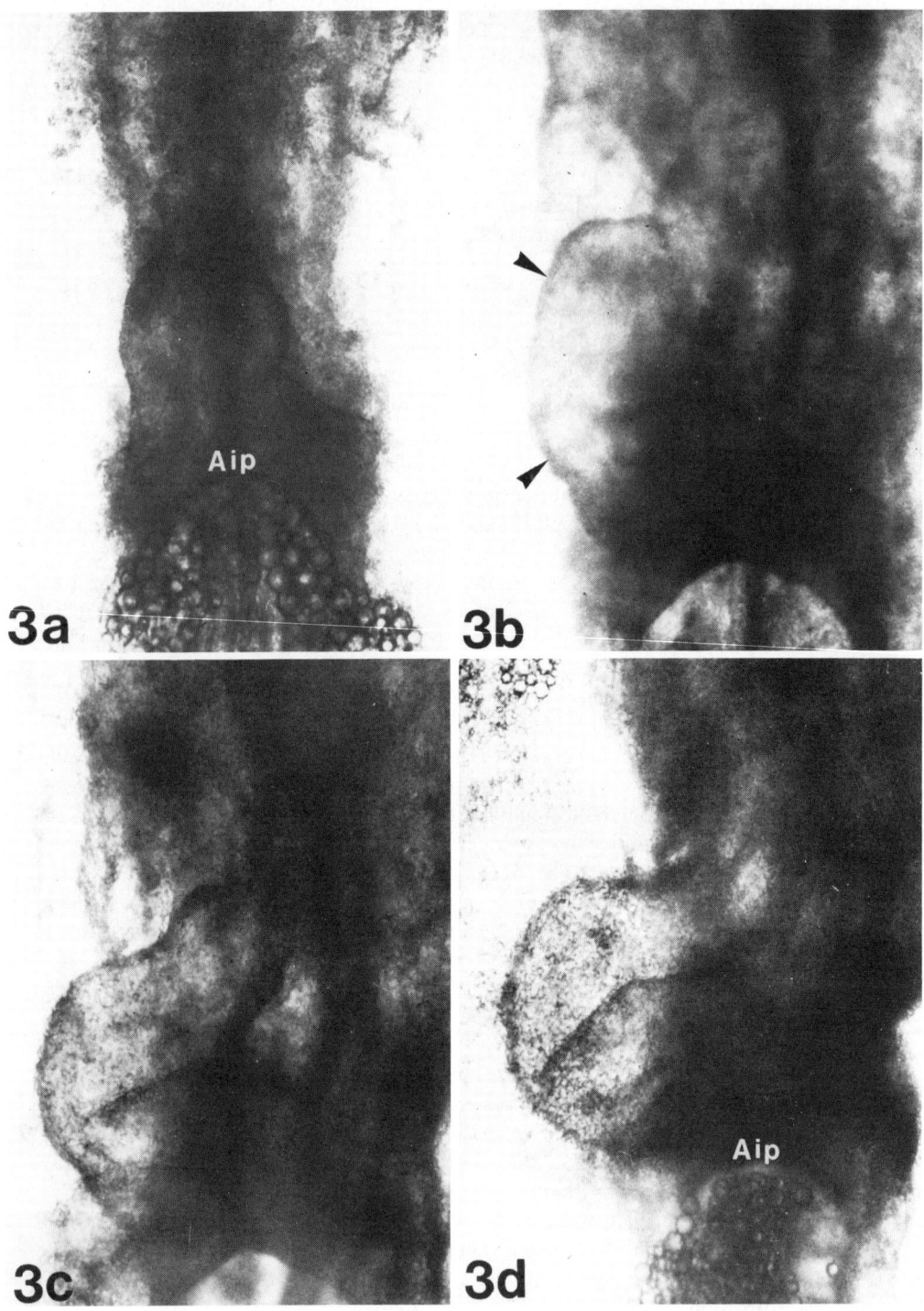

FIGURE 3. This composite illustrates the formation of the cardiac loop. The same embryo has been photographed at several stages while being maintained in New culture; Aip, Anterior intestinal porta. (a) The heart tube bulges outward and begins to rotate toward the right side of the embryo. (b) The heart acquires a C-shaped form. The ventral surface of the heart becomes the convex right side (arrowheads) of the loop. (c) With continued bending, the dorsal mesocardium ruptures, freeing most of the heart from the embryonic body. (d) After rupture of the dorsal mesocardium, the heart twists upon itself and adopts a sigmoid, S-shaped form. (Magnification × 90 [(a) and (b)]; magnification × 80 [(c) and (d)].)

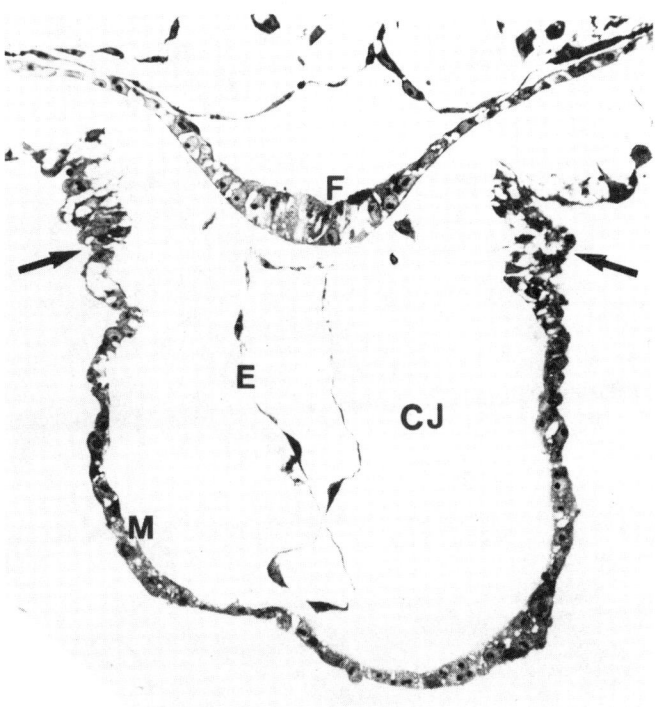

FIGURE 4. Light micrograph showing the histoglogical organization of the tubular heart; transverse semithin section. The outer layer of the heart is formed by the myocardium (M). The endocardium (E) encloses the heart lumen. The cardiac jelly (CJ) occupies the large space interposed between the endocardium and the myocardium. The dorsal mesocardium is indicated by arrows; F, endodermal foregut. (Magnification × 170.) (From Icardo, J. M., *Morphol. Norm. Pathol.*, 7, 43, 1983. With permission.)

THE STRUCTURE OF THE HEART TUBE

The Myocardium

We shall now center our attention on the structure of the heart tube. Throughout loop formation the heart is a very simple structure, being organized into layers (Figure 4). There is an outer layer, or myocardium, and an inner layer, or endocardium. The wide space interposed between the endocardium and the myocardium is occupied by a middle layer of extracellular material called cardiac jelly.[46]

The myocardium is the outermost layer of the heart. It is formed, throughout looping, of a pure population of developing myocytes.[22] The most characteristic feature of these cells at the ultrastructural level is the progressive accumulation of myofibrils in their cytoplasm (Figure 5). Myofibrils initially develop as discrete structures.[22,43] Nascent myofibrils consist of bundles of thick and thin filaments attached to dense bands of Z material (Figure 6). Each of these three elements appears to be synthesized as a separate unit. Electron dense patches of Z material, whether they are associated with the sarcolemma or isolated in the cytoplasm, have been suggested to serve as nucleating centers to which thin and thick filaments become attached.[47-49] However, these studies were carried out within the limits of the electron microscope. This means that a series of events that are involved in myofibrillar development, but occur before the structural appearance of the myofibrils could not be discerned. Only recently has the use of monoclonal antibodies against different myofibrillar components provided new information on myofibrillar assembly. It has been shown that α-

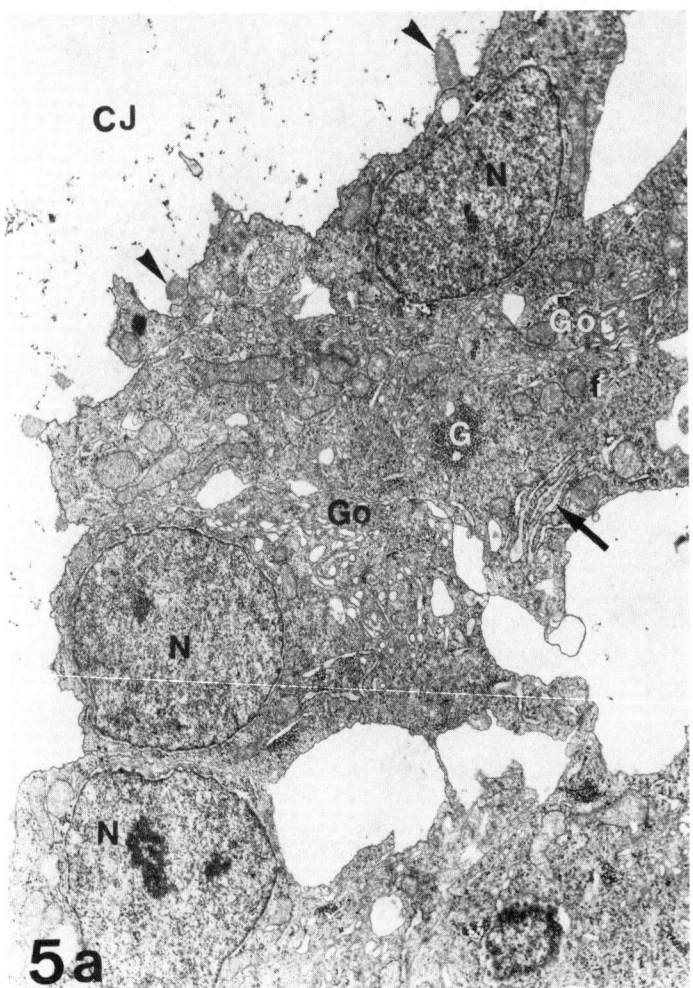

FIGURE 5. TEM micrographs depicting several aspects of the early myocardium. (a) Low-power magnification. Large intercellular spaces appear between the myocytes. These cells show glycogen granules (G), Golgi complexes (Go), granular endoplasmic reticulum filled with amorphous material (arrow), and some myofilaments (f). The cardiac jelly (CJ) presents some electron-dense inclusions. The basal surface of the myocardium shows a developing basal membrane with conspicuous masses of amorphous material (arrowheads). (Magnification × 6100.) (b) High-power magnification of the cytoplasm of an early myocyte. Some myofibrils (M) course parallel to the plane of section, while some others appear to be cut transversely (arrowheads). Large masses of glycogen (G) and numerous mitochondria can be observed. A developing intercalated disk appears (arrow) in the upper left corner. (Magnification × 13,200.) (Figure 5b from Manasek, F. J., *Am. J. Cardiol.*, 25, 149, 1970. With permission.)

actinin is the main component of the dense Z material. Also, the actin and myosin filaments appear to be already organized before the development of the Z patches.[50,51] The incorporation of other elements such as desmin to the nascent myofibrils has also been studied.

From a structural viewpoint, later development of the myofibrils involves a more regular packing of the myofilaments within the myofibril, the thinning and elongation of the Z

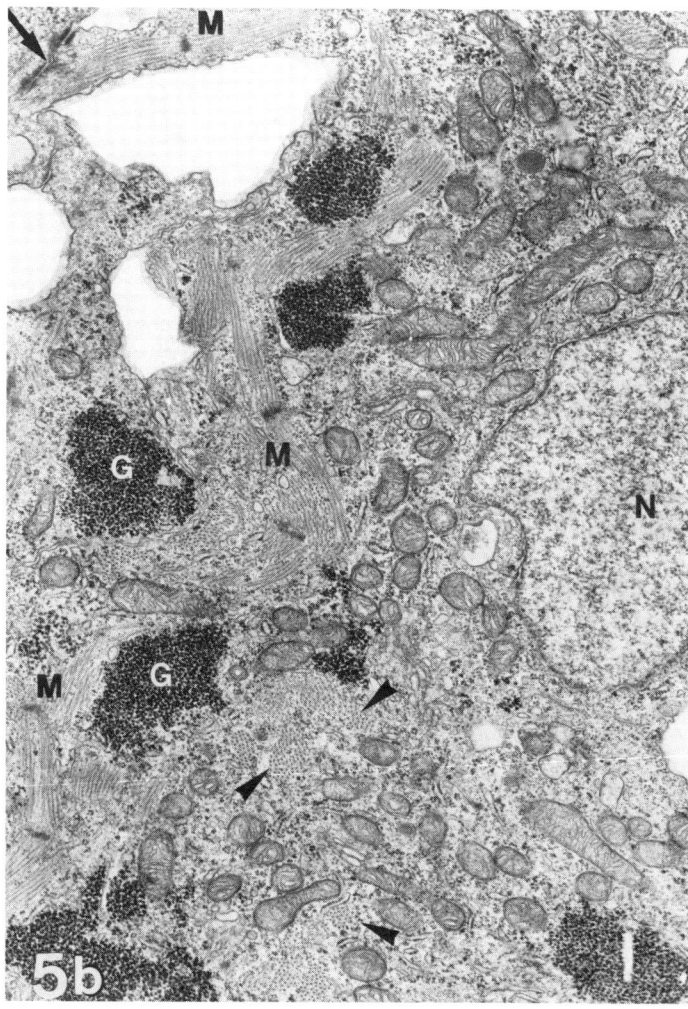

FIGURE 5b.

material, and the progressive addition of sarcomere units (Figure 6). Early myofibrils are only a few sarcomeres in length and most of them appear randomly distributed throughout the cytoplasm. It is only later in development that the myofibrils become aligned, forming parallel bundles typical of mature cardiac muscle.[43] However, the apparently random distribution of the early myofibrils, as shown by transmission electron microscopy, is somewhat misleading. When whole-mount preparations of tubular hearts are examined by means of polarized light microscopy, myofibrils show a predominantly circumferential arrangement.[52] Furthermore, examination of whole hearts with the scanning electron microscope after membrane extraction shows myofibrils to be orderly packed within individual myocytes.[53] The possible significance of this structural arrangement of the early myofibrils will be discussed later.

The synthesis of specific contractile proteins and their assembly into myofibrils is closely related to the initiation of heart function.[54,55] With the progressive increase in functional demands,[56] there is a progressive increase in the number of myofibrils per myocyte and in the contractile capacity of the heart. Overt changes in heart function have been related to changes in the synthetic pattern of contractile proteins,[57,58] especially myosin. Early cardiac

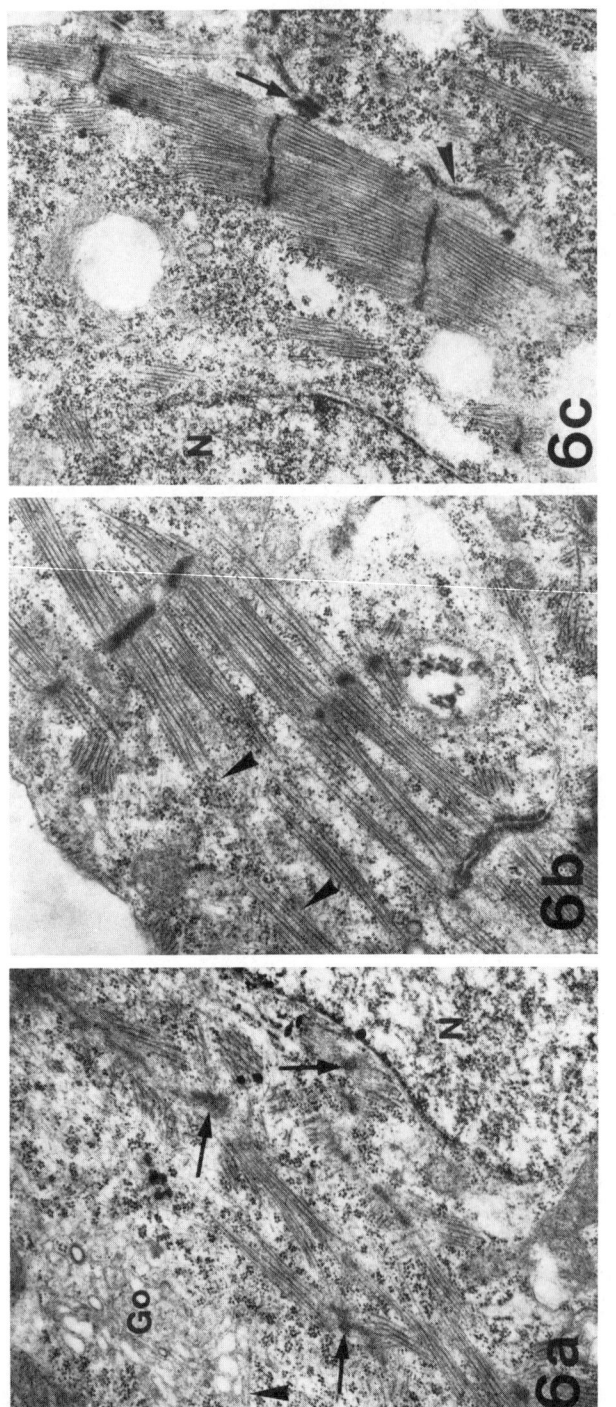

FIGURE 6. TEM micrographs showing several stages of the process of myofibrillogenesis. (a) Groups of thick and thin filaments converge upon Z-band material (arrows); N, nucleus; Go, Golgi complex; arrowhead, microtubular profile. (Magnification × 17,600.) (b) Nascent myofibrils appear more organized. Note their attachment to a developing intercalated disk. A long microtubule (arrowheads) is interposed between the filaments; N, nucleus. (Magnification × 17,900.) (c) More mature myofibrils show a better alignment of filaments and the thinning and elongation of the Z-material. A desmosome (arrow) and a developing intercalated disk (arrowhead) can clearly be observed; N, nucleus. (Magnification × 11,100.)

myocytes express a primordial form of myosin that is also expressed by early skeletal muscle cells.[59] As myocardial cells from the atria and ventricles develop different contractile characteristics, they begin to express different myosin isotypes. Different isomyosins are expressed by the cells of the conduction system,[60] and changes in myosin isoforms can be detected between embryonic, fetal, and adult hearts.[61,62] These changes involve the heavy-chain[63] as well as the light-chain[64] subunits of the myosin molecule. Different isotypes of troponin are also synthesized by the early myocytes.[27,65] The exact relationship between functional changes in the embryonic heart and the proper selection of the different isoforms of contractile proteins is unknown as yet. It is also unclear whether or not these biochemical changes are exclusively related to overt changes in function. For example, the expression of mammalian isomyosins appears to be closely regulated by the levels of thyroid hormone.[66] However, the influence of other factors, such as innervation and the presence of humoral factors(s), on the biochemical status of the embryonic myocytes is largely unknown.

When compared to skeletal muscle cells, cardiac myocytes present a number of seemingly unusual characteristics. Cardiac myocytes are engaged in the production and secretion of large amounts of extracellular material.[36] Rough endoplasmic reticulum, often filled with flocculent material, and well-developed Golgi complexes are conspicuous features of embryonic myocytes.[22,67] Besides their role in the production of extracellular matrix, cardiac muscle cells appear to be involved in a different kind of secretory activity. Mature myocytes (especially right atrial cells) elaborate a number of peptides which are contained in electron-dense granules. These peptides present potent diuretic and natriuretic activities. Therefore, vertebrate hearts can be considered endocrine organs whose secretion products are involved in the control of extracellular fluid volume and blood pressure.[68]

In the heart, commitment to differentiation does not exclude DNA synthesis and mitosis.[69,70] Embryonic myocytes with well-developed myofibrils can often be observed in different phases of the mitotic cycle (Figure 7). Another distinct characteristic of embryonic myocytes is their phagocytic capacity.[71,72] Differentiated muscle cells are able to phagocytose cellular debris (Figure 8) acting as "amateur" phagocytes.

The Endocardium

The endocardium is the innermost layer of the heart; it is located at the interface between the circulating blood and the rest of the heart. The endocardium has long been considered a passive vascular endothelium. Except for its morphogenetic role during cushion tissue formation,[73,74] the possible importance of the endocardium during heart development has largely been overlooked. However, the endocardium does not look like a passive endothelial covering (Figure 9). Different surface features are displayed by the endocardial cells at different stages of development in the different species examined,[75] including man. Several studies[76,77] reveal that endocardial cells show, at least at some stages of development, intense cellular activity, judging by the extension of cell processes and the presence of cell overlapping (Figure 9). Descriptive[78] and experimental[79] studies indicate that endocardial cells are able to change polarity in response of changes in blood flow. Changes in cell morphology have been linked to changes in cellular activities.[80] Hence, it is tempting to suggest that the endocardium may be engaged, at different stages, in various cellular activities. We do not know, however, how these different possible activities reflect (or are related to) different biochemical activities, or how they can be related to morphogenesis. During the stages of cardiac looping, endocardial cells present a dorso-ventral polarity that appears to be the result of active cell behavior.[76] It is, however, difficult to say whether or not endocardial cell polarity is involved in looping, or, rather, whether or not it merely represents a mechanism of adaptation to rapid heart growth.

The Cardiac Jelly

Cardiac jelly, the extracellular matrix of the developing heart, is an acellular, viscous,

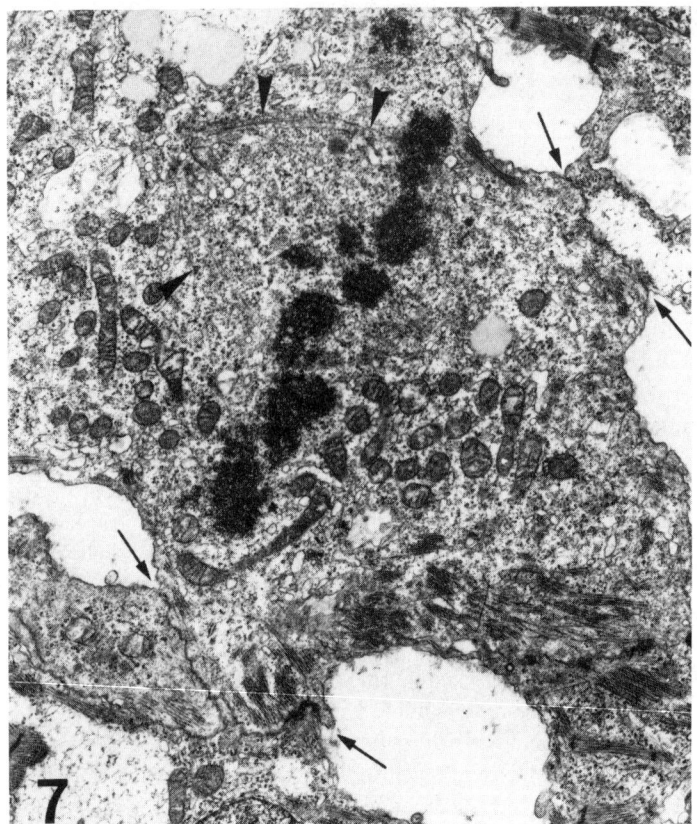

FIGURE 7. TEM micrograph depicting a mitotic myocardial cell. Observe aster microtubules (arrowheads) and the presence of numerous filaments in the cytoplasm. The mitotic myocyte remains attached (arrow) to neighboring myocardial cells. (Magnification × 8100.) (From Manasek, F. J., *J. Cell Biol.*, 37, 191, 1968. With permission.)

sticky gel that accounts for most of the wall thickness of the early heart. Despite its homogeneous aspect under the light microscope (Figure 4), cardiac jelly is a heterogeneous material in terms of both molecular composition and spatial distribution of individual components. In addition to water and ions, cardiac jelly contains the basic components of all connective tissues. Collagen types I, III, and IV, hyaluronate, chondroitins with different degrees of sufation, keratan sulfate, and fibronectin and other glycoproteins have been demonstrated in cardiac jelly using different techniques.[81-85] It is possible, however, that some more molecular species remain unidentified as yet.

The quantitative composition of the cardiac jelly at any given developmental stage is only known imprecisely.[86] Furthermore, adequate attention has seldom been paid to its water and ionic content. We know that early cardiac jelly is rich in hyaluronate and in poorly sulfated chondroitins. Later, its composition changes in both quantitative and qualitative terms: the amounts of highly sulfated chondroitins, keratan sulfate, and collagen increase. Thus, we can speak of a compositional maturation of the cardiac jelly matrix. Cardiac jelly is a heterogeneous gel from the structural viewpoint, too. Under the transmission or electron microscope, the cardiac jelly shows a number of structural inclusions such as fibrillar collagen, nonstriated fibrils, and granular and amorphous masses of electron-dense material (Figure 5). Some of these inclusions have tentatively been identified as filamentous hyalu-

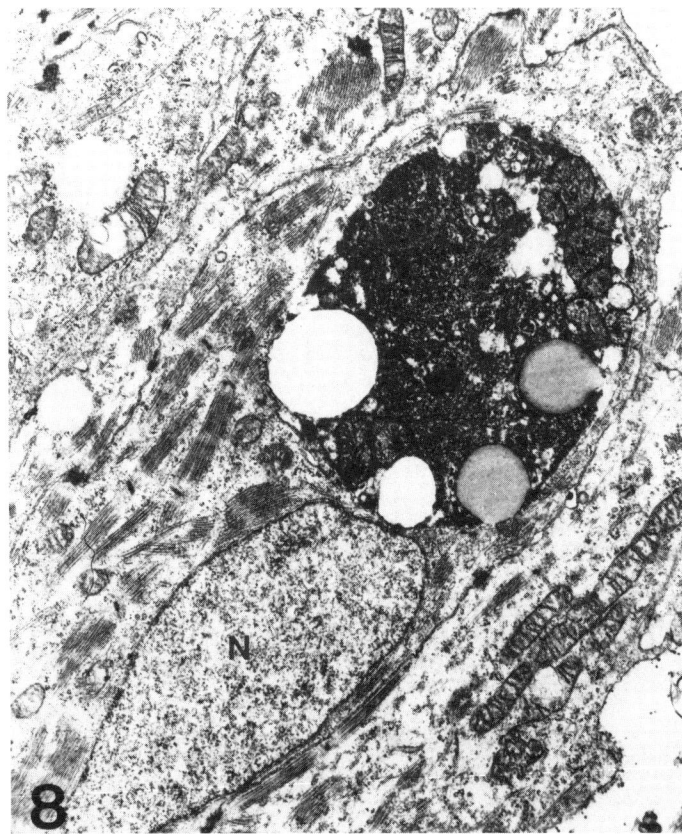

FIGURE 8. TEM micrograph of a myocardial cell showing a phagocytosed cell fragment. Note the numerous myofibrils; N, nucleus. (Magnification × 10,000.)

ronate, granular chondroitin sulfate, and glycoprotein masses.[87,88] The anatomic distribution of such inclusions varies with development, and these variations would be expected to have some influence on heart development.

The area of the dorsal mesocardium shows a high concentration of inclusions, especially of amorphous material, that extends from the endodermal foregut to the endocardium.[89,90] Much of this material appears to be composed of fibronectin.[4] By contrast, the cardiac jelly located between the endocardium and the myocardium shows only a few structural inclusions. However, when this area is observed under the scanning electron microscope, the cardiac jelly appears as a highly organized system of matrical filaments that extend radially from the endocardium toward the myocardium (Figure 10).[91,92] The nature of these filaments is unclear. They do not appear to be composed of collagen or hyaluronate,[93] and they are poorly decorated with antifibronectin antibodies.[4] The density of the radial filaments, as well as the number of lateral connections between them, increases with age. The amount of fibrillar collagen and the number of inclusions also increases with age. Thus, we can then speak of a process of maturation of the cardiac jelly not only from the point of view of its biochemical composition, but also from a structural viewpoint.

The developmental changes observed in the cardiac jelly are not very different from those undergone by other embryonic matrices.[94] The maturation of the cardiac jelly matrix probably reflects the maturation of the cells that produce it (mainly the myocardium). We do not know, however, whether or not the converse is true. Like other extracellular components,[95] cardiac jelly components are probably secreted in the form of monomers, which

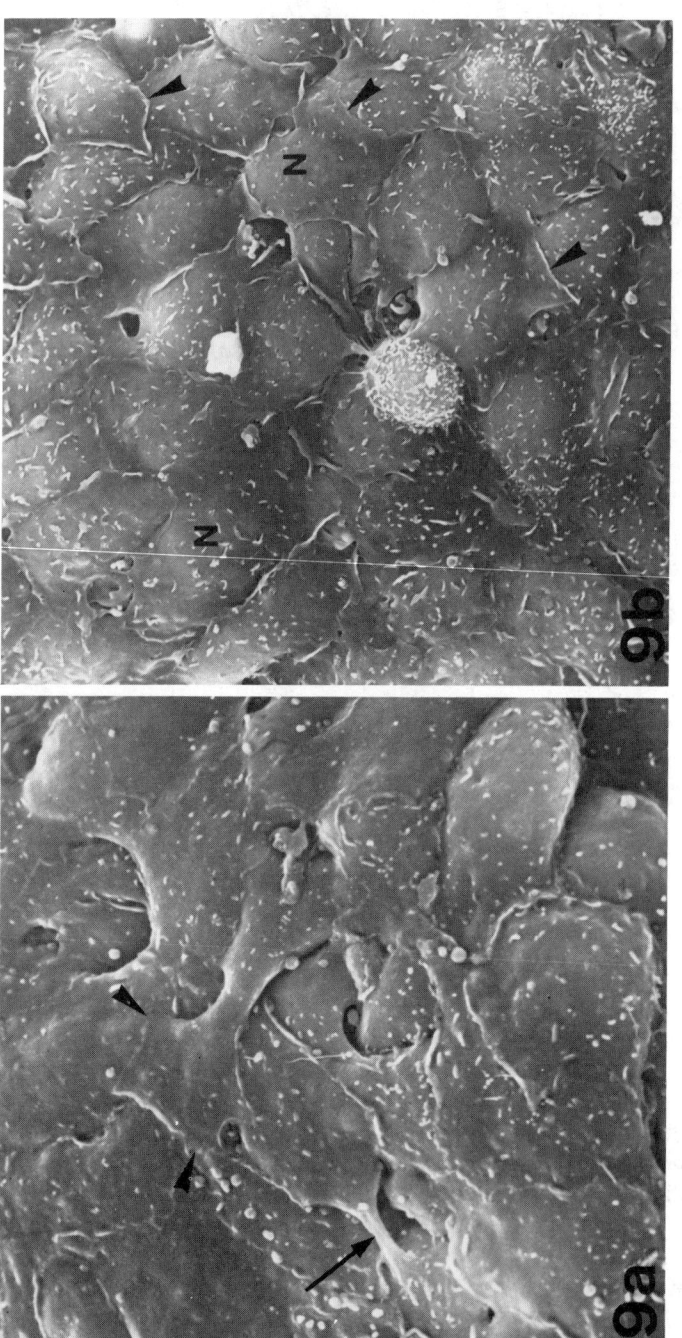

FIGURE 9. SEM micrographs showing different aspects of the embryonic endocardium: (a), (c), and (d) correspond to the same 5-d chicken heart. (a) Endocardial cells on the lateral border of the atrioventricular cushion present cytoplasmic processes such as lamellipodia (arrowheads) and cytoplasmic bridges (arrow). Cell surfaces show small blebs and a few microvilli. (Magnification × 1900.) (b) In the conal cushion of a 4-d chicken heart, the endocardial cells are more regular and show discrete bulging nuclei (N). Note the presence of intense cytoplasm overlapping (arrowheads). A mitotic endocardial cell, showing numerous microvilli and retraction fibers, can be observed in the center of the figure. (Magnification × 1550.) (c) In the center of the atrioventricular cushion, endocardial cells are extremely flattened and polygonal in shape. Cell surfaces are smooth and present small blebs. (Magnification × 1770.) (d) Atrial endocardial cells have irregular contours, rough surfaces, and prominent nuclei (N). Cytoplasmic projections (arrowheads) and intercellular gaps (arrows) can be observed. (Magnification × 1550.)

Part B: Cardiovascular and Respiratory Development 17

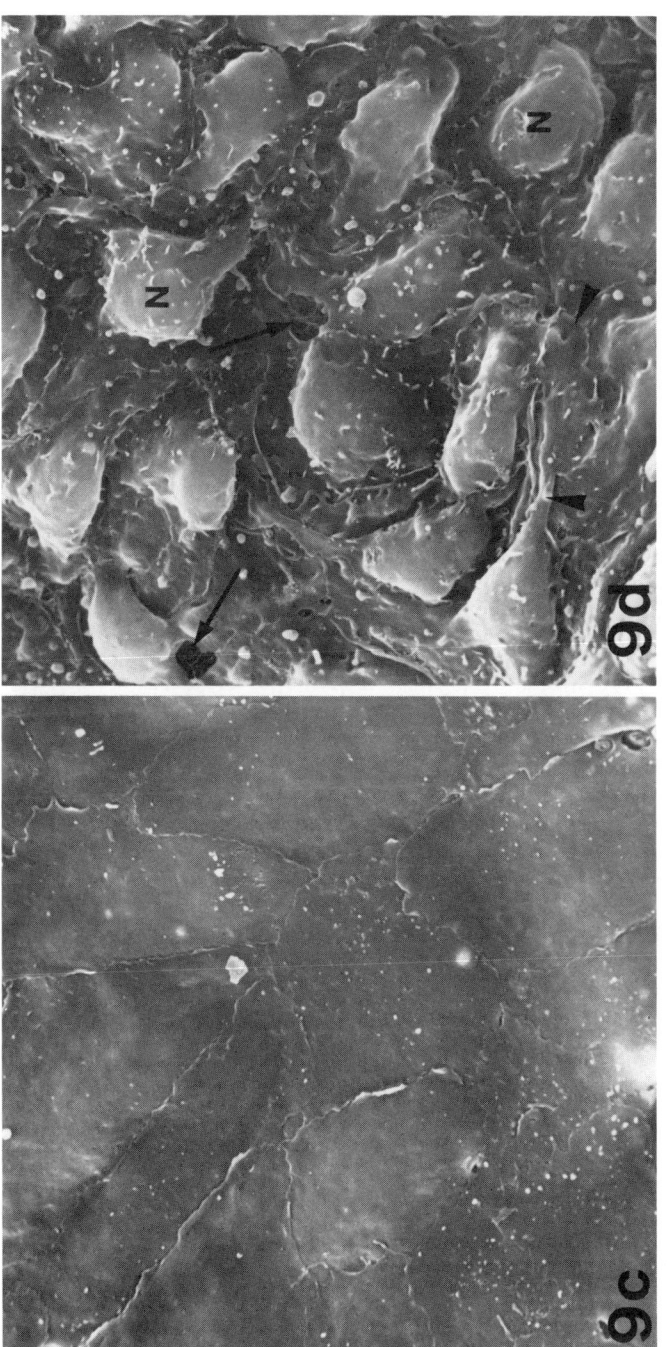

FIGURE 9c.

FIGURE 9d.

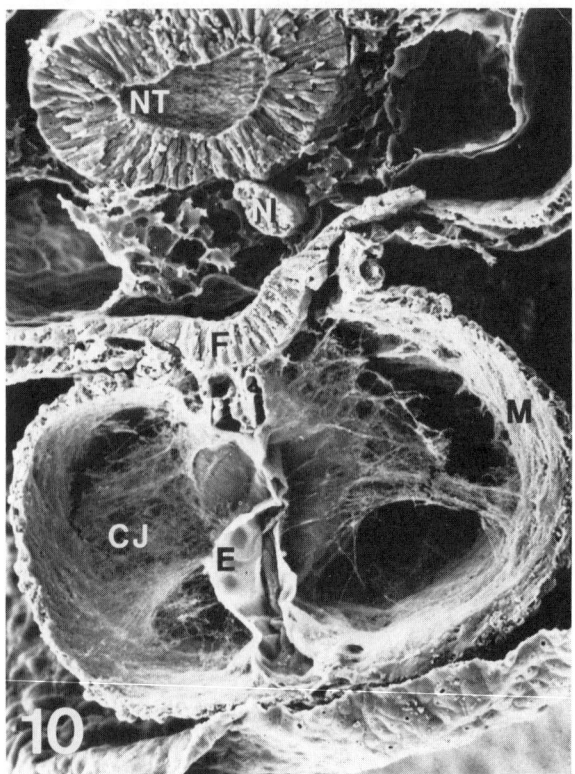

FIGURE 10. SEM micrograph of a transversely fractured heart. The cardiac jelly (CJ) appears to be made up of matrical filaments that extend from the endocardium (E) toward the myocardium (M). The empty spaces probably correspond to soluble material extracted during processing of the specimens; F, endodermal foregut; N, notochord; NT, neural tube. (Magnification × 320.) (From Icardo, J. M., *Morfol. Norm. Pathol.*, 7, 43, 1983. With permission.)

aggregate into large molecular complexes later.[67] Formation of large, three-dimensional molecular aggregates, such as proteoglycan, has been demonstrated in different connective tissues,[96,97] and we would expect diverse proteoglycan complexes to be formed in the cardiac extracellular space. Formation of such complexes would certainly modify the physicochemical properties of the matrix and, in turn, influence (modulate) the biological response of the embryonic myocytes. Although we know practically nothing about the diversity of cardiac proteoglycans, we cannot dismiss their possible developmental significance.[85] In fact, embryonic myocytes in culture have been shown to respond to changes in the environmental concentration of glycosaminoglycans.[98]

Cardiac jelly can be considered from other perspectives. The thickness of the cardiac jelly is essential to heart function. A thick layer of cardiac jelly increases the speed of narrowing of the heart lumen during systole, facilitating embryonic output.[91,99] In addition, cardiac jelly acts as a primitive valvular system. The heart lumen in the atrioventricular canal and in the bulbus cordis is alternatively closed by the rhythmic apposition of cardiac jelly mounds during the cardiac contractile cycle. Thus, the regurgitation of blood is prevented.[42,73] The valve effect of the cardiac jelly is imperfect, however. Some blood regurgitation occurs as is demonstrated by the presence of ventricular diastolic pressure.[100]

We can also consider cardiac jelly as a structural entity with well-defined physical properties. Isolated cardiac jelly is able to maintain its shape and to swell and to shrink in response to changes in the ionic strength of the culture medium.[101] The ability of the cardiac jelly to maintain and to recover its original shape appears to depend upon its glycosaminoglycan (GAG) content.[101] GAG aggregates are polyanionic molecules with huge molecular domains that attract counterions, trap water, and exclude other molecules.[97] Newly synthesized GAGs would increase osmotic pressure, trap more water, and induce volume changes. Thus, the early heart can be considered as a hydrostatically supported structure with an internal pressure that depends on its GAG composition. In fact, the looping heart collapses if testicular hyaluronidase is injected into the cardiac jelly space.[93]

TOWARD A UNIFYING THEORY OF HEART LOOPING

The early heart presents a tubular organization: a broad expanse of extracellular material is enclosed between the endocardium and the myocardium. Because of this concentric organization, deformation of the heart tube during looping involves deformation of the three heart layers. In fact, looping is basically a deformative process.

Most of the requirements for looping appear to be contained within the heart itself.[102] Each of the two areas of the precardiac mesoderm has the capability to loop. This has been shown by the experiments in which bifid hearts are formed (see above). Looping also occurs if the precardiac areas are interchanged,[103] or if they are excised, rotated 180°, and replaced in their original location.[104] This suggests that no preexisting asymmetries exist in the precardiac mesoderm.[15]

All the cellular activities tested to explain modifications of heart shape have failed to explain looping. Asymmetric contribution of cells from the right and left cardiac rudiments does not offer a satisfactory explanation for looping.[105] Differential mitotic rates do not appear to account for looping, either.[106,107] DNA synthesis[15] and cytokinesis[108] appear to be unnecessary. Colchicine studies suggest that looping does not depend on the integrity of the microtubular system, either.[108] Other hypotheses, such as the one based on the deforming effects of the blood flow,[109] have been discarded.

The force necessary to induce heart-shape changes during looping could be generated by the extracellular matrix.[52,86,110] Neither differential secretion rates nor regional differences in matrix accumulation would be needed for the cardiac jelly to produce heart deformation. It would be sufficient to control hydration of the matrix. Newly secreted molecules would produce an increase in the osmotic pressure and an increase of trapped water, inducing the matrix to swell. Matrix swelling would then create an outward pressure. When this outward pressure surpassed the elastic modulus of the myocardium, heart shape would change.

Measurements with servo-null instruments have demonstrated that the interstitial pressure in the cardiac jelly is uniform throughout the heart.[85] Therefore, asymmetric heart deformation can only be achieved if the compliance of the epithelium is different in different regions. During looping, the ventral surface of the heart expands more rapidly than the area of the dorsal mesocardium. The surface features of ventral myocytes are consistent[110] with great cell deformability, i.e., high epithelial compliance. These characteristics are not shared by the myocytes located dorsally. Furthermore, the cardiac jelly in the area of the dorsal mesocardium has more structural inclusions. This could mean that this area of the heart is more resistant to deformation. Although internal matrix pressure may provide the driving force for heart bulging and bending, we still need a system to regulate the extent and the quality of the deformation. In hydrostatically supported organisms this sort of control is effected by helical winding patterns of extracellular fibers.[111] In the heart, such a system has not been found in the extracellular milieu. However, it could reside in the myocardium.

There appears to be a close relationship between myofibrillogenesis and looping. Myofibrillogenesis accompanies looping, and looping is altered when the process of myofibrillogenesis is disrupted.[37] Examination of the myocardium *in situ* with the scanning electron microscope[53] has shown regional differences in the orientation and density of myofibrils, as well as changes in their arrangement with age. Although a helical winding pattern of fibers could not be demonstrated, the structural arrangement of the myofibrils could be consistent with such a system. It has been postulated that changes in myofibrillar pattern (density, angle, etc.) may provide the myocardium with the physical properties (anisotropy) sufficient to regulate the deforming pressure exerted by the cardiac jelly. Hence, the myocardium could control heart deformation.

The above model relates phenotype expression, biosynthetic activities, and structure to heart shape and heart shape changes. The model does not explain why the heart always loops to the right, nor does it take into account the possible involvement of extracardiac factors in looping.[103,112,113] However, it provides us with a testable hypothesis. This model proposes that when any of the biological parameters cited above is modified, looping will be abnormal. Direct and indirect evidence cited in this chapter lend support to the main points of this proposal.

Rotation of the early heart to the right is essential to normal heart development. If the heart loops to the left, situs inversus anomalies result. A strain of mice is now available in which 50% of the individuals develop diverse types of heterotaxia.[114] The hearts of these mice have been shown to loop to the left in 50% of the cases.[115] Thus, we have an animal model for testing any of the intra- and extracellular factors presumably involved in the regulation of looping. Only the work is yet to be done.

ACKNOWLEDGMENTS

The authors wish to thank Dr. F. J. Manasek for kindly providing Figures 5 and 7, Dr. J. M. Hurle for Figure 4, and Dr. M. Lafarga for Figure 8. The expert photographic assistance of A. de la Hoz is also acknowledged. This work was supported by Comisión Asesora de Investigación Cientifica y Técnica (CAICYT) Grant No. 2985/83.

REFERENCES

1. **Patten, B. M.**, The development of the heart, in *Pathology of the Heart and Blood Vessels*, Gould, S. E., Ed., Charles C. Thomas, Springfield, IL, 1968, 20.
2. **Stalsberg, H. and Dehaan, R. L.**, The precardiac areas and formation of the tubular heart in the chick embryo, *Dev. Biol.*, 19, 128, 1969.
3. **Le Douarin, G. H.**, Analyse experimentale des premiers stades du developpement cardiaque chez les vertebres superieurs, *Annee Biol.*, 13, 43, 1974.
4. **Icardo, J. M. and Manasek, F. J.**, Fibronectin distribution during early chick embryo heart development, *Dev. Biol.*, 95, 19, 1983.
5. **Linask, K. K. and Lash, J. W.**, Precardiac cell migration: fibronectin localization at mesoderm-endoderm interface during directional movement, *Dev. Biol.*, 114, 87, 1986.
6. **Linask, K. K. and Lash, J. W.**, A role for fibronectin in the migration of avian precardiac cells. I. Dose-dependent effects of fibronectin antibody, *Dev. Biol.*, 129, 315, 1988.
7. **Ojeda, J. L. and Hurle, J. M.**, Cell death during the formation of the tubular heart of the chick embryo, *J. Embryol. Exp. Morphol.*, 33, 523, 1975.
8. **Ojeda, J. L. and Hurle, J. M.**, Establishment of the tubular heart. Role of cell death, in *Mechanisms of Cardiac Morphogenesis and Teratogenesis*, Pexieder, T., Ed., Raven Press, New York, 1981, 101.
9. **Dehaan, R. L., Williams, E. H., Ypey, D. L., and Clapham, D. E.**, Intercellular coupling of embryonic heart cells, in *Mechanisms of Cardiac Morphogenesis and Teratogenesis*, Pexieder, T., Ed., Raven Press, New York, 1981, 299.
10. **Manasek, F. J. and Kulikowski, R. R.**, Myocardial filopodia during early heart development, *Scanning Electron Microsc.*, 2, 281, 1981.

11. **Lofberg, J.**, Apical surface topography of invaginating and noninvaginating cells. A scanning-transmission study of amphibian neurulae, *Dev. Biol.*, 36, 311, 1974.
12. **Thiery, J.-P., Duband, J. L., Rutishauer, U., and Edelman, G. M.**, Cell adhesion molecules in early chick embryogenesis, *Proc. Natl. Acad. Sci. U.S.A.*, 79, 6737, 1982.
13. **Goss, C. M.**, Double hearts produced experimentally in rat embryos, *J. Exp. Zool.*, 72, 33, 1935.
14. **Dehaan, R. L.**, Morphogenesis of the vertebrate heart, in *Organogenesis*, DeHaan, R. L. and Ursprung, H., Eds., Holt, Reinhart & Winston, New York, 1965, 377.
15. **Lacktis, J. W.**, An Experimetnal and Descriptive Study of Heart Morphogenesis in the Chick, Ph.D. thesis, University of Chicago, Chicago, IL, 1981.
16. **Folkman, J., Taylor, S., and Spillberg, C.**, The role of heparin in angiogenesis, in *Development of the Vascular System*, Nuggent, J. and O'Connor, M., Eds., Pitman, London, 1983, 132.
17. **McAuslan, B. R., Reilly, W., and Hannan, G. N.**, Inducers of neovascularisation: criteria for definition of a putative direct acting angiogenic factor, in *Progress in Microcirculation Research II*, Courtice, F. C., Garlick, D. G., and Perry, M. A., Committee in Postgraduate Medical Education, University of New South Wales, Sidney, 1984, 3.
18. **Kuettner, K. E. and Pauli, B. U.**, Inhibition of neovascularization by a cartilage factor, in *Development of the Vascular System*, Nuggent, J. and O'Connor, M., Eds., Pitman, London, 1983, 163.
19. **Wiens, D. and Spooner, B. S.**, Actin isotype biosynthetic transitions in early cardiac organogenesis, *Eur. J. Cell Biol.*, 30, 60, 1983.
20. **Woodroofe, M. N. and Lemanski, L. F.**, Two actin variants in developing axolotl heart, *Dev. Biol.*, 82, 172, 1981.
21. **Ebert, J. D.**, An analysis of the synthesis and distribution of the contractile protein, myosin, in the development of the heart, *Proc. Natl. Acad. Sci. U.S.A.*, 39, 333, 1953.
22. **Manasek, F. J.**, Embryonic development of the heart. I. A light and electron microscopic study of myocardial development in the early chick embryo, *J. Morphol.*, 125, 329, 1968.
23. **Wiens, D., Sullins, M., and Spooner, B. S.**, Precardiac mesoderm differentiation in vitro. Actin isotype synthetic transitions, myofibrillogenesis, initiation of heart beat, and the possible involvement of collagen, *Differentiation*, 28, 62, 1984.
24. **Hirota, A., Sakai, T., Fujii, S., and Kamino, K.**, Initial development of conduction pattern of spontaneous action potential in early embryonic precontractile chick heart, *Dev. Biol.*, 99, 517, 1983.
25. **Kucera, P. and De Ribaupierre, Y.**, *In situ* recording of the mechanical behaviour of cell in the chick embryo, in *Embryonic Development*, Part B, Burger, M. M. and Weber, R., Eds., Alan R. Liss, New York, 1982, 433.
26. **Patten, B. M. and Kramer, T. C.**, The initiation of contraction in the embryonic chick heart, *Am. J. Anat.*, 53, 349, 1933.
27. **Shimada, Y. and Toyota, N.**, Troponin types in cardiac and skeletal muscles *in vivo* and *in vitro*: an immunofluorescence microscopic study, in *Congenital Heart Disease, Causes and Processes*, Nora, J. and Takao, A., Eds., Futura, New York, 1984, 133.
28. **Lee, W. H.**, The glycogen content of various tissues of the chick embryo, *Anat. Rec.*, 110, 465, 1951.
29. **Icardo, J. M.**, Heart anatomy and developmental biology, *Experientia*, 44, 910, 1988.
30. **Chacko, S. and Joseph, X.**, The effect of 5-bromodeoxyuridine (BrDU) on cardiac muscle differentiation, *Dev. Biol.*, 40, 340, 1974.
31. **Spooner, B. S., Cohen, H. I., and Faubion, J.**, Development of the embryonic mammalian pancreas: the relationship between morphogenesis and cytodifferentiation, *Dev. Biol.*, 61, 119, 1977.
32. **Bacon, R. L.**, Self-differentiation and induction in the heart of amblyostoma, *J. Exp. Zool.*, 98, 87, 1945.
33. **Orts-Llorca, F.**, Influence of the endoderm on heart differentiation during the early stages of development of the chicken embryo, *Wilheim Roux Arch. Entwicklungsmech. Org.*, 154, 533, 1963.
34. **Copenhaver, W. M.**, Heart, blood vessels, blood, and entodermal derivatives, in *Analysis of Development*, Willier, B. H., Weiss, P. A., and Hamburger, V., Eds., Hafner, New York, 1971, 440.
35. **Hay, E. D.**, Cell-matrix interaction in the embryo: cell shape, cell surface, cell skeletons, and their role in differentiation, in *The Role of Extracellular Matrix in Development*, Trelstad, R. L., Ed., Alan R. Liss, New York, 1984, 1.
36. **Manasek, F. J., Reid, M., Vinson, W., Seyer, J., and Johnson, R.**, Glycosaminoglycans synthesis by the early embryonic chick heart, *Dev. Biol.*, 35, 332, 1973.
37. **Manasek, F. J.**, Heart development: interactions involved in cardiac morphogenesis, in *The Cell Surface in Animal Embryogenesis and Development*, Poste, G. and Nicolson, G. L., Eds., North-Holland, Amsterdam, 1976, 545.
38. **Wartiovaara, J. and Vaheri, A.**, Fibronectin and early mammalian embryogenesis, in *Development in Mammals*, Johnson, M. H., Ed., Elsevier/North-Holland, Amsterdam, 1980, 233.
39. **Castro-Quezada, A., Nadal-Ginard, B., and De La Cruz, M. V.**, Experimental study of the formation of the bulboventricular loop in the chick, *J. Embryol. Exp. Morphol.*, 27, 623, 1972.
40. **De La Cruz, M. V., Sanchez Gomez, C., Arteaga, M. M., and Arguello, C.**, Experimental study of the development of the truncus and the conus in the chick embryo, *J. Anat.*, 123, 661, 1977.

41. **Kramer, T. C.,** The partitioning of the truncus and conus and the formation of the membranous portion of the interventricular septum in the human heart, *Am. J. Anat.,* 71, 343, 1942.
42. **Streeter, G. L.,** Developmental horizons in human embryos. Description of age group XI, 13 to 20 somites, and age group XII, 21 to 29 somites, *Carnegie Inst. Washington Contrib. Embryol.,* 30, 211, 1942.
43. **Manasek, F. J.,** Histogenesis of the embryonic myocardium, *Am. J. Cardiol.,* 25, 149, 1970.
44. **Gros, D., Mocquard, J. P., Schrevel, J., and Challice, C. E.,** Assembly of gap junctions in developing mouse cardiac muscle, in *Mechanisms of Cardiac Morphogenesis and Teratogenesis,* Pexieder, T., Ed., Raven Press, New York, 1981, 285.
45. **Le Douarin, G. H. and Renaud, D.,** Differentiation of cellular electrical properties in the developing embryonic chick heart, in *Mechanisms of Cardiac Morphogenesis and Teratogenesis,* Pexieder, T., Ed., Raven Press, New York, 1981, 317.
46. **Davis, C. L.,** Development of the human heart from its first appearance to the stage found in embryos of 20 paired somites, *Carnegie Inst. Washington Contrib. Embryol.,* 19, 245, 1927.
47. **Hay, E. D.,** The fine structure of differentiating muscle in the salamander tail, *Z. Zellforsch. Mikrosk. Anat.,* 59, 6, 1963.
48. **Markwald, R. R.,** Distribution and relationship of precursor Z material to organizing myofibrillar bundles in embryonic rat and hamster ventricular myocytes, *J. Mol. Cell. Cardiol.,* 5, 341, 1973.
49. **Saetersdal, T., Engedal, H., Lie, R., and Mykeblust, R.,** On the origin of Z-band material and myofilaments in myoblasts from the human atrial wall, *Cell Tissue Res.,* 207, 21, 1980.
50. **Tokuyasu, K. T. and Maher, P. A.,** Immunocytochemical studies of cardiac myofibrillogenesis in early chick embryos. I. Presence of immunofluorescence actitin spots in premyofibril stages, *J. Cell Biol.,* 105, 2781, 1987.
51. **Tokuyasu, K. T. and Maher, P. A.,** Immunocytochemical studies of cardiac myofibrillogenesis in early chick embryos. II. Generation of α-actinin dots within actitin spots at the time of the first myofibril formation, *J. Cell Biol.,* 105, 2795, 1987.
52. **Nakamura, A., Kulikowski, R. R., Lacktis, J. W., and Manasek, F. J.,** Heart looping: a regulated response to deforming forces, in *Etiology and Morphogenesis of Congenital Heart Disease,* van Praagh, R. and Takao, A., Eds., Futura, New York, 1980, 81.
53. **Manasek, F. J., Isobe, Y., Shimada, Y., and Hopkins, W.,** The embryonic myocardial cytoskeleton, interstitial pressure, and the control of morphogenesis, in *Congenital Heart Disease, Causes and Processes,* Nora, J. and Takao, A., Eds., Futura, New York, 1984, 359.
54. **Lindner, E.,** Myofibrils in the early development of the chick embryo hearts as observed with the electron microscope, *Anat. Rec.,* 136, 234, 1960.
55. **Challice, C. E. and Edwards, G. A.,** On the micromorphology of the developing ventricular muscle, in *The Specialized Tissues of the Heart,* Paes de Carvalho, A., de Mello, W. C., and Hofman, B. F., Eds., Elsevier/North-Holland, Amsterdam, 1961, 44.
56. **Clark, E. B.,** Functional aspects of cardiac development, in *Growth of the Heart in Health and Disease,* Zak, R., Ed., Raven Press, New York, 1984, 81.
57. **Litten, R. Z., III, Martin, B. J., Low, R. B., and Alpert, N. R.,** Altered myosin isozyme patterns from pressure-overloaded and thyrotoxic hypertrophied rabbit hearts, *Circ. Res.,* 50, 856, 1982.
58. **Gorza, L., Mercadier, J. J., Schwartz, K., Thornell, L. E., Sartore, S., and Schiaffino, S.,** Myosin types in the human heart. An immunofluorescence study of normal and hypertrophied atrial and ventricular myocardium, *Circ. Res.,* 54, 694, 1984.
59. **Sweeney, L. J., Clark, W. A., Jr., Umeda, P. K., Zak, R., and Manasek, F. J.,** Immuofluorescence analysis of the primordial myosin detectable in embryonic striated muscle, *Proc. Natl. Acad. Sci. U.S.A.,* 81, 797, 1984.
60. **Gorza, L., Saggin, L., Sartore, S., and Ausoni, S.,** An embryonic-like myosin heavy chain is transiently expressed in nodal conduction tissue of the rat heart, *J. Mol. Cell. Cardiol.,* 20, 931, 1988.
61. **Thornell, L. E., Forsgren, S., Gorza, L., Sartore, S., and Schiaffino, S.,** Differentiation of fiber types in cardiac muscle, in *Congenital Heart Disease. Causes and Processes,* Nora, J. and Takao, A., Eds., Futura, New York, 1984, 157.
62. **Sweeney, L. J.,** Contractile protein expression in embryonic heart development, in *Cardiac Morphogenesis,* Ferrans, V. J., Rosenquist, G. C., and Weinstein, C., Eds., Elsevier, New York, 1985, 78.
63. **Evans, D., Miller, J. B., and Stockdale, F. E.,** Developmental patterns of expression and coexpression of myosin heavy chains in atria and ventricles of the avian hearts, *Dev. Biol.,* 127, 376, 1988.
64. **Kawashima, M., Nabeshima, Y.-I., Obinata, T., and Fujii-Kuriyama, Y.,** A common myosin light chain is expressed in chicken embryonic skeletal, cardiac and smooth muscles and in brain continuously from embryo to adult, *J. Biol. Chem.,* 262, 14,408, 1987.
65. **Sabry, M. A. Dhoot, G. K.,** Identification and changes in the expression of troponin T isoforms in developing avian and mammalian heart, *J. Mol. Cell. Cardiol.,* 21, 85, 1989.
66. **Izumo, S., Nadal-Ginard, B., and Mahdavi, V.,** All members of the myosin heavy chain multigene family respond to thyroid hormone in a highly tissue-specific manner, *Science,* 231, 597, 1986.

67. **Icardo, J. M., Ojeda, J. L., and Hurle, J. M.**, A quantitative study of the position of the Golgi apparatus in the developing chick myocardium. Effects of colchicine administration, *Acta Anat.*, 117, 152, 1983.
68. **Forssmann, W. G., Scheuermann, D. W., and Alt, J., Eds.**, *Functional Morphology of the Endocrine Heart*, Steinkopff Verlag, Darmstadt, 1989.
69. **Manasek, F. J.**, Mitosis in developing cardiac muscle, *J. Cell Biol.*, 37, 191, 1986.
70. **Rumyantsev, P. P. and Snigirevskaya, E. S.**, The ultrastructure of differentiating cells of the heart muscle in the state of mitotic division, *Acta Morphol. Acad. Sci. Hung.*, 16, 271, 1968.
71. **Gardfield, R. E., Chacko, S., and Blose, S.**, Phagocytosis by muscle cells, *Lab. Invest.*, 33, 418, 1975.
72. **Hurle, J. M., Lafarga, M., and Ojeda, J. L.**, Cytological and cytochemical studies of the necrotic area of the bulbus of the chick embryo heart: phagocytosis by developing myocardial cells, *J. Embryol. Exp. Morphol.*, 41, 161, 1977.
73. **Patten, B. M., Kramer, T. C., and Barry, A.**, Valvular action in the embryonic chick heart by localized apposition of endocardial masses, *Anat. Rec.*, 102, 299, 1948.
74. **Markwald, R. R., Fitzharris, T. P., and Manasek, F. J.**, Structural development of endocardial cushions, *Am. J. Anat.*, 148, 85, 1977.
75. **Pexieder, T.**, Prenatal development of the endocardium: a review, *Scanning Electron. Microsc.*, 2, 223, 1981.
76. **Icardo, J. M., Ojeda, J. L., and Hurle, J. M.**, Endocardial cell polarity during the looping of the heart in the chick embryo, *Dev. Biol.*, 90, 203, 1982.
77. **Icardo, J. M.**, Changes in endocardial cell morphology during development of the endocardial cushions, *Anat. Embryol.*, 179, 443, 1989.
78. **Hurle, J. M. and Colvee, E.**, Changes in the endothelial morphology of the developing semilunar heart valves, *Anat. Embryol.*, 167, 67, 1983.
79. **Icardo, J. M.**, Endocardial cell arrangement. Role of hemodynamics, *Anat. Rec.*, 225, 150, 1989.
80. **Bell, P. B. and Revel, J.-P.**, Scanning electron microscope application to cells and tissue in culture, in *Biomedical Research Applications of Scanning Electron Microscopy*, Vol. 2, Hodges, G. M. and Hallowes, R. C., Eds., Academic Press, London, 1980, 1.
81. **Markwald, R. R. and Adams Smith, W. N.**, Distribution of mucosubstances in the developing rat heart, *J. Histochem. Cytochem.*, 29, 896, 1972.
82. **Manasek, F. J.**, The extracellular matrix: a dynamic component of the developing embryo, *Curr. Top. Dev. Biol.*, 10, 35, 1975.
83. **Manasek, F. J.**, Macromolecules of the extracellular compartment of embryonic and mature heart, *Circ. Res.*, 38, 331, 1976.
84. **Icardo, J. M.**, Ultrastructure and function of the cardiac jelly: a review, *Morfol. Norm. Pathol.*, 7, 43, 1983.
85. **Manasek, F. J., Icardo, J. M., Nakamura, A., and Sweeney, L.**, Cardiogenesis: developmental mechanisms and embryology, in *The Heart and Cardiovascular System*, Fozzard, H. A., Haber, E., Jennings, R. B., Katz, A. M., and Morgan, H. E., Eds., Raven Press, New York, 1986, 965.
86. **Manasek, F. J. and Nakamura, A.**, Forces and deformations: their origins and regulation in early development, in *Cardiac Morphogenesis*, Ferrans, V. J., Rosenquist, G., and Weinstein, C., Eds., Elsevier, New York, 1985, 126.
87. **Markwald, R. R., Fitzharris, T. P., Bank, H., and Bernanke, D. H.**, Structural analysis of cell:matrix interaction during morphogenesis of atrioventricular cushion tissue, *Dev. Biol.*, 69, 634, 1978.
88. **Markwald, R. R., Fitzharris, T. P., and Adams Smith, W. N.**, Morphologic recognition of complex carbohydrates in embryonic cardiac extracellular matrix, *J. Histochem. Cytochem.*, 27, 1171, 1979.
89. **Johnson, R. C., Manasek, J. F., Vinson, W., and Seyer, J.**, The biochemical and structural demonstration of collagen during early heart development, *Dev. Biol.*, 36, 252, 1974.
90. **Hurle, J. M. and Ojeda, J. L.**, Cardiac jelly arrangement during the formation of the tubular heart of the chick embryo, *Acta Anat.*, 98, 444, 1977.
91. **Nakamura, A. and Manasek, F. J.**, Cardiac jelly fibrils: their distribution and organization, in *Morphogenesis and Malformation of the Cardiovascular System*, Rosenquist, G. C. and Bergsma, D., Eds., Alan R. Liss, New York, 1978, 229.
92. **Hurle, J. M., Icardo, J. M., and Ojeda, J. L.**, Compositional and structural heterogenicity of the cardiac jelly of the chick embryo tubular heart: a TEM, SEM and histochemical study, *J. Embryol. Exp. Morphol.*, 562, 211, 1980.
93. **Nakamura, A. and Manasek, F. J.**, An experimental study of the relation of cardiac jelly to the shape of the early chick embryo heart, *J. Embryol. Exp. Morphol.*, 65, 235, 1981.
94. **Toole, B. P.**, Developmental role of hyaluronate, *Connect. Tissue Res.*, 10, 93, 1982.
95. **Kimura, J. H., Hardingham, T. E., and Hascall, V. C.**, Assembly of newly synthesized proteoglycan and link protein into aggregates in cultures of chondrosarcoma chondrocytes, *J. Biol. Chem.*, 255, 7134, 1980.
96. **Hascall, V. C. and Hascall, G. K.**, Proteoglycans, in *Cell Biology of Extracellular Matrix*, Hay, E. D., Ed., Plenum Press, New York, 1981, 39.

97. **Kuettner, K. E. and Kimura, J. H.,** Proteoglycans: an overview, *J. Cell. Biochem.*, 27, 327, 1985.
98. **Manasek, F. J., Lacktis, J. W., Aiton, J., and Lieberman, M.,** Synthesis and distribution of glycopeptides and glycosaminoglycans in cultures of embryonic heart cells, in *Mechanisms of Cardiac Morphogenesis and Teratogenesis,* Pexieder, T., Ed., Raven Press, New York, 1981, 181.
99. **Barry, A.,** The functional significance of the cardiac jelly in the tubular heart of the chick embryo, *Anat. Rec.,* 102, 289, 1948.
100. **Paff, G. H., Boucek, R. J., and Gutten, G. S.,** Ventricular blood pressure and competency of valves in the early embryonic chick heart, *Anat. Rec.,* 151, 119, 1965.
101. **Nakamura, A. and Manasek, F. J.,** Experimental studies of the shape and structure of isolated cardiac jelly, *J. Embryol. Exp. Morphol.,* 43, 167, 1978.
102. **Butler, J. K.,** An Experimental Analysis of Cardiac Loop Formation in the Chick, M.A. thesis, University of Texas, Austin, 1952.
103. **Del Rio, J. S.,** Influence of extrinsic factors on the development of the bulboventricular loop of the chick embryo, *J. Embryol. Exp. Morphol.,* 31, 199, 1974.
104. **Orts-Llorca, F. and Jimenez Collado, J.,** Determination of heart polarity (arteriovenous axis) in the chicken embryo, *Wilhelm Roux Arch. Entwicklungsmech. Org.,* 158, 147, 1967.
105. **Stalsberg, H.,** The origin of heart symmetry: right and left contributions to the early chick embryo heart, *Dev. Biol.,* 19, 109, 1969.
106. **Sissman, N. J.,** Cell multiplication rates during development of the primitive cardiac tube in the chick embryo, *Nature (London),* 210, 504, 1966.
107. **Stalsberg, H.,** Regional mitotic activity in the precardiac mesoderm and differentiating heart tube in the chick embryo, *Dev. Biol.,* 20, 18, 1969.
108. **Icardo, J. M. and Ojeda, J. L.,** Effects of colchicine on the formation and looping of the tubular heart of the embryonic chick, *Acta Anat.,* 119, 1, 1984.
109. **Manasek, F. J. and Monroe, R. G.,** Early cardiac morphogenesis is independent of function, *Dev. Biol.,* 27, 584, 1972.
110. **Manasek, F. J., Kulikowski, R. R., Nakamura, A., Nguyenphuc, Q., and Lacktis, J. W.,** Early heart development: a new model of cardiac morphogenesis, in *Growth of the Heart in Health and Disease,* Zak, R., Ed., Raven Press, New York, 1984, 105.
111. **Wainwright, S. A., Biggs, W. D., Currey, J. D., and Gosline, J. M.,** *Mechanical Designs in Organisms,* John Wiley & Sons, New York, 1976.
112. **Orts-Llorca, F.,** Les facteurs determinants de la morphogenese et la differentiation cardiaque, *Bull. Assoc. Anat.,* 23, 1, 1964.
113. **Lepori, N. G.,** Research on heart development in chick embryos under normal and experimental conditions, *Monit. Zool. Ital.,* 1, 159, 1967.
114. **Layton, W. M., Jr.,** Heart malformation in mice homozygous for a gene causing situs inversus, in *Morphogenesis and Malformation of the Cardiovascular System,* Rosenquist, G. C. and Bergsma, D., Eds., Alan R. Liss, New York, 1978, 277.
115. **Layton, W. M., Jr. and Manasek, F. J.,** Cardiac looping in early iv/iv mouse embryos, in *Etiology and Morphogenesis of Congenital Heart Disease,* Van Praagh, R. and Takao, A., Eds., Futura, New York, 1980, 109.

LATE HEART EMBRYOLOGY. THE MAKING OF AN ORGAN

Jose M. Icardo, M. Angeles Fernandez-Teran, and Jose L. Ojeda

INTRODUCTION

After completion of looping the heart undergoes a series of profound morphogenetic changes. As a result, it progressively acquires the adult configuration. Many of these morphologic changes are well known, and most of them have frequently been reviewed. Thus, we will try not to engage the reader into extensive descriptions. Only those shape changes that have been less studied over the years, or those that are not easily found in current textbooks, will be emphasized. Since maturation of the different tissues and structures continues until advanced postnatal stages, development of the heart will be followed, when appropriate, into the postnatal period. Emphasis will also be put on the different aspects of cell behavior and on the different developmental mechanisms that are involved in late heart development and growth.

EXTERNAL SHAPE CHANGES

The external modifications of heart shape after looping are shown in Figure 1. These modifications have been reviewed several times in recent years,[1-5] and they will not be reviewed again (for a brief description, see the legend of Figure 1).

At the stages represented in Figure 1 the heart is growing rapidly, increasing several times in size and mass. The increase in mass of any organ is accomplished by two basic mechanisms: increase in cell number (hyperplasia) and increase in cell mass (hypertrophy). Muscle cells are continuously engaged in the synthesis of myofibrillar elements. The cytoplasm of the cardiac myocytes soon becomes filled with myofibrils that progressively assume a parallel arrangement following the direction of the main cell axis (Figures 2 and 3). Thus, myocyte mass increase constitutes a basic mechanism of heart growth; however, the exact importance of cell hypertrophy during the developmental stages studied here has yet to be assessed.

Cardiac myocytes are also able to divide (see the preceding chapter). Although the mitotic activity of these cells declines with age, the rapid expansion of both atria and ventricles (see Figure 1) appears to be determined, at least in part, by the presence of centers of active cell proliferation.[6,7] Myocardial mitotic activity is also high in the cono-truncus,[8,9] suggesting that mitosis plays an important role in the lengthening of this region of the heart.

Heart growth as described above does not stop at the stages represented in Figure 1, nor does it stop at birth. The process of sarcomerogenesis is still going on several months after birth.[10] In postnatal hearts, myocytes with developing myofibrils can be observed side by side with myocytes whose cytoplasm is fully packed with well-oriented, mature myofibrils. The capability of cardiac myocytes to enter into the mitotic cycle continues also in early postnatal periods.[11-12] Later, the capability to undergo mitosis is lost, except for some atrial cells.[13,14] However, mature cardiac muscle cells are still able to synthesize DNA.[15] DNA synthesis in mature myocytes leads either to endoreplication, resulting in different degrees of ploidy, or to amitotic divisions, resulting in multinucleation.[16] Polyploidy is a very common feature in human hearts,[17] while other mammals have mostly binucleated diploid myocytes.[18,19]

Hypertrophy and hyperplasia are not the only mechanisms of growth during heart development. Different types of nonmyocardial cells will be incorporated into the developing heart (see below). Although nonmuscle cells do not hypertrophy, they present higher mitotic

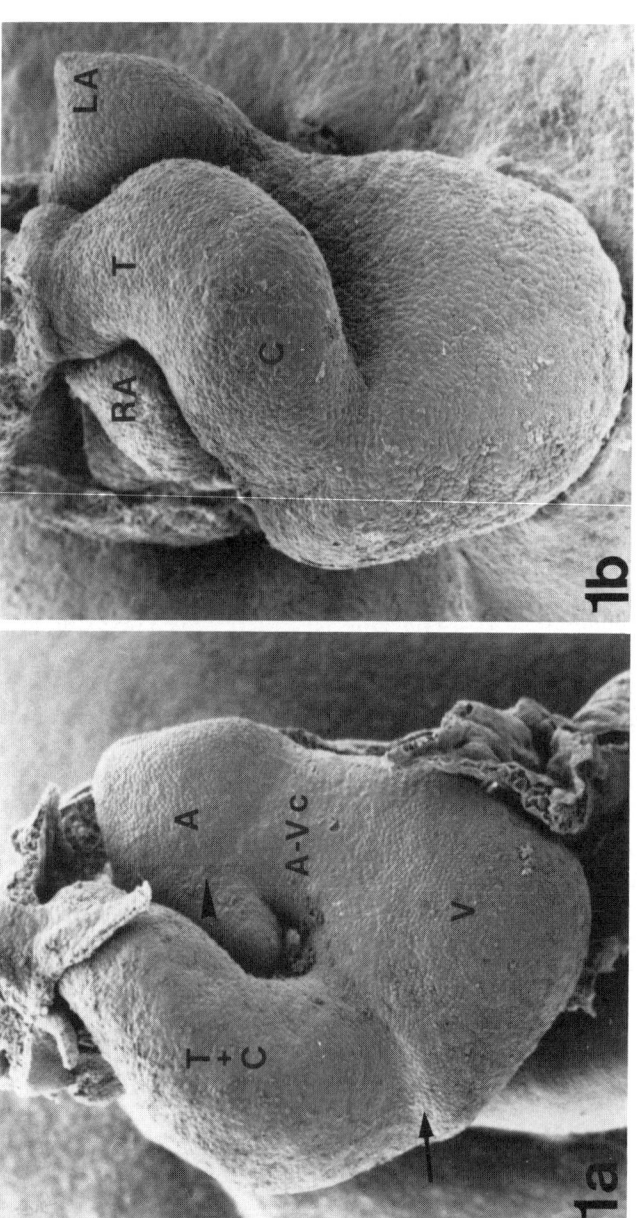

FIGURE 1. This series of scanning electron micrographs depicts the external modifications of heart shape after looping (chicken embryos, ventral views): A, primitive atrium; Ao, aorta; A-Vc, atrioventricular canal; C, conus; LA, left atrium; LV, left ventricle; P, Pulmonary artery; RA, right atrium; RV, right ventricle; T, truncus; V, primitive ventricle. (a) $3\frac{1}{2}$ d of incubation. The bulbus cordis (T + C) is located to the right of the primitive atrium. The ventricle occupies a caudal position, being separated from the atrium by the AV canal. The bulboventricular sulcus (arrow) separates the bulbus from the ventricle. Arrowhead indicates the interatrial sulcus; the partitioning of the primitive atrium is beginning. (Magnification × 100.) (b) Fourth day of incubation. The bulbus cordis bends ventrally and undergoes a relative displacement to the left, occupying a middle position between the two developing atria. The truncus and conus now appear as two clearly distinct regions. The left atrium is bigger than the right atrium. The ventricle is becoming more trapezoidal. (Magnification × 125.) (c) Fifth day of incubation. The two atria now present a similar size. A deep interventricular sulcus is clearly observed between the right and left ventricles. The apex of the heart is already formed by the left ventricle. The rough aspect of the heart surface is due to the collapse of the subepicardial space. (Magnification × 57.) (d) The heart has nearly assumed the adult configuration. The proximal segments of the aorta and the pulmonary artery are already formed, and the external separation of the arterial trunks is beginning. (Magnification × 63.)

Part B: Cardiovascular and Respiratory Development 27

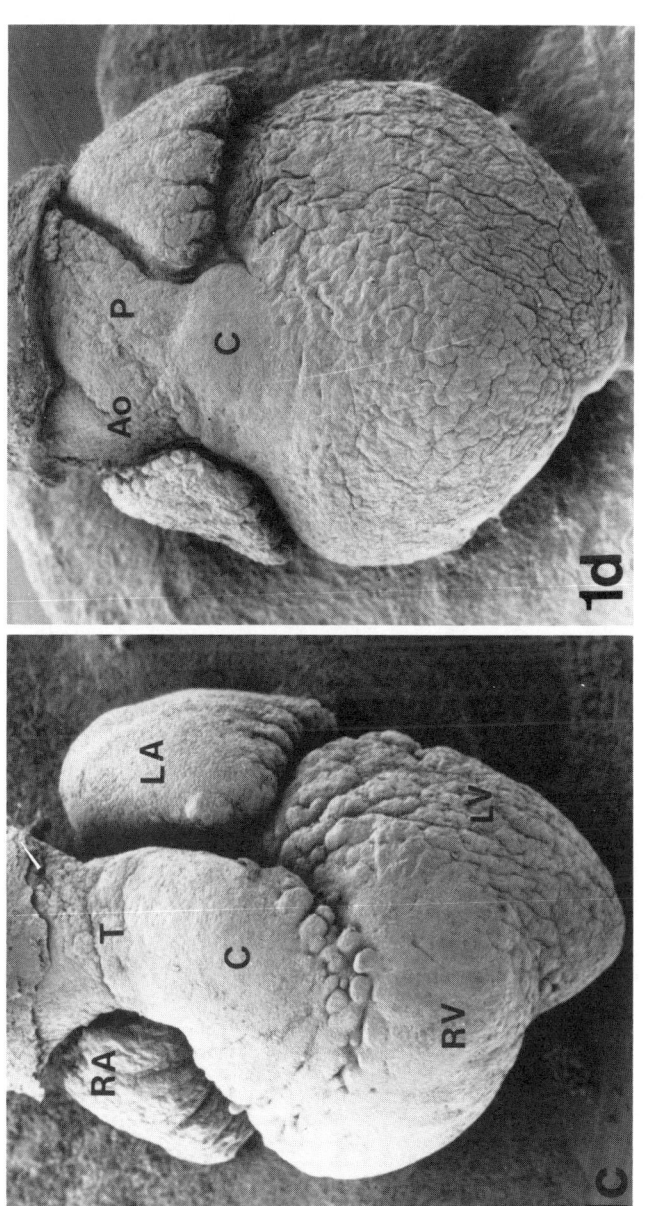

FIGURE 1c.

FIGURE 1d.

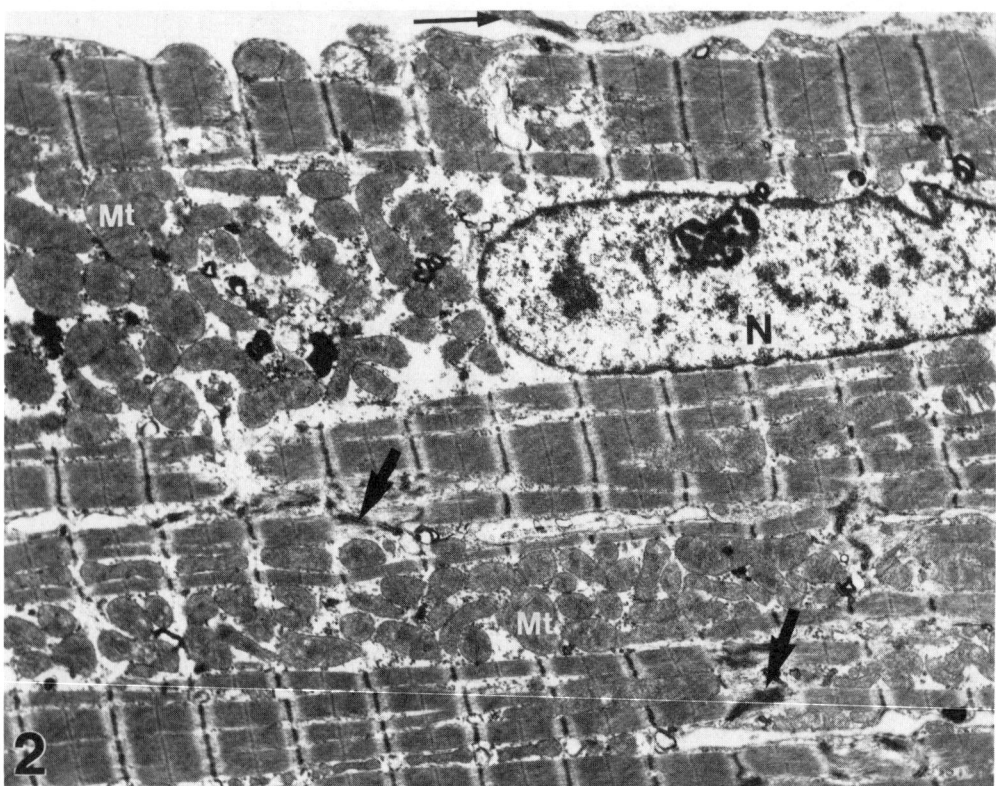

FIGURE 2. Transmission electron micrograph depicting a panoramic view of several cardiac myocytes in the adult rabbit. Myofibrils appear oriented in parallel following the direction of the main cell axis. Note the attachment of the myofibrils to the intercalated disks (thick arrows). Numerous mitochondriae (Mt) can be observed in the perinuclear region and in between the myofibrils. A blood vessel (thin arrow) appears in the upper part of the picture; N, nucleus. (Magnification × 5800.)

rates than the myocytes and their number increases progressively. In the adult heart they represent about 80% of the total cell number.[20,21] Although in the developing heart the incorporation of nonmuscle cells does not represent probably a significant part of the total increase in heart mass, their presence must be taken into account. Another developmental process that helps to explain the rapid enlargement of the atrial region has yet to be considered. Part of the venous system that empties into the atria is incorporated into the expanding atrial walls.[22,23] The smooth area that can be observed on the inner surface of the adult atria, and about 50% of the definitive interatrial septum, is derived from this process of venous incorporation.

The possible involvement of other mechanisms (hormones, growth factors, cell degradation products, etc.) in the control of heart growth has recently been reviewed.[24] The role of heart function as a stimulus for growth will be treated below. The morphogenetic role of cell death should also be emphasized. Opposite to the concept of organ growth by mitosis, cell death appears as a paradox of development. However, programmed cell death plays an essential role in the remodeling and in the fine shape tuning of different embryonic organs and structures.[25] The importance of cell death during heart development has been stressed earlier.[26-28]

INNER SHAPE CHANGES

Concomitant with the changes in its outer surface, the heart also experiences a process of internal remodeling. Although these changes will be treated separately, they constitute a

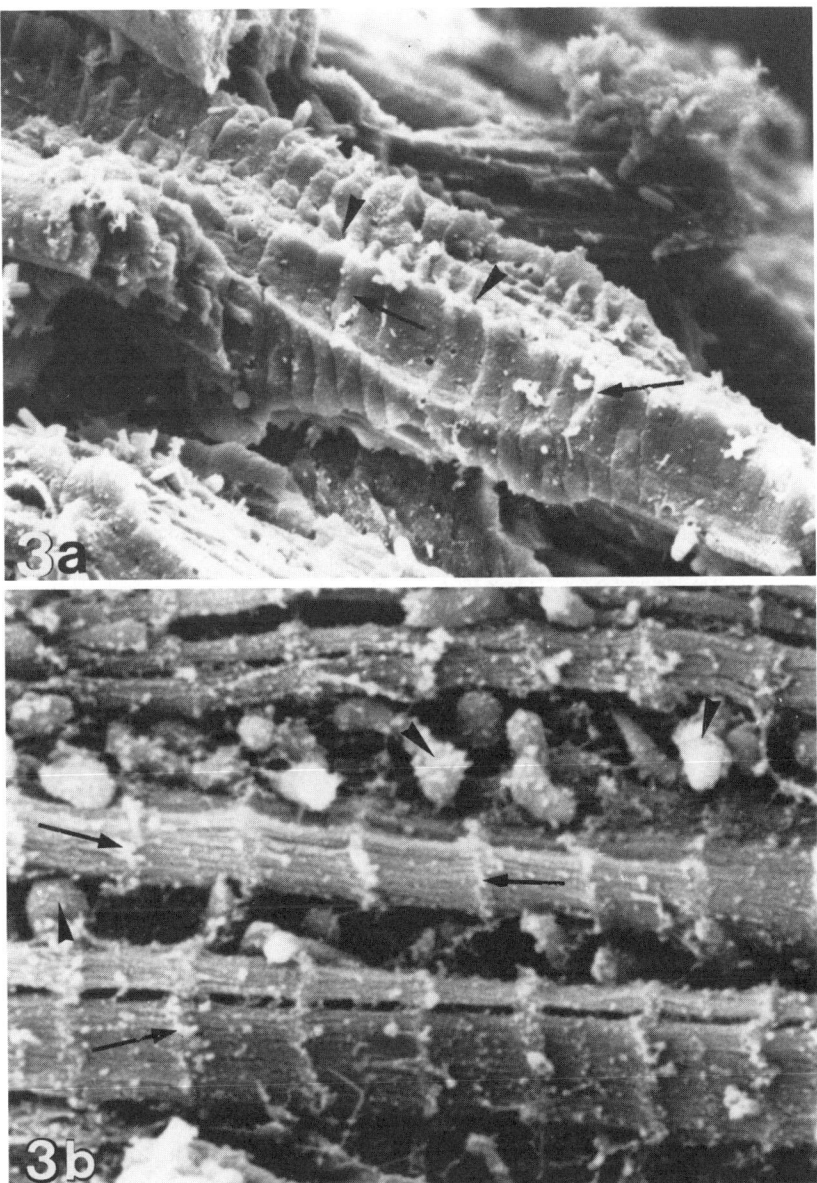

FIGURE 3. Scanning electron micrographs depicting the external and internal appearance of the cardiac myocytes in the rabbit. (a) Surface view of a myocardial cell isolated from neighboring myocytes after digestion with ClH. The sarcolemma protrudes over the Z-bands (arrows) giving to the outer surface of the myocyte a characteristic banding pattern. Arrowheads indicate the surface pits that correspond to the T-tubules. (Magnification × 6700.) (b) This fracture shows the internal structure of a cardiac myocyte. The myofibrils appear arranged along the main cell axis. The location of the Z-bands is indicated by arrows. The space between the myofibrils is loaded with mitochondriae (arrowheads). Some microfilaments (probably intermediate filaments) can be observed in the vicinity of the Z-bands. (Magnification × 18,000.)

perfectly synchronized continuum. Trabeculation and formation of the endocardial cushions constitute a first step in heart partitioning. The development of the distinct septa and valve apparatuses gradually divides the single heart lumen into separate compartments. Only minor shape modifications will be necessary after birth for the heart to adapt to normal physiologic

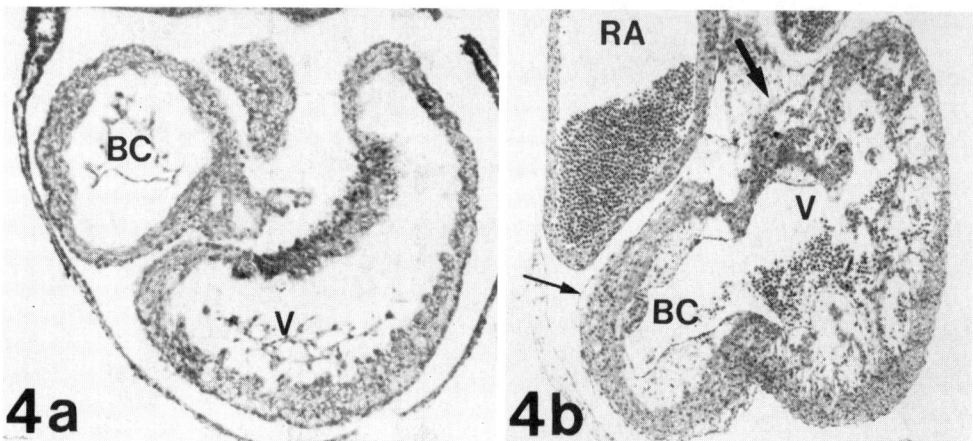

FIGURE 4. Light micrographs depicting different phases of the development of the heart in the human embryo: BC, bulbus cordis; RA, right atrium; V, ventricle. (a) Cross section of the heart at the cardiac loop stage. The plane of section passes through the bulbus cordis and the ventricle. Note in both the bulbus cordis and the ventricle the wide layer of cardiac jelly interposed between the endocardium and the myocardium. (Kyoto Collection, 1845-H12, Slide 5, Section 7.) (b) The structure of the heart tube has changed considerably in this section (compare to a). The process of trabeculation has started in the ventricle; note the decrease in thickness of the cardiac jelly. In the bulbus cordis, the layer of cardiac jelly remains thick and it has been invaded by mesenchyme. No cardiac jelly can be observed in the right atrium. The epicardium is indicated by arrows. (Kyoto Collection, 1127-H14, Slide 6, Section 3.) (From Icardo, J. M., in *Growth of the Heart in Health and Disease*, Zak, R., Ed., Raven Press, New York, 1984, 41. With permission.)

and hemodynamic demands. The chronological relationship of the main events of cardiac morphogenesis in different species has already been established.[29]

Trabeculation

At the end of looping the mycoardium is a compact tissue layer. However, the compact aspect of the myocardium is soon to change. In the ventricular region large intercellular spaces appear between the myocytes, dividing the myocardium into two distinct zones: an inner, spongy layer, and an outer, compact layer.[30] While these changes are occurring in the myocardium the endocardium begins to extend discrete evaginations that reach the basal surface of the myocardium. Then, the basal surface of the myocardium is disrupted and the endocardium penetrates into the myocardial wall[31] (Figure 4).

Despite its apparent simplicity, trabeculation is far from being a simple process. It is still unknown why the endocardium, at a fixed developmental time, presents an invasive behavior. Is this activity genetically programmed, or does it occur in response to changes in the physicochemical properties of the microenvironment? Cardiac jelly becomes progressively thinner during trabeculation. At these stages hyaluronidase[32,33] and plasminogen activator (see Reference 5) activity can be detected in the ventricular area. Although its relation to trabeculation is unclear, enzyme activity may disrupt the supramolecular organization of the preexisting cardiac jelly, thus facilitating the progression of the endocardial outgrowths. It should be stressed at this point that the reduction in thickenss of the cardiac jelly does not necessarily mean a reduction in the total amount of extracellular material; the ventricle is growing rapidly, so the total amount of extracellular material may actually increase. Another unanswered question is why the myocardium becomes spongy and opens up to receive the endocardium. The kind of interaction that the endocardium establishes with the myocardium throughout invasion is also unclear. Frequent contacts between the two tissue layers can be observed. However, these contacts appear to be transient structures,[31,34] and they may only serve to facilitate endocardial progression during the process of invasion.

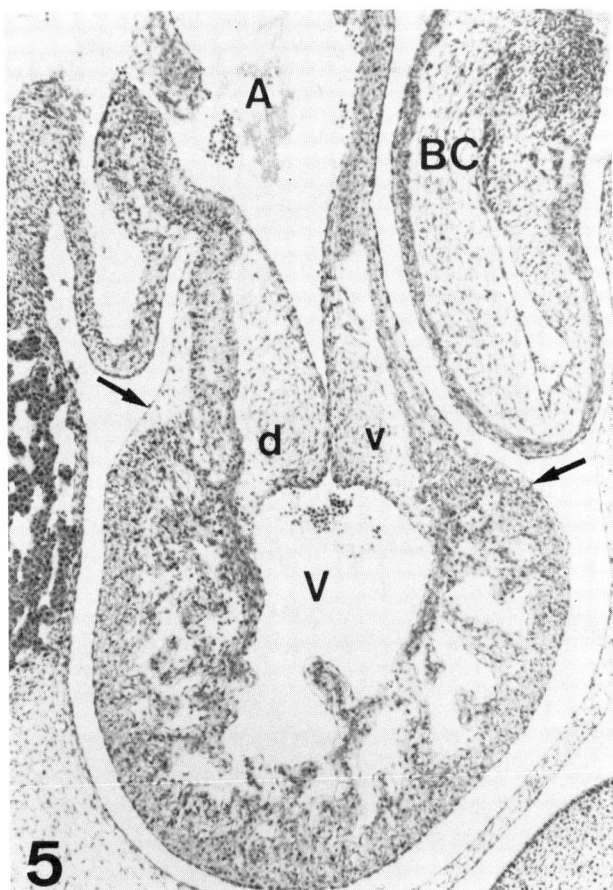

FIGURE 5. Light micrograph depicting a longitudinal section of the heart of a human embryo. The section passes through the atrioventricular canal. No cardiac jelly can be observed in the atrial region (A). Only a thin layer of cardiac jelly remains between the endocardium and the myocardial core of the ventricular trabeculae. The trabeculae adopt a radial orientation. The cardiac jelly has accumulated in the bulbus cordis (BC) and in the atrioventricular canal, and it has been invaded by mesenchyme: d and v, dorsal and ventral endocardial cushions; V, Ventricle. (Kyoto Collection, 1290-H14, Slide 6, Section 3.) (From Icardo, J. M., in *Growth of the Heart in Health and Disease*, Zak, R., Ed., Raven Press, New York, 1984, 41. With permission.)

Trabeculation starts at the ventricular apex. As development progresses and the ventricular chambers expand centrifugally, the endocardium penetrates deeper into the myocardium. Initially, trabeculation seems to be a nonstructured process. However, as the ventricles expand centrifugally, the trabeculae adopt a definite radial orientation[31,35] (Figure 5). Later in development the ventricular walls undergo a process of compaction, and most of the intertrabecular spaces disappear.[36] The remaining trabeculae are transformed into the trabeculae carnae observed in the adult heart. Some of the intertrabecular spaces (sinusoids) persist, eventually being transformed into the Thebesian vessels.

Without trabeculation the ventribular wall would become too thick and compact to perform its contractile functions efficiently. Trabeculation has another important advantage. The intertrabecular spaces permit the easy access of oxygen and nutrients to the ventricular cells; they perform as a capillary system until development of the coronary vascular bed.

The Endocardial Cushions

In the ventricle, the thickness of cardiac jelly is greatly reduced during trabeculation. In the atrial region the thickness of cardiac jelly is also reduced, and the myocardium and the endocardium become progressively closer to each other (Figures 4 and 5). However, cardiac jelly accumulates in the atrioventricular canal and in the bulbus cordis, forming the endocardial cushions (Figures 4 and 5). The most probable explanation for the accumulation of cardiac jelly in these areas is that of a continued secretion. The suggestion that the blood flow acts like a meandering river, displacing the cardiac jelly from areas of high hemodynamic shear to areas of low hemodynamic shear,[37,38] appears to be incorrect. The displacement of preexisting cardiac jelly is unlikely due to the structural integration of the components of the extracellular material.

Endocardial cushions appear as large expanses of acellular cardiac jelly between the endocardium and the myocardium. Soon after cushion formation, endothelial cells limiting the cushions develop migratory appendages and are seeded into the matrix. Then, they are transformed into mesenchyme.[39,40] Cushion mesenchymal (CM) cells migrate from the endocardium toward the myocardium using the cardiac jelly as substratum (Figure 6). As the CM cells populate the cardiac jelly the composition of the extracellular matrix is modified. Whereas the premigratory matrix is very rich in hyaluronate, the postmigratory matrix is richer in sulfated glycosaminoglycans.[41-43] The structural arrangement of the matrical elements is also modified. Concomitant with CM cell migration, the radial orientation of the cardiac jelly fibrils is lost (Figure 6) (see, however, References 42 and 43). After completion of cell migration, the cushions adopt the appearance of any other embryonic mesenchyme (Figures 5 and 6).

The mechanisms that trigger the migratory activity of the endocardial cells are unknown. It has been suggested that the changes in the composition and arrangement of the cardiac jelly glycosaminoglycans control cushion cell migration.[42,43] Furthermore, macromolecules secreted by the myocardium could guide the migration of these cells.[44] However, the exact role played by the cardiac jelly matrix is unclear as yet. For example, contact guidance and chemotaxis, proposed to explain the directionality of CM cell migration,[42,43] have not been proved adequately. On the contrary, recent *in vitro*[45] and *in vivo*[46] evidence suggests that the CM cells may be able to control their own migratory behavior. Once the migratory process has started, the directionality of CM cell migration may be enforced by a mechanism of contact inhibition of movement. Differences in adhesiveness between the cells and their substratum may push CM cells from areas of high cell density (near the endocardium) toward available cell-free space (near the myocardium).[46]

It is important to note that, at least in the truncal cushions, the origin of the CM cells appears to be twofold. A stream of migrating cells invades the cardiac jelly following a cranio-caudal direction. These cells come from the occipital neural crest,[47] and they appear to be confined largely to the truncus.[48] Although the quantitative importance of this process of migration is not yet clear, it has a profound morphogenetic significance. Ablation of the neural crest between somites 1 and 3 suppresses cell migration, resulting in anomalies of truncal septation.[49]

The interactions that the CM cells establish with the matrical components of the cardiac jelly appear to be of prime importance in the process of migration.[45] CM cells have large amounts of fibronectin associated with their surfaces.[46] This glycoprotein is actively synthesized by the cushion cells themselves, as demonstrated by *in situ* hybridization techniques.[50] It has been proposed that fibronectin could be involved in the adhesion of the CM cells to their substratum. To test this suggestion, minute amounts of fibronectin, antifibronectin antibodies, and peptide fragments of fibronectin have been injected into the cushions. After injection, the migration of the CM cells stops.[134] Although the effects of these injections are short lived, and the CM cells seem to be able to overcome the experimental manipulation, these results have important embryological and clinical implications.

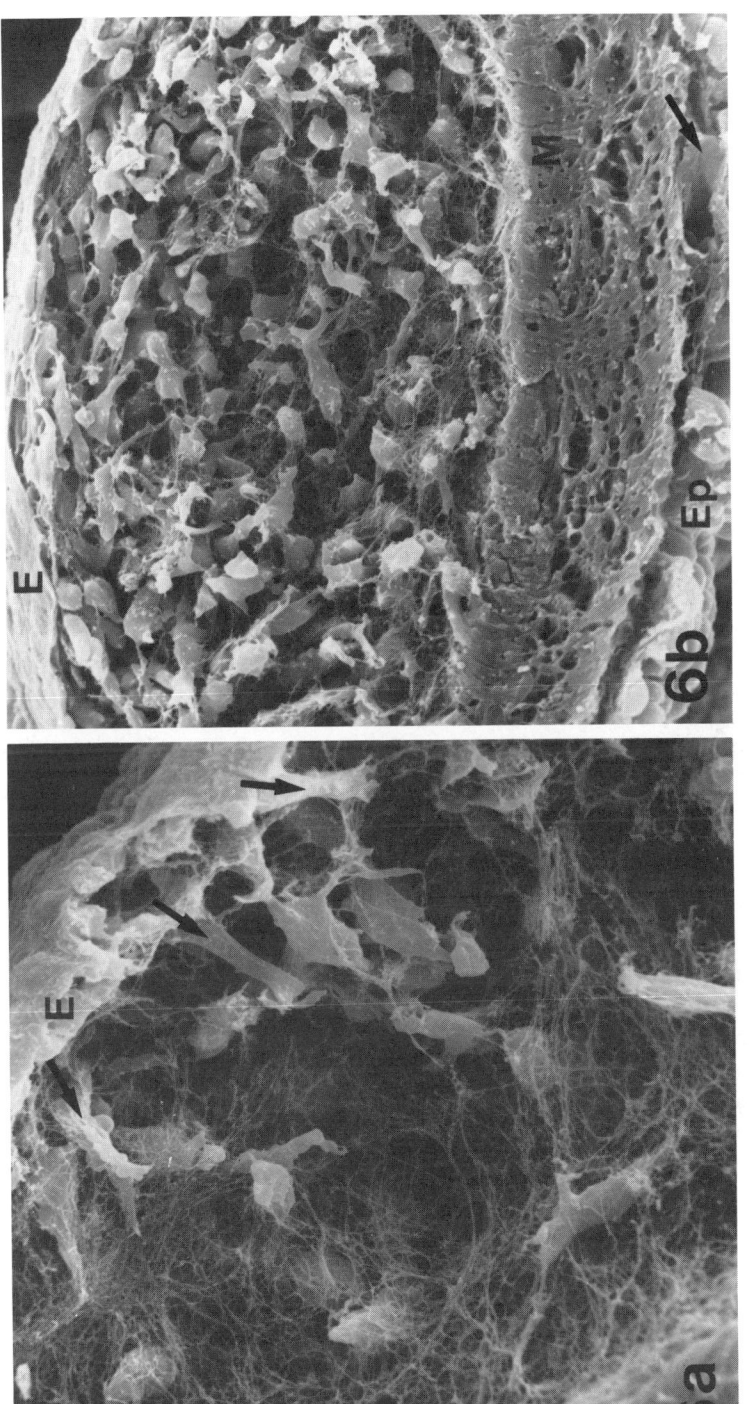

FIGURE 6. Scanning electron micrographs showing the development of the endocardial cushions: E, endocardium; Ep, epicardium; M, myocardium. (a) Several endocardial cells (arrows) are migrating into the cardiac jelly where they will be transformed into mesenchyme. Mesenchymal cells appear enmeshed in a three-dimensional network of matrical fibers. Cardiac jelly fibrils do not present a radial orientation. Note the condensation of the fibrillar material around the mesenchymal cells. (Magnification × 880.) (b) After completion of cell migration, the cushions adopt the appearance of any other embryonic mesenchyme. Note the intercellular spaces of the myocardium. A vascular channel (arrow) can be observed in the subepicardial space. (Magnification × 710.)

Persistent defects in cell-matrix interaction, whether they are due primarily to the cells or to the matrix, may lead to defects in CM cell migration. Therefore, defects in cell adhesion may result in cushion maldevelopment. Endocardial cushions are the forerunners of septal and valvular structures, and, needless to say, cushion defects will result in congenital heart disease. Infants with Down's syndrome very often have severe cardiac anomalies that are attributable to cushion maldevelopment. Heart fibroblasts from these patients present increased adhesiveness when compared to controls.[51] If the CM cells are more adhesive than normal, a smaller number of cells will populate the cushions. Hence, cushion development will be arrested and cardiac anomalies will result. This hypothesis has been tested in computer simulations employing different values of cell adhesiveness.[52] The success of these simulations allows the development of heart malformations to be predicted and opens up a new approach to understanding the complex interactions that occur during both normal and abnormal cardiogenesis.

Heart Septation

The heart septa arise independently in the different regions of the heart: atrium, ventricle, atrioventricular (AV) canal, and cono-truncus. With continued development the different septa meet on top of the interventricular septum. Then, the merging of all septa brings about the division of the single heart lumen into two independent circulatory channels.

Septation of hollow organs takes place by two basic mechanisms.[53] The first one consists of the formation of opposite masses of tissue (cushions) that grow toward each other and fuse. The second one consists of the rapid growth of one area together with the slower growth of the area adjacent. This results in formation of an eccentric communication that is later closed by proliferation of the neighboring tissue. The AV canal and the cono-truncus are septated by formation of endocardial cushions. The primitive atrium and ventricle are septated by the second mechanism.[2]

The division of the AV canal is carried out by formation of two cushions: the dorsal and ventral endocardial cushions. Initially the cushions project into the lumen, later growing toward each other and fusing.[54] Fusion of the dorsal and ventral endocardial cushions divides the lumen of the AV canal into two independent apertures: the right and left AV orifices (Figures 7 and 8).

Cushion fusion is still a poorly understood process. It does not appear to involve mechanisms of cell death or cell fusion. Endocardial cells from opposite cushions become adherent, form transient junctional structures, and break if traction is applied.[55] The possible importance of cell surface molecules[56] responsible for mechanisms of cell recognition and adhesion during cushion fusion remains to be explored. Concomitant with the process of fusion, the endocardial cells that line the cushions become loose and are transformed into mesenchyme.

The *interatrial septum* first appears as a thin myocardial partition arising from the dorsocephalic wall of the atrium. This partition, the septum primum, grows downward toward the AV canal and fuses with the dorsal and ventral endocardial cushions (Figures 7 and 8). Before this process of fusion is completed an opening is formed in the cranial aspect of the septum primum. This opening, the foramen secundum, permits blood communication between the two atria.

Several mechanisms appear to be involved in formation of the foramen secundum. The endocardium covering the septum primum invades the myocardial core of the septum, disrupting the tissue and forming small holes. The small holes coalesce to form a big aperture,[1] which is the foramen. Cell death is also observed in this area, but it appears to play only a minor role during foramen formation.[57] The suggestion that the foramen opens in response to increasing blood pressure appears to be incorrect, since the foramen opens in spite of the absence of hemodynamic pressure.[58]

The septation of mammalian atria is completed by formation of yet another septum, the

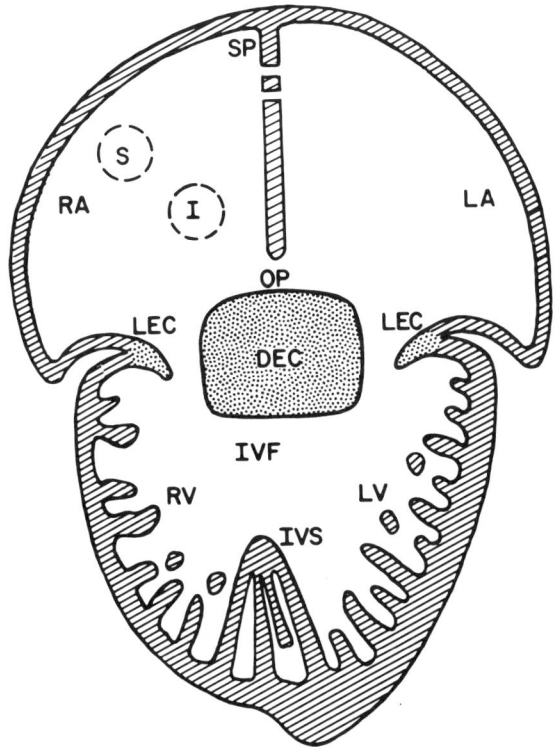

FIGURE 7. Diagram depicting a frontal section of the heart showing the formation of the interatrial (SP, septum primum) and interventricular (IVS) septa. The septum primum grows downward toward the AV canal. The foramen delimited between the free border of the septum and the endocardial cushions of the atrioventricular canal is the ostium primum (OP). The dorsal endocardial cushion (DEC) has not yet merged with the ventral cushion. The interventricular septum develops by coalescence of the primitve trabeculae. The openings of the inferior (I) and superior (S) cava veins into the right atrium are represented by broken circles: LA, left atrium; LEC, lateral endocardial cushion; LV, left ventricle; RA, right atrium; RV, right ventricle. (Modified from Icardo, J. M., in *Growth of the Heart in Health and Disease*, Zak, R., Ed., Raven Press, New York, 1984, 41. With permission.)

septum secundum, that develops to the right of the septum primum[57,59] (Figure 8). The septum secundum acts as a flap-like valve, permitting the shunt of blood from the right to the left atrium, but not in the opposite direction. Since no pulmonary circulation is established at this time, the interatrial communication assures the blood supply for the left side of the heart. Furthermore, it protects the lungs from massive blood affluence. Apposition of the septum primum and septum secundum after birth closes the interatrial shunt, resulting in formation of the fossa ovalis.

The *interventricular septum* forms at the apex of the ventricle by coalescence of the primitive trabeculae (Figures 7 and 8). The intertrabecular septum develops toward the right side of the AV canal. The opening located between the free margin of the septum and the AV canal is the interventricular foramen. This foramen permits the communication between the left ventricle and the cono-truncus, being later used to connect the aorta with the left ventricle.

The septation of the *conus and truncus* takes place by formation of two sets of cushions, the conal and truncal cushions, that develop along a spiraling course[4,60-62] (Figure 9). While the cono-truncal cushions are still developing, the region of the heart connecting the cono-truncus with the aortic arches becomes a distinct region: the truncoaortic sac. Division of the truncoaortic sac is accomplished by formation of a spur of mesenchymal tissue, the

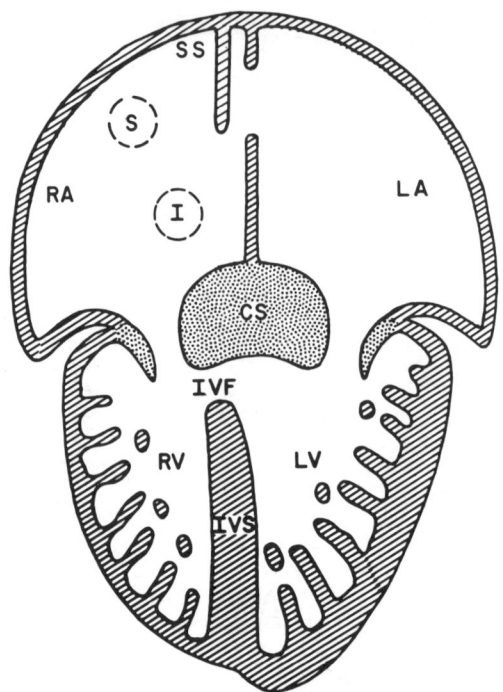

FIGURE 8. Diagram depicting a frontal section of the heart at a later stage than that shown in Figure 7. The dorsal and ventral endocardial cushions have already merged with each other, forming the cushion septum (CS). The septum primum has merged with the dorsal and ventral endocardial cushions and the ostium primum has been closed. A new foramen, the ostium secundum, has formed in the dorsal aspect of the septum primum. Another septum, the septum secundum (SS), is developing to the right of the septum primum. The interventricular septum (IVS) has grown toward the right side of the atrioventricular canal. The interventricular foramen (IVF) communicates the two ventricular chambers. The openings of the inferior (I) and superior (S) cava veins into the right atrium are represented by broken circles: LV, left ventricle; RV, right ventricle. (Modified from Icardo, J. M., in *Growth of the Heart in Health and Disease,* Zak, R., Ed., Raven Press, New York, 1984, 41. With permission.)

aortopulmonary septum, which arises between aortic arches IV and VI.[63] This septum is a horseshoe-shaped structure, anchored into the truncus by two limbs of densely arranged mesenchymal cells. Formation of the aortopulmonary septum and fusion of the truncal and conal ridges divides the whole outflow tract area into two separate (aortic and pulmonic) channels. The spiraling trajectory of the septum accounts for the spiraling trajectory of the aorta and pulmonary artery.

Concomitant with this process of septation, the distal part of the truncus is transformed into the proximal part of the aorta and the pulmonary artery.[64] The mechanisms of myocardial regression in the truncus are not yet clear. Several hypotheses, such as myocardial cell death,[27] myocardial dedifferentiation,[65] and translocation of the entire septation complex toward the ventricle,[66] have been proposed.

It is interesting to note that the mesenchymal cells involved in forming the aortopulmonary septum, and in the formation of the tunica media of the aorta and pulmonary artery, are intensely fluorescent for fibronectin.[67] Fibronectin appears to be a selective marker for these mesenchymal cells, permitting a clear visualization of the different processes of cell rearrangement which occur during truncal morphogenesis. These cells are also positive for desmin,[68] for elastin precursors,[69] and for F-actin.[70] It is worth noting that the staining for fibronectin disappears (or becomes very reduced) after completion of truncus septation. The staining for elastin precursors and for desmin also disappears in the aortopulmonary septum

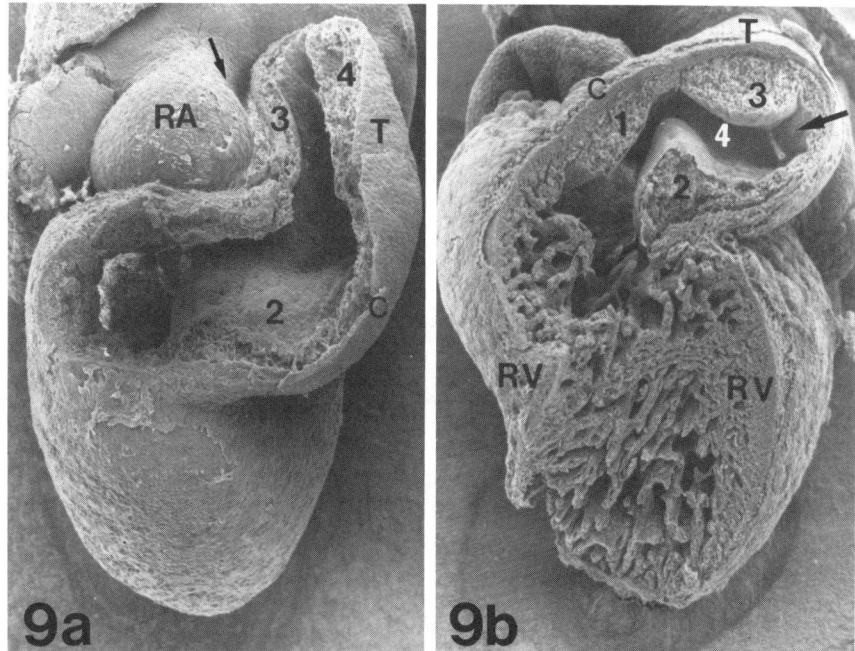

FIGURE 9. Scanning electron micrographs depicting the development of the truncal and conal cushions: 1, dextrodorsal conal ridge; 2, sinistroventral conal ridge; 3, sinistroventral truncal ridge; 4, dextrodorsal truncal ridge; C, conus; RA, right atrium; RV, right ventricle; T, truncus. The right wall of the conotruncus and part of the right ventricular wall have been dissected away (chicken embryos, left segment of the specimen, seen from the right). (a) Fourth day of incubation. The truncal cushions are more developed than the conal cushions. The dextrodorsal conal cushion has been removed. Arrow indicates the interatrial sulcus. (Magnification × 66.) (b) Fifth day of incubation. The four cono-truncal cushions are well developed. Note their spiraling trajectory. Arrow indicates one of the intercalated valve swellings that contributes to the formation of the semilunar valves. (Magnification × 55.) (From Icardo, J. M., in *Growth of the Heart in Health and Disease,* Zak, R., Ed., Raven Press, New York, 1984, 41. With permission.)

after completion of septation. The striking reduction in fibronectin staining has been related to the acquisition by the mesenchymal cells of mature phenotypes (fibroblast, smooth muscle). It should be stressed that the expression of fibronectin appears to be switched on and off intermittently during heart development. These changes in the expression of fibronectin can be correlated precisely to a number of distinct developmental events.[46,67,71]

After septation of the trunco-conus, the divided conus lies on top of the interventricular septum. Then, the growth and merging of the surrounding mesenchymal tissue isolates the pulmonary from the aortic area, forming the pulmonary and the aortic infundibulum. This area of growing mesenchymal tissue constitutes the origin of the membranous portion of the adult interventricular septum (see, for more details, References 4, 60, and 72).

Heart Valves

The *semilunar valves* develop at the trunco-conal junction from three pairs of small tubercles of cushion tissue.[73] During the process of truncus septation, each channel (aortic and pulmonic) receives one of each pair of the valve swellings. Then, the valve swellings are excavated in the proximal direction and adopt a pocket-like form. The space located between the arterial face of the valve leaflets and the arterial wall constitutes the primordium of the Valsalva sinus. Close interactions of both the mesenchyme and the endocardium with the extracellular matrix of the valve anlage appear to be involved in the process of valve excavation.[74,75] Developmental maturation of the semilunar valve tissue continues until advanced postnatal periods.[76]

The *AV valves* have a dual origin. They arise in part from the mesenchymal tissue that surrounds the AV orifices and in part from a sleeve of myocardium that remains attached to the ventricular muscle by a few trabeculae.[72] The contribution of different areas of cushion tissue to valve formation (see, for example, Reference 77) may help to explain the origin of some congenital valve anomalies. With development, the myocardial sleeve is replaced by connective tissue and the trabeculae are transformed into the chordae tendineae of the adult valves. The endocardium appears to be actively involved in the early formation of the chordae tendineae.[78]

Adult AV valve leaflets present a core of connective tissue with some muscular, vascular, and neural elements. Interweaving of muscle cells with the connective core may increase the physical resistance of the valve leaflets, preventing excessive ballooning toward the atria during ventricular systole.[79]

THE EPICARDIUM. THE SUBEPICARDIAL TISSUE

Concomitant with all the shape changes that we have just described, new cell types are being added to the developing heart. Addition of these cell types increases the complexity of cellular interactions and makes it more difficult to analyze the development of the embryonic heart.

Formation of the epicardium (the prospective visceral pericardium) begins with migration over the dorsal surface of the heart of cells that originate in the region of the sinus venosus.[80-82] The epicardial cells migrate upon the outer surface of the myocardium, forming a cohesive epithelial sheet (Figures 4 and 5). Epicardial cells along the free edges of the advancing epicardium extend filopodial and lamellipodial projections that make contact with the naked myocardial surface (Figure 10). Progressively, the entire myocardial surface is invested with the epicardium (Figure 10). At the time of epicardial migration the apical surface of the myocardium presents a thin layer of fibronectin that can be demonstrated by immunofluorescent staining.[46] Epicardial cells, unlike cushion mesenchymal cells, do not have fibronectin associated with their surfaces. However, they may rely on the myocardial fibronectin to assure the process of migration.

After completion of epicardial migration, a thin layer of extracellular material develops at the epicardial-myocardial interface.[80] This space soon enlarges and is invaded by a population of mesenchymal cells (Figure 11). Subepicardial cells originate most probably by delamination of the epicardium, being subsequently transformed into mesenchyme. However, it has also been suggested that these cells may arise from the region of the septum transversum.[82] Subepicardial cells are enmeshed in a three-dimensional network of matrical fibers (Figure 11). The exact composition of this extracellular material is still unknown, but collagen fibrils and filamentous and amorphous material can be demonstrated by transmission and scanning electron microscopy (Figure 11). Mesenchymal cells in the subepicardium are spindle-shaped or stellated, frequently presenting long prolongations that establish contact with other subepicardial cells, with the myocardium and with the extracellular material (Figure 11).

THE CORONARY VESSELS

The development of the coronary vascular bed is still a matter of discussion. It appears to start in the subepicardial space. Groups of mesenchymal cells organize to form vessels of small caliber (Figure 11) that interconnect with each other and form an extense subepicardial plexus. Angiogenic cords[33,83] and free erythrocytes can be observed in the subepi-

Part B: Cardiovascular and Respiratory Development 39

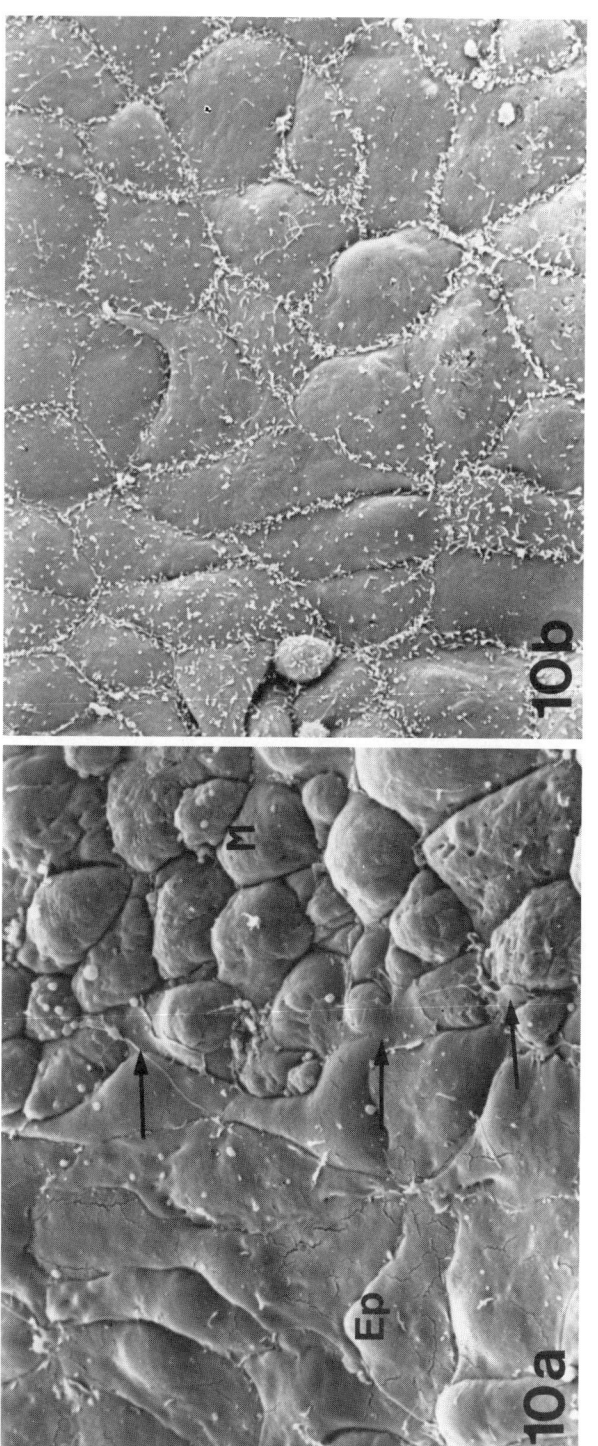

FIGURE 10. Scanning electron micrographs depicting the formation of the epicardial layer: Ep, epicardium; M, myocardium. (a) The apical surface of the myocardium is being covered by a sheet of flattened cells. Compare the flattened appearance of the epicardial cells to the rounded, bulging appearance of the myocytes. Epicardial cells located at the free edge of the advancing epicardium extend lamellipodial (arrows) projections over the naked myocardial surface. (Magnification × 1800.) (b) The entire myocardial surface has been covered by the epicardium. Epicardial cells are flattened and present numerous microvilli at the cell periphery. (Magnification × 1860.)

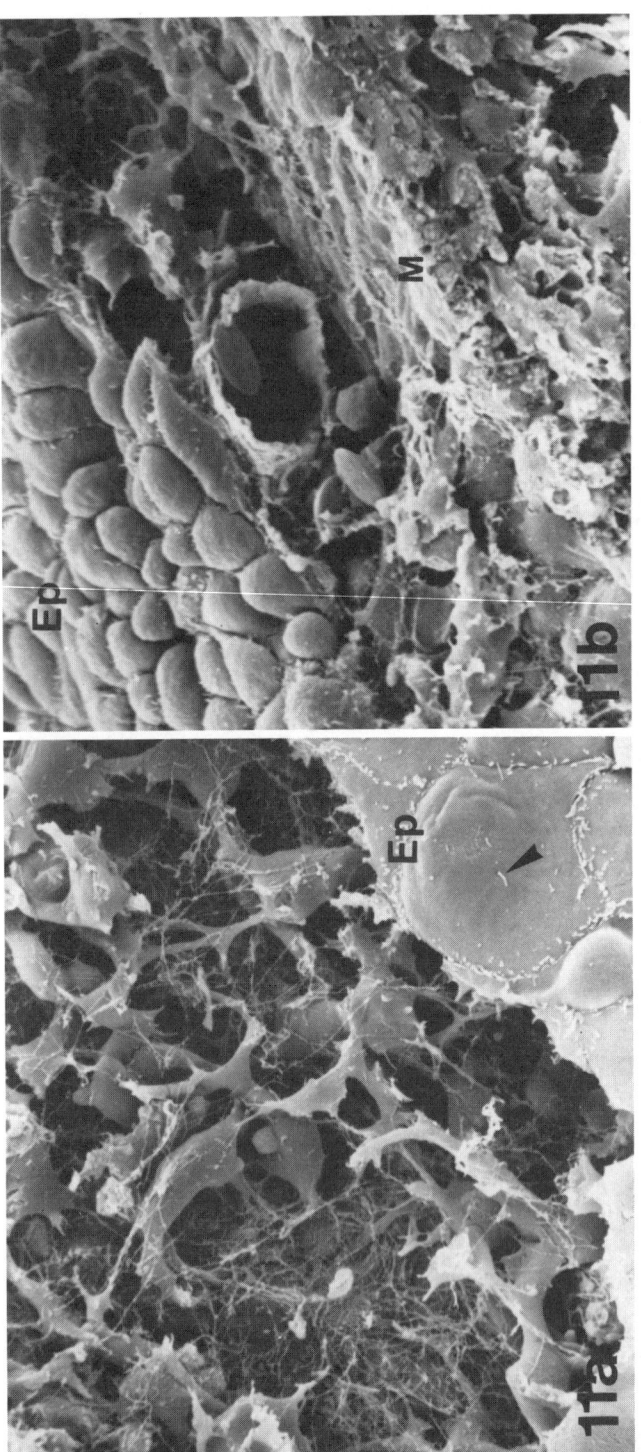

FIGURE 11. Scanning electron micrographs depicting several aspects of the development of the subepicardial space: Ep, epicardium; M, myocardium. (a) The epicardium has partially been removed. The subepicardial space appears occupied by mesenchymatous cells. These cells are spindle-shaped or stellated and appear enmeshed in a three-dimensional network of matrical fibers. No orientation of the matrical fibrils can be discerned. The apical surface of the myocardium appears beneath the subepicardial tissue. An epicardial cell shows a solitary cilium (arrowhead). (Magnification × 1975.) (b) This picture illustrates the aspect of the subepicardial space at a later stage than that shown in (a). The subepicardial cells appear densely arranged. The formation of the subepicardial vascular plexus has already started: a cross section of a vascular channel can be observed in the center of the picture. (Magnification × 1800.)

cardial space. It has been suggested that free blood cells may have an inductive role by generating vessels around them.[33] The subepicardial plexus develops vascular sprouts that invade the myocardium. Some of these sprouts connect directly with the intertrabecular spaces.

The origin of the myocardial capillaries is still under discussion. They are considered to be formed by sprouting from the subepicardial plexus,[83,84] directly from the endocardium,[85] or *in situ*, by formation of angiogenic cords.[86] It is clear, however, that both systems, the intertrabecular sinusoids and the capillaries, coexist in the myocardium for some time. It is only later in development, when the capillary bed is well developed, that the myocardial trabeculae coalesce and most of the intertrabecular spaces disappear.[35] Myocardial vessels are surrounded by pericytes and smooth muscle cells. These cells probably come from the subepicardial mesenchyme, migrating into the myocardium as individual cells.[86] The main trunks of the coronary arteries appear to develop as multiple offshoots from the aortic[87] and pulmonic[88] Valsalva sinuses. The vascular offshoots connect with the subepicardial plexus. Blood flow through these nascent vascular channels has been implicated in the establishment of the definitive arterial trunks.[88] The multiple origin of the coronary arteries, and their connection with the subepicardial plexus, can explain most of the anatomical variations observed in the origin and course of the main coronary vessels. While the myocardial capillaries are still developing, the cardiac veins arise by endothelial sprouting from the region of the sinus venosus. Cardiac veins connect with the subepicardial plexus and with the myocardial capillary bed.

Myocardial capillaries are the substitute for a phylogenetic system, the sinusoids, that is clearly unable to provide for the metabolic needs of myocytes in higher vertebrates. It should be emphasized that proliferation of heart capillaries is very intense at early postnatal periods.[89] Furthermore, diverse stimuli can induce capillary growth in adult hearts.[90] The mechanisms that regulate the development of the coronary vascular bed are still unknown. A recent paper[91] has shown that injection of protamine (an inhibitor of angiogenesis) into young rats significantly decreases capillarization of the myocardium. This suggests that capillary growth in the heart may be governed by the same angiogenic factors that govern the proliferation of capillaries in other organs.

The development of the lymphatic capillaries of the heart is still controversial. Budding lymphatic trunks have been shown extending toward the base of the heart and spreading outwards from there.[92] However, it is still unclear whether or not formation of the entire capillary bed occurs by invasive proliferation.[93] Some of the small vessels situated in the subepicardium may develop later into lymphatic capillaries. The adult anatomy of the lymphatic system, and the circulation of the lymph flow, have been extensively reviewed.[94]

DEVELOPMENTAL ROLE OF HEMODYNAMICS

Classically, blood flow has been considered a prime factor in heart shaping and development. The presence of two independent bloodstreams, spiraling around each other,[37] was suggested to determine not only the site for development of the heart septa, but also to be involved in the patterning of vessels derived from the aortic arches. The fact that heart malformations can be produced by diverse manipulations that are assumed to alter blood flow has been taken as confirmation of the theories cited above. However, the concept of hemodynamic molding is still under discussion. For example, blood flow does not appear to be necessary for heart looping.[95] Furthermore, it is actually unclear whether or not the streaming blood follows the patterns classically described.[96] Mechanical interference with cardiogenesis can influence other parameters in addition to the blood flow. For example, constriction of the cono-truncus may stop the migration into the heart of the cells derived from the neural crest, resulting by itself in cardiac malformations. Pharmacologic agents may affect primarily the vascular bed,[97] resulting only secondarily in heart maldevelopment. Also, selection of the vessels derived from the aortic arches probably involves more than simple blood flow. In none of these experiments have the exact modifications of the blood flow been monitored. We do not mean to say that cardiac anomalies could not be produced by the direct effects of turbulent flow; however, it is very difficult to separate these effects from those derived from the altered hemodynamic load.[98]

It seems probable that the questions which must be answered are those related to the ways in which the heart tissue responds to blood flow. Concomitant with the development of the intra- and extraembryonic vasculature, the volume of circulating blood increases considerably. The stroke volume, the heart rate, and the cardiac output also increase.[99,100] The heart responds to increasing functional demands by adjusting the muscle mass to its hemodynamic load.[24] Thus, myocyte hypertrophy appears to be a first effect of the increasing work load. This fact is well documented in adult hearts.[17]

Cardiac myocytes not only change size in response to functional demands; the expression of contractile proteins,[101] together with the speed of contraction and the muscle efficiency,[102,103] change as well. The electrical properties of the sarcolemma also undergo changes.[104] Research into the nature of the regulatory mechanisms that transform physical forces into biological responses is just beginning; we do not yet know exactly how all these different parameters of the embryonic heart are influenced by the changing work load. Nevertheless, it appears that the response of the embryonic heart to an increased work load is similar to that of the mature heart. Preliminary reports indicate that, when the heart afterload is increased, the growth of the embryonic ventricles is accelerated.[105]

The endocardium is in direct contact with the streaming blood. Hence, it should be the first structure to be influenced by hemodynamic changes. However, the biological responses of the embryonic endocardium to the changing blood flow have been largely unexplored. One of the first parameters that could be explored is whether or not endocardial cells are orientated in the direction of the blood flow, as are normal endothelial cells.[106] Endocardial cells in the portion of the atria that funnels into the AV canal appear massively aligned in the direction of the blood flow.[107] Cell alignment with the blood flow occurs also in the ventricular face of the semilunar valve leaflets.[71,108] It has been suggested that shear stress is responsible for the orientation of the endocardial cells along the flow paths. To test whether or not changes in blood flow can modify endocardial cell orientation, the left vitellin vessels of chicken embryos have been severed during the third day of incubation. This procedure increases the amount of blood directed toward the right atrium. Under these conditions, the endocardial cells located in the vicinity of the opening of the sinus venosus change shape from polygonal to elongated and align themselves in the direction of the increased blood flow.[107] These experiments strongly suggest that endocardial cells are able to respond to modifications of the hemodynamic forces. However, the response of the endocardium under normal flow conditions is difficult to evaluate. Furthermore, the mechanisms by which these stimuli are translated into regulators of shape are still speculative.

The cardiac extracellular matrix undergoes drastic modifications concomitant with the increase in cardiac workload. For example, the amount of collagen increases considerably in late fetal periods. Collagen fibrils organize progressively into a network of fine bundles (struts) that course from myoctye to myocyte and from myocyte to capillaries. Development of this complex fibrillar network is completed in the first postnatal days[109] when both the blood pressure and the cardiac output experience a significant increase.[110] This collagen network appears to maintain the capillaries open during systolic contraction, ensuring uninterrupted blood flow.[111] It is difficult to determine whether or not the changes in the organization of the extracellular material (for example, an increase in the synthesis of collagen, and the orientation of the collagen fibrils in the direction of lines of stress) are a direct effect of the increased workload. It is clear, however, that these changes represent modifications of the extracellular matrix that maintain the shape and integrity of the heart wall under increased luminal pressure.[109]

THE CONDUCTION SYSTEM. THE INNERVATION OF THE HEART

The conduction system of the heart is made up of specialized myocardial cells that present morphologic and electric characteristics different from those of the primarily working cells.[112] These cells group to form the sinoatrial node, the internodal tracts, the AV node, and the His-Purkinje system. The location and the histologic characteristics of these structures have been extensively studied.[112-114]

From an embryologic standpoint it is difficult to determine when the conduction system develops. Adult-like electrocardiographic recordings can be obtained from looping hearts[115] well before an anatomic conduction system is established. Furthermore, development of the conduction system is not completed until after birth.[116] The mode of propagation of the electrical impulses in the absence of specialized tissue is still uncertain.[117] Several techniques[116,118,119] have been used to demonstrate the emergence of the conducting cells under the light microscope. With the transmission electron microscope, cells of the conduction system present a paler cytoplasm and a smaller number of myofibrils than the working myocytes.[120] The paucity of myofibrils in these cells has been related to the fact that they are surrounded by connective tissue that may protect them from the normal working stress.[121] Morphologic[120] and immunohistochemical[122] evidence suggests that the cells of the conduction system develop from the bulk of the cardiac myocytes, following a specialized differentiation path.

The heart is innervated by both adrenergic and cholinergic nerves. Adrenergic nerve trunks start growing along the coronary arteries involving, at term, most of the myocardial structures.[123] The vagus nerve supplies the cholinergic innervation. Most of the cardiac ganglion cells (parasympathetic, postganglionic) come from the occipital neural crest.[48] The autonomic system establishes a balanced influence upon the pacemaking cells. This balance, however, is not attained until the postnatal period, when the innervation is completed.[89]

CONCLUSIONS

In these two chapters we have reviewed the developmental anatomy of the embryonic heart. Little attention has been focused on gross anatomic changes. Instead, in an attempt to understand the complex biological interrelations that result in heart shape, mechanisms have been emphasized whenever possible. Certainly, the discussion of the possible developmental mechanisms has not been exhausted. Many questions have been raised in areas where our knowledge is most imperfect, especially at the cellular and subcellular levels. Perhaps the main question we have to ask now is how the expression of the genomic information is regulated through development. On the contrary, we could ask whether or not there is a gene (or gene cluster) directly responsible for heart shape.

The discovery of the "homeo box" gene, responsible for segmentation in *Drosophila*,[124] appears to indicate that morphogenesis is a more or less direct transcript of the genomic information. Furthermore, gene sequences similar to the homeo box gene have been shown to occur in several mammals, including man.[125] However, the possible organizing role of these sequences in humans remains to be clarified.

In a few cases, single gene defects have been held responsible for cardiac anomalies,[126] and we can assume that the normal gene sequence regulates normal morphogenesis. In most cases, however, chromosome aberrations result in complex syndromes in which congenital heart disease is just a part of the syndrome. The presence of cardiac anomalies in these cases appears to be due to multifactorial inheritance.[127] We have seen above how a failure of mesenchymal cells to adhere to their substratum may be amplified within the developmental "cascade", so as to result in congenital heart disease.

In many cases the chromosomal change precedes the developmental change. In other cases, however, the chromosomal change may result from altered development.[128] The genome is very sensitive to the influence of factors that operate both from the inside and from the outside of the cells. The expression of "heat-shock" proteins[129] and the changes in the expression of myosin isoforms in response to changes in workload[101] are clear examples of external factors modifying the expression of the genome. Changes in the interaction between the cells and their substratum, changes in cell shape, physical tension applied upon the cell surfaces, etc., may act as initiators of structural changes that may be sensed by intracellular organelles (i.e., cytoskeleton) via transmembrane components.[130] For example, regenerating aortic endothelial cells undergo structural modifications that may be translated into biochemical changes.[131] In isolated myocytes, the force of contraction appears to be a direct stimulus for myosin synthesis and cell growth.[132] The way in which this information is translated into biochemical processes remains unidentified as yet. However, the cytoplasm appears to play a key role as a signal transducer. In fused heterokaryons, nonmuscle nuclei can be forced to synthesize muscle proteins by muscle cytoplasm.[133] Thus, although in a still undefined way, nuclear gene expression appears to be subjected to modulation by the cytoplasm. The nature of the stimuli responsible for structural modifications at the cell and tissue level and the ways in which these structural modifications are translated into gene regulators are the questions we shall have to answer in the near future.

ACKNOWLEDGMENTS

The authors are indebted to Dr. A. Nakamura for his invaluable expertise with the microinjection techniques cited in this paper. Thanks are also due to Dr. K. Hoshino, who made available the embryos of the Kyoto Collection, and to Dr. F. J. Manasek, who photographed the specimens. Experimental work reported in this chapter was supported by a grant from the Spanish-American Committee for the Scientific and Technological Cooperation (CCB-8402-010) and Comisión Asesora de Investigación Cientifica y Técnica (C.A.Y.C.I.T.) Grant No. 2985/83.

REFERENCES

1. **Patten, B. M.,** The development of the heart, in *Pathology of the Heart and Blood Vessels,* Gould, S. E., Ed., Charles C Thomas, Springfield, IL, 1968, 20.
2. **Van Mierop, L. H. S. and Netter, F. H.,** Embryology, in *CIBA Journal,* Vol. 5, CIBA Pharmaceutical, Summit, NJ, 1969, 112.
3. **Steding, G. and Seidl, E.,** Contribution to the development of the heart. I. Normal development, *Thorac. Cardiovasc. Surg.,* 28, 386, 1980.
4. **Icardo, J. M.,** The growing heart: an anatomical perspective, in *Growth of the Heart in Health and Disease,* Zak, R., Ed., Raven Press, New York, 1984, 41.
5. **Manasek, F. J., Icardo, J. M., Nakamura, A., and Sweeney, L. J.,** Cardiogenesis: developmental mechanisms and embryology, in *The Heart and Cardiovascular System,* Fozzard, H. A., Haber, E., Jennings, R. B., Katz, A. M., and Morgan, H. E., Eds., Raven Press, New York, 1986, 965.
6. **Goertler, K.,** Die Stoffwechseltopographie des embryonalen Huhnerherzens und hire Bedeutung fur die Entstehung angeborener Herzfehler, *Verh. Dtsch. Ges. Pathol.,* 40, 181, 1957.
7. **Rychter, Z. and Rychterova, V.,** Angio- and myoarchitecture of the heart wall under normal and experimentally changed morphogenesis, in *Mechanisms of Cardiac Morphogenesis and Teratogenesis,* Pexieder, T., Ed., Raven Press, New York, 1981, 431.
8. **Thompson, R. P. and Fitzharris, T. P.,** Morphogenesis of the truncus arteriosus of the chick embryo heart: the formation and migration of mesenchymal tissue, *Am. J. Anat.,* 154, 545, 1979.

9. **Paschoud, N. and Pexieder, T.**, Patterns of proliferation during the organogenetic phase of heart development, in *Mechanisms of Cardiac Morphogenesis and Teratogenesis*, Pexieder, T., Ed., Raven Press, New York, 1981, 431.
10. **Legato, M. J.**, Ultrastructural changes during normal growth in the dog and rat ventricular myofiber, in *Developmental and Physiological Correlates of Cardiac Muscle*, Lieberman, M. and Sano, T., Eds., Raven Press, New York, 1975, 249.
11. **Zak, R.**, Cell proliferation during cardiac growth, *Am. J. Cardiol.*, 31, 211, 1973.
12. **Schmid, G. and Pfitzer, P.**, Mitoses and binucleated cells in perinatal human hearts, *Virchows Arch. B*, 48, 59, 1985.
13. **Rumyantsev, P. P.**, DNA synthesis and nuclear division in embryonal and postnatal histogenesis of myocardium (autoradiographic study), *Fed. Proc. Fed. Am. Soc. Exp. Biol.*, 24, T899, 1965.
14. **Bugaisky, L. and Zak, R.**, Cellular growth of cardiac muscle after birth, *Tex. Rep. Biol. Med.*, 39, 123, 1979.
15. **Nag, A. C., Carey, T. R., and Cheng, M.**, DNA synthesis in rat heart cells after injury and the regeneration of myocardia, *Tissue Cell*, 15, 597, 1983.
16. **Rumyantsev, P. P.**, Interrelations of the proliferation and differentiation processes during cardiac myogenesis and regeneration, *Int. Rev. Cytol.*, 51, 187, 1977.
17. **Ferrans, V. J.**, Cardiac hypertrophy: morphological aspects, in *Growth of the Heart in Health and Disease*, Zak, R., Ed., Raven Press, New York, 1984, 187.
18. **Oberpriller, J. O., Ferrans, V. J., and Carroll, R. J.**, Changes in DNA content, number of nuclei and cellular dimensions of young rat atrial myocytes in response to left coronary artery ligation, *J. Mol. Cell. Cardiol.*, 15, 31, 1983.
19. **Clubb, F. J., Jr. and Bishop, S. P.**, Formation of binucleated cells in the neonatal heart. An index for growth hypertrophy, *Lab. Invest.*, 50, 571, 1984.
20. **Zak, R.**, Development and proliferative capacity of cardiac muscle cells, *Circ. Res.*, 34-35 (Suppl. II), 17, 1974.
21. **Katzberg, A., Farmer, B., and Harris, B.**, The predominance of binucleation in isolated rat heart myocytes, *Am. J. Anat.*, 149, 489, 1977.
22. **Quiring, D. P.**, The development of the sino-atrial region of the chick heart, *J. Morphol.*, 55, 81, 1933.
23. **Romanoff, A. L.**, *The Avian Embryo*, Macmillan, New York, 1960.
24. **Zak, R.**, Factors controlling growth, in *Growth of the Heart in Health and Disease*, Zak, R., Ed., Raven Press, New York, 1984, 165.
25. **Glucksmann, A.**, Cell death in normal development, *Arch. Biol. (Liege)*, 76, 419, 1965.
26. **Pexieder, T.**, Cell death in the morphogenesis and teratogenesis of the heart, *Adv. Anat. Embryol. Cell Biol.*, 51, 1, 1975.
27. **Hurle, J. M., Lafarga, M., and Ojeda, J. L.**, Cytological and cytochemical studies of the necrotic area of the bulbus of the chick embryo heart: phagocytosis by developing myocardial cells, *J. Embryol. Exp. Morphol.*, 41, 161, 1977.
28. **Okamoto, N., Akimoto, N., Satow, Y., Hidaka, N., and Miyabara, S.**, Role of cell death in conal ridges of developing human heart, in *Mechanisms of Cardiac Morphogenesis and Teratogenesis*, Pexieder, T., Ed., Raven Press, New York, 1981. 127.
29. **Sissman, N. J.**, Developmental landmarks in cardiac morphogenesis: comparative chronology, *Am. J. Cardiol.*, 25, 141, 1970.
30. **Kurkiewicz, T. O.**, O histogenezie miesnia sarcowego zwierzat kregowych, *Bull. Acad. Sci. Cracovie*, 148, 1909.
31. **Icardo, J. M. and Fernandez-Teran, M. A.**, Morphologic study of ventricular trabeculation in the embryonic chick heart, *Acta Anat.*, 130, 264, 1987.
32. **Orkin, R. W. and Toole, B. P.**, Hyaluronidase activity and hyaluronate content of the developing chick embryo heart, *Dev. Biol.*, 66, 308, 1978.
33. **Nakamura, A.**, Cardiac hyaluronidase activity of chick embryos at the time of endocardial cushion formation, *J. Mol. Cell. Cardiol.*, 12, 1239, 1980.
34. **Tokuyasu, K. T.**, Development of myocardial circulation, in *Cardiac Morphogenesis*, Ferrans, V. J., Rosenquist, G., and Weinstein, C., Eds., Elsevier, New York, 1985, 226.
35. **Ben-Shachar, G., Arcilla, R. A., Lucas, R. V., and Manasek, F. J.**, Ventricular trabeculations in the chick embryo heart and their contribution to ventricular and muscular septal development, *Circ. Res.*, 57, 759, 1985.
36. **Patten, B. M.**, The development of the ventricular wall and its blood supply, in *Cardiac Development with Special Reference to Congenital Heart Disease*, Jaffee, O. C., Ed., University of Dayton Press, Dayton, OH, 1970, 1.
37. **Minot, C. S.**, On a hitherto unrecognized form of blood circulation without capillaries in the organs of vertebrata, *Proc. Boston Soc. Nat. Hist.*, 29, 185, 1900.
38. **Barthel, H.**, *Missbildungen des Menschieblen Herzens*, Georg Thieme Verlag, Stuttgart, 1960.

39. **Markwald, R. R., Fitzharris, T. P., and Manasek, F. J.,** Structural development of endocardial cushions, *Am. J. Anat.,* 148, 85, 1977.
40. **Kinsella, M. G. and Fitzharris, T. P.,** Origin of cushion tissue in the developing chick heart: cinematographic recordings of in situ formation, *Science,* 207, 1359, 1980.
41. **Markwald, R. R., Fitzharris, T. P., Bank, H., and Bernanke, D. H.,** Structural analysis on the matrical organization of glycosaminoglycans in developing endocardial cushions, *Dev. Biol.,* 62, 292, 1978.
42. **Markwald, R. R., Fitzharris, T. P., Bolender, D. L., and Bernanke, D. H.,** Structural analysis of cell:matrix interaction during morphogenesis of atrioventricular cushion tissue, *Dev. Biol.,* 69, 634, 1979.
43. **Markwald, R. R., Funderberg, F. M., and Bernanke, D. H.,** Glycosaminoglycans: potential determinants in cardiac morphogenesis, *Tex. Rep. Biol. Med.,* 39, 253, 1979.
44. **Krug, E. L., Runyan, R. B., and Markwald, R. R.,** Protein extracts from early embryonic hearts initiate cardiac endothelial cytodifferentiation, *Dev. Biol.,* 112, 414, 1985.
45. **Fitzharris, T. P. and Markwald, R. R.,** Cellular migration through the cardiac jelly matrix: a stereoanalysis by high-voltage electron microscopy, *Dev. Biol.,* 92, 315, 1982.
46. **Icardo, J. M. and Manasek, F. J.,** An indirect immunofluorescence study of the distribution of fibronectin during the formation of the cushion tissue mesenchyme in the embryonic heart, *Dev. Biol.,* 101, 336, 1984.
47. **Kirby, M. L., Gale, T. F., and Stewart, D.,** Neural crest cells contribute to aorticopulmonary septation, *Science,* 220, 1059, 1983.
48. **Taylor, I. M., Stewart, P. A., and Jeffery, T. J.,** The neural crest contribution to aortico-pulmonary septation, *Anat. Rec.,* 211, 193A, 1985.
49. **Kirby, M. L., Aronstam, R. S., and Buccafusco, J. J.,** Changes in cholinergic parameters associated with failure of conotruncal septation in embryonic chick hearts, *Circ. Res.,* 56, 392, 1985.
50. **French-Constant, C. and Hynes, R. O.,** Patterns of fibronectin gene expression and splicing during cell migration in chicken embryos, *Development,* 104, 369, 1988.
51. **Wright, T. C., Destrempes, M., Orkin, R., and Kurnitt, D. M.,** Increased adhesiveness of Down syndrome fetal fibroblasts in vitro, *Proc. Natl. Acad. Sci. U.S.A.,* 81, 2426, 1984.
52. **Kurnitt, D. M., Aldridge, J. F., Matsuoka, R., and Matthysse, S.,** Increased adhesiveness of trisomy 21 cells and atrioventricular canal malformations in Down's syndrome: a stochastic model, *Am. J. Med. Genet.,* 20, 385, 1985.
53. **Langman, J. and Van Mierop, L. H. S.,** Development of the cardiovascular system, in *Heart Disease in Infants, Children and Adolescents,* Moss, A. J. and Adams, F. H., Eds., Williams & Wilkins, Baltimore, 1968, 3.
54. **Hay, D. A. and Low, F. N.,** The fusion of dorsal and ventral endocardial cushions in the embryonic chick heart: a study in fine structure, *Am. J. Anat.,* 133, 1, 1972.
55. **Hay, D. A.,** Development and fusion of the endocardial cushions, in *Morphogenesis and Malformations of the Cardiovascular System,* Rosenquist, G. C. and Bergsma, D., Eds., Alan R. Liss, New York, 1978, 69.
56. **Edelman, G. M.,** Cell adhesion molecules, *Science,* 219, 450, 1983.
57. **Morse, D. E., Rogers, C. S., and McCann, P. S.,** Atrial septation in the chick and rat: a review, *J. Submicrosc. Cytol.,* 16, 259, 1984.
58. **Hendrix, M. J. C., Brailey, J. L., Brown, B. J., and Sorrentino, J. M.,** The development of the atrial septum in absence of hemodynamic pressure, *Anat. Rec.,* 205, 80a, 1983.
59. **Igarashi, H.,** Scanning electron microscopic study on the formation of the atrial septum in rat embryos, *Acta Anat. Nippon.,* 59, 28, 1984.
60. **De Vries, P. A. and Saunders, J. B. de C. M.,** Development of the ventricles and spiral outflow tract in the human heart, *Carnegie Inst. Washington Contrib. Embryol.,* 37, 87, 1962.
61. **Van Mierop, L. H. S., Alley, R. D., Kausel, H. W., and Stranahan, A.,** The anatomy and embryology of endocardial cushion defects, *J. Thorac. Cardiovasc. Surg.,* 43, 71, 1962.
62. **Pexieder, T.,** Development of the outflow tract of the embryonic heart, in *Morphogenesis and Malformations of the Cardiovascular System,* Rosenquist, G. C. and Bergsma, D., Eds., Alan R. Liss, New York, 1978, 29.
63. **Los, J. A.,** Cardiac septation and development of the aorta, pulmonary trunk, and pulmonary veins: previous work in the light of recent observations, in *Morphogenesis and Malformations of the Cardiovascular System,* Rosenquist, G. C. and Bergsma, D., Eds., Alan R. Liss, New York, 1978, 109.
64. **Challice, C. E. and Viragh, S.,** The embryologic development of the mammalian heart, in *Ultrastructure of the Mammalian Heart,* Challice, C. E. and Viragh, S., Eds., Academic Press, New York, 1973, 91.
65. **Arguello, C., De La Cruz, M. V., and Sanchez, C.,** Ultrastructural and experimental evidence of myocardial cell differentiation into connective tissue cells in embryonic chick heart, *J. Mol. Cell. Cardiol.,* 10, 307, 1978.
66. **Thompson, R. P., Wong, Y.-M. M., and Fitzharris, T. P.,** A computer graphic study of cardiac truncal septation, *Anat. Rec.,* 206, 207, 1983.
67. **Icardo, J. M.,** Distribution of fibronectin during the morphogenesis of the truncus, *Anat. Embryol.,* 171, 193, 1985.

68. **Sumida, H., Nakamura, H., Akimoto, N., Okamoto, N., and Satow, Y.,** Desmin distribution in the cardiac outflow tract of the chick embryo during aortico-pulmonary septation, *Arch. Histol. Jpn. (Niigata, Jpn.),* 50, 525, 1987.
69. **Rosenquist, T. H., McCoy, J. R., Waldo, K. L., and Kirby, M. L.,** Origin and propagation of elastogenesis in the developing cardiovascular system, *Anat. Rec.,* 221, 860, 1988.
70. **Sumida, H., Ashcraft, R. A. and Thompson, R. P.,** Cytoplasmic stress fibers in the developing heart, *Anat. Rec.,* 223, 82, 1989.
71. **Icardo, J. M. and Manasek, F. J.,** Fibronectin distribution during early chick embryo heart development, *Dev. Biol.,* 95, 19, 1983.
72. **Van Mierop, L. H. S.,** Morphological development of the heart, in *Handbook of Physiology,* Sect. 2, Vol. 1, American Physiological Society, Bethesda, MD, 1979, 1.
73. **Kramer, T. C.,** The partitioning of the truncus and conus and the formation of the membranous portion of the interventricular septum in the human heart, *Am. J. Anat.,* 71, 343, 1942.
74. **Maron, B. J. and Hutchins, G. M.,** The development of the semilunar valves in the human heart, *Am. J. Pathol.,* 74, 331, 1974.
75. **Hurle, J. M., Colvee, E., and Blanco, A. M.,** Development of mouse semilunar valves, *Anat. Embryol.,* 160, 83, 1980.
76. **Colvee, E. and Hurle, J. M.,** Maturation of the extracellular material of the semilunar heart valves in the mouse: a histochemical analysis of collagen and mucopolysaccharides, *Anat. Embryol.,* 162, 343, 1981.
77. **Garcia-Pelaez, I., Diaz-Gongora, C., and Arteaga, M.,** Contribution of the superior atrioventricular cushion to the left ventricular infundibulum. Experimental study on the chick embryo, *Acta Anat.,* 118, 224, 1984.
78. **Morse, D. E., Hamlett, W. C., and Noble, C. W., Jr.,** Morphogenesis of chordae tendinae. I. Scanning electron microscopy, *Anat. Rec.,* 210, 629, 1984.
79. **Cooper, T., Napolitano, L. M., Fitzgerald, M. J. T., Moore, K. E., Daggett, W. M., Willman, V. L., Sonnenblick, E. H., and Hanlon, C. R.,** Structural basis of cardiac valvular function, *Arch. Surg.,* 93, 767, 1966.
80. **Manasek, F. J.,** Embryonic development of the heart. II. Formation of the epicardium, *J. Embryol. Exp. Morphol.,* 22, 333, 1969.
81. **Shimada, Y. and Ho, E.,** Scanning electron microscopy of the embryonic chick heart: formation of the epicardium and surface structure of the four heterotypic cells that constitute the embryonic heart, in *Etiology and Morphogenesis of Congenital Heart Disease,* van Praagh, R. and Takao, A., Eds., Futura, New York, 1980, 63.
82. **Viragh, S. and Challice, C. E.,** The origin of the epicardium and the embryonic myocardial circulation in the mouse, *Anat. Rec.,* 201, 157, 1981.
83. **Hirakow, R.,** Development of the cardiac blood vessels in staged human embryos, *Acta Anat.,* 115, 220, 1983.
84. **Licata, R. H.,** Coronary circulation: embryology, in *Blood Vessels and Lymphatics,* Abramson, D. I., Ed., Academic Press, New York, 1962, 258.
85. **Voboril, Z. and Schlieber, T. H.,** Uber die Entwicklung der GefaBversorgung der Rattenherzens, *Z. Anat. Entwicklungs gesch.,* 129, 24, 1969.
86. **Manasek, F. J.,** The ultrastructure of embryonic myocardial blood vessels, *Dev. Biol.,* 26, 42, 1971.
87. **Aikawa, E. and Kawano, J.,** Formation of coronary arteries sprouting from the primitive aortic sinus wall of the chick embryo, *Experientia,* 38, 816, 1982.
88. **Conte, G. and Pellegrini, A.,** On the development of the coronary arteries in human embryos, stages 14-19, *Anat. Embryol.,* 169, 209, 1984.
89. **Rakusan, K.,** Postnatal development of the heart, in *Hearts and Heart-Like Organs,* Vol. 1, Bourne, G. H., Ed., Academic Press, New York, 1980, 301.
90. **Hudlicka, O.,** Growth of capillaries in skeletal and cardiac muscle, *Circ. Res.,* 50, 451, 1982.
91. **Rakusan, K. and Turek, Z.,** Protamine inhibits capillary formation in growing rat hearts, *Circ. Res.,* 57, 393, 1985.
92. **Rychter, Z., Jelinek, R., Klika, E., and Antalikova, L.,** Development of the lymph bed in the wall of the chick embryo heart, *Physiol. Bohem.,* 20, 533, 1971.
93. **Klika, E., Rychter, Z., Antalikova, L., and Zajicora, A.,** Mesenchyma and development of lymphatic capillary bed, in *Lymphology: Proceedings of the 6th International Congress,* Malek, P., Bartos, V., Weissleder, H., and Witte, M., Eds., Georg Thieme Verlag, Stuttgart, 1979, 9.
94. **Miller, A. J.,** *Lymphatics of the Heart,* Raven Press, New York, 1982.
95. **Manasek, F. J. and Monroe, R. G.,** Early cardiac morphogenesis is independent of function, *Dev. Biol.,* 27, 584, 1972.
96. **Yoshida, H., Manasek, F. J., and Arcilla, R. E.,** Intracardiac flow patterns in early embryonic life. A reexamination, *Circ. Res.,* 53, 363, 1983.
97. **Clark, E. D., Hu, N., and Dooley, J. B.,** The effect of isoproterenol on cardiovascular function in the stage 24 chick embryo, *Teratology,* 31, 41, 1985.

98. **Zak, R.,** Contractile function as a determinant of muscle growth, in *Cell and Muscle Motility,* Vol. 1, Dowben, R. M. and Shay, J. W., Eds., Plenum Press, New York, 1981, 1.
99. **Faber, J. J., Green, T. J., and Thornburg, K. L.,** Embryonic stroke volume and cardiac output in the chick, *Dev. Biol.,* 41, 14, 1974.
100. **Clark, E. D.,** Functional aspects of cardiac development, in *Growth of the Heart in Health and Disease,* Zak, R., Ed., Raven Press, New York, 1984, 81.
101. **Alpert, N. R.,** *Myocardial Hypertrophy and Failure,* Raven Press, New York, 1983.
102. **Mansier, P., Schwartz, K., Lelievre, L., Moalic, J. M., Charlemagne, D., Samuel, J. L., Rappaport, L., and Swynghedaw, B.,** New trends in biology of cardiac overload: plasma membranes, enzymes, cytoskeleton proteins and in vivo traduction of RNA, in *Cardiac Adaptation to Hemodynamic Load, Training and Stress,* Jacob, R., Gulch, R. W., and Kissling, G., Eds., Steinkopff Verlag, Darmstadt, 1983, 94.
103. **Hirzel, H. O., Tuchschmid, C. R., Schneider, J., Krayenbuehl, H. P., and Schaub, M. C.,** Relationship between myosin isoenzyme composition, hemodynamics, and myocardial structure in various forms of human cardiac hypertrophy, *Circ. Res.,* 57, 729, 185.
104. **Gulch, R. W. and Mohrmann, J.,** Alterations in electrical properties of rat myocardium accompanying different models of cardiac hypertrophy, in *Cardiac Adaptation to Hemodynamic Load, Training and Stress* Jacob, R., Gulch, R. W., and Kissling, G., Eds., Steinkopff Verlag, Darmstadt, 1983, 174.
105. **Clark, E. D.,** Ventricular function and cardiac growth in the chick embryo, in *Cardiac Morphogenesis,* Ferrans, V. J., Rosenquist, G. C., and Weinstein, C., Eds., Elsevier, New York, 1985, 238.
106. **Reidy, M. A. and Langille, B. L.,** The effect of local blood flow patterns on endothelial cell morphology, *Exp. Mol. Pathol.,* 32, 276, 1980.
107. **Icardo, J. M.,** Endocardial cell arrangement. Role of hemodynamics, *Anat. Rec.,* 225, 150, 1989.
108. **Hurle, J. M. and Colvee, E.,** Changes in the endothelial morphology of the developing semilunar heart valves. A TEM and SEM study in the chick, *Anat. Embryol.,* 167, 67, 1983.
109. **Borg, T. K. and Caufield, J. B.,** The collagen matrix of the heart, *Fed. Proc., Fed. Am. Soc. Exp. Biol.,* 40, 2037, 1981.
110. **Hopkins, F., McCutcheon, E. P., and Wekstein, D. R.,** Post-natal changes in rat ventricular function, *Circ. Res.,* 32, 685, 1973.
111. **Caufield, J. B. and Borg, T. K.,** The collagen network of the heart, *Lab. Invest.,* 40, 364, 1979.
112. **Viragh, S. and Challice, C. E.,** The impulse generation and conduction system of the heart, in *Ultrastructure of the Mammalian Heart,* Challice, C. E. and Viragh, S., Eds., Academic Press, New York, 1973, 43.
113. **Truex, R. C. and Smythe, M. Q.,** Comparative morphology of the cardiac conduction tissue in animals, *Ann. N.Y. Acad. Sci.,* 127, 18, 1965.
114. **Lev, M.,** The conduction system, in *Pathology of the Heart and Blood Vessels,* Gould, S. E., Ed., Charles C Thomas, Springfield, IL, 1968, 180.
115. **Paff, G. H., Boucek, R. J., and Klopfenstein, H. S.,** Experimental heart-block in the chick embryo, *Anat. Rec.,* 149, 217, 1965.
116. **James, T. N.,** Cardiac conduction system: fetal and postnatal development, *Am. J. Cardiol.,* 25, 213, 1970.
117. **Lieberman, M.,** Physiologic development of impulse conduction in embryonic cardiac tissue, *Am. J. Cardiol.,* 25, 279, 1970.
118. **Viragh, S. and Challice, C. E.,** The development of the conduction system in the mouse heart. I. The first A-V conduction pathway, *Dev. Biol.,* 56, 382, 1977.
119. **Marino, T. A. and Severdia, J.,** The early development of the A-V node and bundle in the ferret heart, *Am. J. Anat.,* 167, 299, 1983.
120. **Manasek, F. J.,** Electron microscopic contributions to conduction development, in *Proc. Conduction Development Conf.,* National Institutes of Health, Bethesda, MD, 1970, 3.
121. **Marino, T. A., Kent, R. L., and Cooper, G.,** IV, Relationship between hemodynamic load and cardiac cell differentiation, in *Cardiac Morphogenesis,* Ferrans, V. J., Rosenquist, G. C., and Weinstein, C., Eds., Elsevier, New York, 1985, 208.
122. **Forsgren, S., Strehler, E., and Thornell, L.-E.,** Differentiation of Purkinje fibres and ordinary ventricular and atrial myocytes in the bovine heart: an immuno- and enzyme histochemical study, *Histochem. J.,* 14, 929, 1982.
123. **Lebowitz, E. A., Novick, J. S., and Rudolph, A. M.,** Development of myocardial sympathetic innervation in the fetal lamb, *Pediatr. Res.,* 6, 887, 1972.
124. **Gehring, W. J.,** The homeo box: a key to the understanding of development?, *Cell,* 40, 3, 1985.
125. **Hart, C. P., Awgulewitsch, A., Fainsod, A., McGinnis, W., and Ruddle, F. H.,** Homeo box gene complex on mouse chromosome 11: molecular cloning, expression in embryogenesis, and homology to a human homeo box locus, *Cell,* 43, 9, 1985.
126. **Nora, J. J. and Nora, A. H.,** The genetic contribution to congenital heart disease, in *Congenital Heart Disease. Causes and Processes,* Nora, J. J. and Takao, A., Eds., Futura, New York, 1985, 3.

127. **Nora, J. J. and Takao, A., Eds.,** *Congenital Heart Disease. Causes and Processes,* Futura, New York, 1985.
128. **Bostock, C. J.,** Chromosomal changes associated with changes in development, *J. Embryol. Exp. Morphol.,* Suppl. 83, 7, 1984.
129. **Lindquist, S.,** Heat shock proteins — a comparison of *Drosophila* and yeast, *J. Embryol. Exp. Morphol.,* 83, 147, 1984.
130. **Ingber, D. E. and Jamieson, J. D.,** Cells as integrity structures: architectural regulation of histodifferentiation by physical forces transduced over basement membrane, in *Gene Expression During Normal and Malignant Differentiation,* Anderson, L. C., Gahmberg, C. G., and Ekblom, P., Eds., Academic Press, London, 1985, 13.
131. **Huttner, I., Walker, C. and Gabbiani, G.,** Aortic endothelial cell during regeneration. Remodeling of cell junctions, stress fibers, and stress fiber-membrane attachment domains, *Lab. Invest.,* 53, 287, 1985.
132. **McDermott, P., Daood, M., and Klein, I.,** Contraction regulates myosin synthesis and myosin content of cultured heart cells, *Am. J. Physiol.,* 249, H763, 1985.
133. **Blau, H. M., Pavlath, G. K., Hardeman, E. C., Chiu, C.-P., Silverstein, L., Webster, S. G., Miller, S. C., and Webster, C.,** Plasticity of the differentiated state, *Science,* 230, 758, 1985.
134. **Icardo, J. M., Nakamura, A., Fernandez-Teran, M. A., and Manasek, F. J.,** Effects of injecting fibronectin and antifibronectin antibodies on cushion mesenchyme cell migration. An in vivo study, *Development,* submitted.

DEVELOPMENTAL CHANGES IN ELECTRICAL ACTIVITY AND ION CHANNELS IN THE HEART

Nicholas Sperelakis

INTRODUCTION

The heart is the first organ to become functional during embryogenesis. Important physiological, electrophysiological, contractile, pharmacological, biochemical, and ultrastructural changes occur in myocardial cells during the embryonic development of avian and mammalian hearts. In many mammalian hearts, some of the changes extend into the early postnatal period. These changes affect and determine the functional behavior and properties of the heart at each stage of development and differentiation. Therefore, it is the purpose of this article to review and summarize many of these changes in properties. For example, striking changes occur in the electrical properties of ventricular myocardial cells during embryonic development of chick heart. Because of space limitations, this article will focus primarily on the changes in the properties of the ion channels that occur during development of chick heart. The reader is referred to several review-type articles that summarize other developmental changes as well as go into greater detail.[1-6]

RESTING POTENTIAL AND K$^+$ PERMEABILITY

The resting potential (E_m), measured by intracellular microelectrodes, of the ventricular portion of the chick and rat hearts increases during embryonic development.[7-13] In embryonic chick heart, the greatest changes occur between days 2 and 7, and thereafter the increase is smaller. For example, in a 2-d-old heart, the mean resting E_m is about -40 mV, and this increases to about -51 mV on day 3. The resting potential is close to -80 mV by day 12, nearly the final adult value. The large increase in resting E_m during the first few days is due mainly to an increase in K$^+$ permeability (P_K), although there is also an increase in $[K]_i$ and in K$^+$ equilibrium potential (E_K). However, larger values of resting E_m have been reported for young hearts.[14]

The relationship between resting E_m and external K$^+$ concentration ($[K]_o$) was determined for embryonic hearts of different ages.[10,12,15] $[K]_i$, estimated by extrapolation of the curves to zero potential, varied between 125 mM (for 2-d hearts) and 155 mM (for 14 to 20-d hearts). The data points for the 3-d hearts most closely fitted the theoretical curve (calculated from the Goldman constant-field equation) for a P_{Na}/P_K ratio of 0.2; those for the 5-d hearts fitted the curve for a P_{Na}/P_K of 0.1, and those for the 15-d hearts most closely fitted between the 0.05 and 0.01 curves. These data suggest that the P_{Na}/P_K ratio is very high in young hearts and may account for the low measured resting E_m; i.e., the low resting E_m is not due to a greatly lower $[K]_i$ and E_K. Only a small increase in the calculated E_K occurs during development. Thus, in the young hearts, the resting E_m is far from E_K due to the high P_{Na}/P_K ratio. In this respect, the myocardial cells in young embryonic hearts resemble sinoatrial nodal cells in adult hearts.

In old embryonic chick or adult hearts, the E_m vs. log $[K]_o$ curve is nearly linear above 10 mM K$^+$, with a slope approaching the theoretical 61 mV per decade (from the Nernst equation). If the slope were exactly 61 mV per decade, then E_m would be equal to E_K, and the membrane would be completely K$^+$ selective. The slope for hearts 7 to 20 d old was 51 to 53 mV per decade, whereas the average slopes (curves continually bend) for 4-, 3-, and 2-d hearts were 46, 40, and 30 mV per decade, respectively. Similar values for $[K]_i$ and slope were found for embryonic chick atrial cells at various stages of development.[16]

The input resistance (r_{in}) of the ventricular cells, determined from steady-state voltage/current curves, is high (13 MΩ) in young 2-d-old hearts and rapidly declines over the next few days, reaching the final adult value of about 4.5 MΩ by day 14. If the average cell size and the degree of electrical coupling between the cells remain unchanged, the high r_{in} of the young hearts would suggest that membrane resistivity (R_m) is very high, consistent with a low K$^+$ conductance and P_K. These results suggest that the P_{Na}/P_K ratio is high in young hearts because P_K is low and not because P_{Na} is high. Consistent with a low P_K is the finding that the chronaxie of young hearts (2-d-old) is about fourfold higher than that of 9- to 16-d hearts.[11] This indicates that the membrane time constant is fourfold higher in young hearts and, for a constant membrane capacitance, membrane resistivity must be fourfold higher. Consistent with the conclusions from these electrical studies, Carmeliet et al.[17] have reported, on the basis of ^{42}K flux measurements, that P_K is about twofold to threefold lower in 6- to 8-d hearts than in 18- to 20-d hearts. The values for the P_K coefficients were 13.2×10^{-8} cm/s for 7-d hearts and about 27.5×10^{-8} cm/s for 19-d hearts. From the constant-field equation, the P_{Na}/P_K ratios calculated to be 0.018 for the 19-d hearts and 0.037 for the 7-d hearts; P_{Na} did not change during development (constant at about 0.50×10^{-8} cm/s).

Since flattening of the resting potential vs. log $[K]_o$ curve at lower $[K]_o$ levels is much more prominent for young hearts, i.e., they are depolarized less by a given increment in $[K]_o$, young hearts should be, and are, less affected by elevation of $[K]_o$.[12,18] This is true for both inhibition of automaticity, as well as for loss of excitability of the ventricle to electrical stimulation.[12]

The young ventricular cells are not hyperpolarized by acetylcholine (ACh), even though a large hyperpolarization is theoretically possible because the resting E_m is much below E_K.[12] Therefore, ACh does not greatly increase P_K in ventricular cells. The atrial cells of young hearts are slightly depolarized by ACh, but slightly hyperpolarized in Na$^+$-free medium, suggesting that ACh increases both Na$^+$ conductance and K$^+$ conductance.[16,19,20]

In single-channel recordings using a cell-attached patch clamp, it was found that the inwardly rectifying K$^+$ channel was missing in young 3-d-old hearts, but present in older hearts (Figure 1). Since this K$^+$ channel is important in determining the resting potential, this finding is consistent with the low recorded resting potential and high input resistance in the young cells.

(Na,K)-ATPASE ACTIVITY AND ELECTROGENIC PUMP POTENTIAL

The specific activity of the (Na,K)-ATPase is low in young embryonic chick hearts and rises during development.[21,22] The average value on day 4 is about 35% of that on day 16 (7.4 ± 0.7 mol P_i/h/mg protein). This enzyme activity is highest on day 20, and the adult level is about equal to that on embryonic day 16. Thus, while P_K is increasing during development, and hence the outward passive leak of K$^+$ and inward leak of Na$^+$ (due to the increased electrochemical driving forces), the capability of the Na$^+$-K$^+$ pump is increasing correspondingly, thus compensating for the increased demand on the pump due to increased cation leak. However, the pumping capacity of the young hearts must be sufficient to maintain the relatively high $[K]_i$ and low $[Na]_i$ already present in the young cells.

It has been reported that the calculated $[K]_i$ levels may actually decrease during development.[10,17,23] The electrophysiologic data from resting E_m vs. log $[K]_o$ curves[12] indicate that $[K]_i$ is about 125 mM on day 2 and that it increases gradually to about 155 mM on days 14 to 20. A high $[K]_i$ value (145 mM) was reported for chick atrial cells on day 4.[16] Thus, $[K]_i$ is already high in young hearts, and so the cardiac cells must actively transport cations before day 2.

Tissue electrolyte analyses in chick embryonic hearts (ventricles) indicate that the total

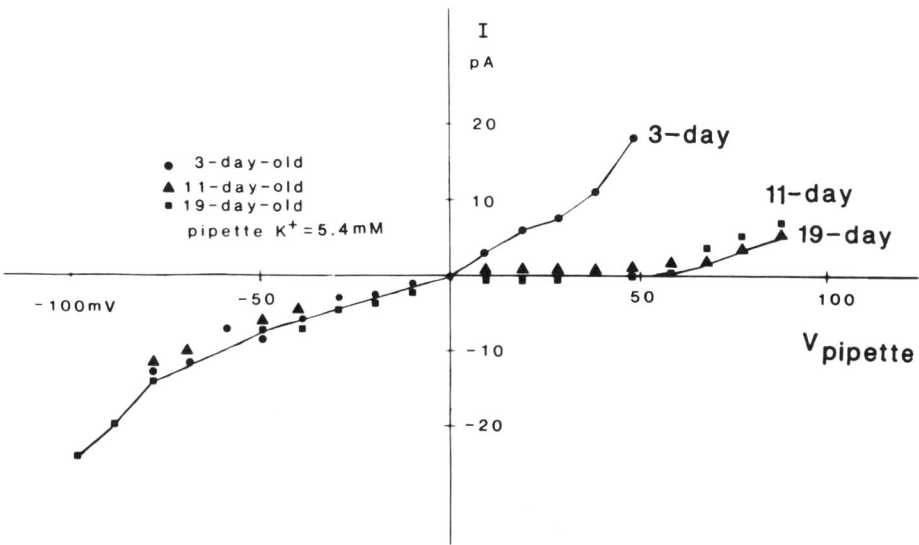

FIGURE 1. Absence of the inward-rectifying K$^+$ channel in young (3-d-old) embryonic chick hearts. To characterize the age-related changes in K$^+$ channel-kinetic properties, cell-attached patch clamp experiments were conducted on embryonic chick myocytes prepared from three different embryonic stages: 3, 11, and 19 d old. Shown is the I-V curve of steady-state K$^+$ currents attained 1 min after changes in pipette potential. Data plotted are the mean values (n = 4 to 6) of the channel current amplitude for each age embryonic heart. As can be seen, the inward-going rectification was substantial at potentials higher than 0 mV for 11- and 19-d-old hearts, but it was absent in 3-d-old hearts. (From Sada, H., Bkaily, G., and Sperelakis, N., unpublished observations.)

tissue content of Na$^+$ in young hearts is very high and that it decreases gradually during development of the heart;[13,24] however, there is a great amount of bound Na$^+$ in young hearts, especially in the extracellular mucopolysaccharide cardiac jelly.[25] The data from electrophysiologic studies[12,13] indicated that the thermodynamically active free intracellular Na$^+$ concentration must not be very high, because the Na$^+$-dependent action potentials (APs) already overshoot to +11 mV in day 2 hearts; the overshoot increases rapidly, reaching +28 mV by day 8; the latter value is the same as the adult value. Others also reported that [Na]$_i$ in the chick heart did not change much during development.[10,17,23] For example, an [Na]$_i$ value of 16 mM was reported for 7-d embryonic chick hearts.[17] Thus, there seems little doubt that the free [Na]$_i$ is relatively low in young hearts.

The energy-dependent pump located in the cell membrane, which maintains the ionic gradients for Na$^+$ and K$^+$ ions across the cell membrane, becomes electrogenic when the ratio of Na$^+$ ions pumped out to K$^+$ ions pumped in is greater than 1; i.e., the pump directly produces a potential (V$_{ep}$) which contributes to the measured resting potential, usually between 2 and 15 mV, depending on the type of cell and the physiologic conditions.[26] Electrogenic Na$^+$ pump potentials have been demonstrated in various tissues of the heart[27-31] and in cultured embryonic chick myocardial cells.[32] Pelleg et al.[33] observed an electrogenic pump potential of a few millivolts in cultured embryonic heart cell reaggregates derived from early (3-d-old) and late (16-d-old) stages of development that were subjected to overdrive stimulation. When automatic heart cells are driven at a faster rate than their intrinsic rate, upon termination of the drive there is a transient pause followed by a gradual recovery to the predrive firing rate. This phenomenon of overdrive suppression of automaticity is caused by a small transient hyperpolarization of a few millivolts due to stimulation of an electrogenic Na$^+$ pump potential, presumably resulting from an increase in [Na]$_i$ and in [K]$_o$ during the drive. These studies indicate that electrogenic transport is present in early stages of ontogenesis, and that this ability is retained *in vitro*.

AUTOMATICITY

Requirements for automaticity include (1) a low Cl^- conductance (g_{Cl}), as is generally true for myocardial cells, and (2) a low K^+ conductance (g_K). A low g_K enhances membrane apparent inductance in series with the inward-rectifying anomalous K^+ channel having a negative slope conductance region, and tends to cause oscillations in E_m. The low g_K also is responsible for the low resting potential (moves the resting E_m farther from E_K), and places E_m in the region that can support pacemaker oscillations. Pronounced changes in automaticity of the ventricular cells also occur during development, as could be predicted from the changes in P_K. The incidence of hyperpolarizing afterpotentials and pacemaker potentials is very high (about 90%) in the young hearts, and this incidence decreases to 0% in the old embryonic hearts.[12] In an excised portion of the ventricle, the incidence of pacemaker potentials is 100% for embryos up through day 10, whereas the incidence is 0% on day 12 or older. These results indicate that the ventricular myocardial cells possess automaticity capability when they are young, but this capability diminishes as the cells age.

However, old ventricular cells again become automatic when they were trypsin dispersed and placed into monolayer cell culture. For example, ventricular myocardial cells dispersed from 16-d-old chick embryos and cultured as monolayers usually revert toward the young embryonic state with respect to their electrical properties, including gain of automaticity.[12] When the cells are allowed to reaggregate into small spheres, however, they often retain their highly differentiated electrical properties, including lack of automaticity.[34,35] The gain in automaticity of cultured cells appears to reflect a decrease in P_K.[36-38] In some cases, isolated single ventricular cells in culture have such a low P_K that they become depolarized too far and do not normally exhibit automaticity or excitability.[39] However, if these cells are hyperpolarized by intracellular application of current pulses, spontaneous APs and contractions occur.

ACTION POTENTIAL CHANGES

The APs of the cells of intact chick hearts undergo sequential changes during development *in situ*.[12,15,40] There is a progressive increase in maximal rate of rise ($+\dot{V}_{max}$) and overshoot of the AP, as well as an increase in resting potential. The overshoot averaged $+11$ mV on day 2 and increased progressively over the next few days, reaching the maximal value of about $+28$ mV by day 8. The duration (at 50% repolarization) was hardly changed during development, the average value being 110 ms. The time course of the increase in $+\dot{V}_{max}$ was not parallel to the increase in resting E_m, the increase in resting E_m preceding the increase in $+\dot{V}_{max}$ by several days. Thus, in young hearts, it was not unusual to find a cell with a large resting potential, but with a low $+\dot{V}_{max}$.

Young (2 to 3 d *in ovo*) myocardial cells possess slowly rising (10 to 40 V/s) APs preceded by pacemaker potentials (Figure 2). Hyperpolarization does not greatly increase the rate of rise, thus indicating that the low $+\dot{V}_{max}$ is not due to inactivation of fast Na^+ channels at the low resting potential, but rather to a low density of fast Na^+ channels. Excitability is not lost until the membrane is depolarized to less than -20 mV, thus indicating the preponderance of slow channels. The AP upstroke in young hearts is generated by Na^+ influx through tetrodotoxin (TTX)-insensitive and Mn^{2+}-insensitive slow Na^+ channels, as indicated by the dependence of the AP overshoot and rate of rise on $[Na]_o$. The slope of overshoot as a function of $[Na]_o$ approaches the theoretical 61 mV per decade at the lower $[Na]_o$ concentrations.

Kinetically fast Na^+ channels are substantial in number by day 5. At this time, $+\dot{V}_{max}$ is about 50 to 70 V/s. During this intermediate stage of development (from day 5 to day 7), a large number of slow Na^+ channels still coexist with the fast Na^+ channels in the

Part B: Cardiovascular and Respiratory Development   55

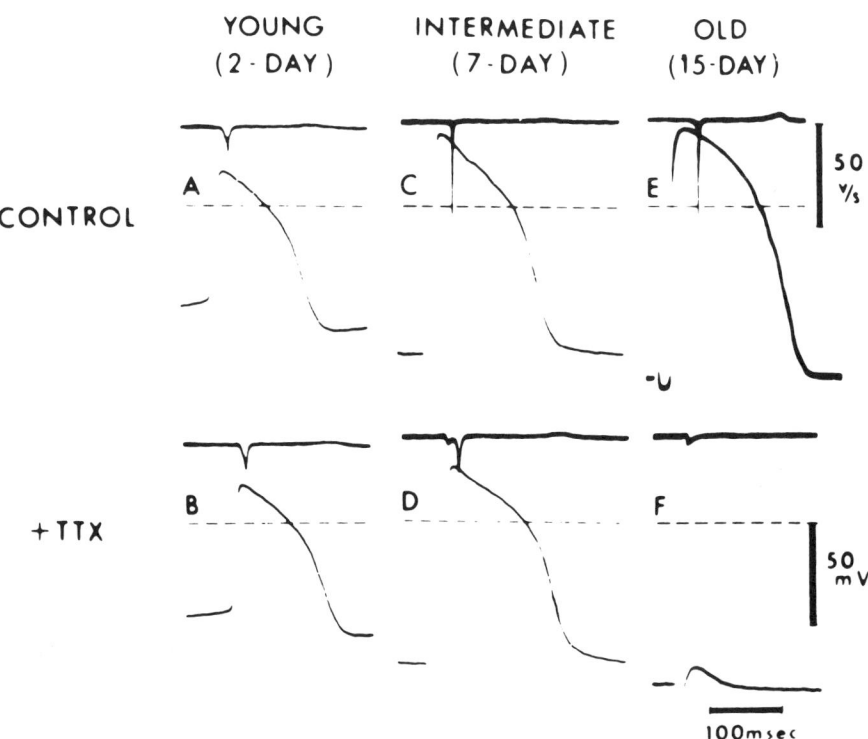

FIGURE 2. Characteristics of the action potentials in intact embryonic chick hearts at different stages of development. (A and B) Intracellular recordings from a 2-d-old heart before (A) and 20 min after (B) the addition of tetrodotoxin (TTX) (20 µg/ml). (C and D) Recordings from a 7-d-old heart before (C) and 2 min after (D) the addition of TTX (2 µg/ml). Note depression of the rate of rise in D. (E and F) Records from a 15-d-old heart prior to (E) and 2 min after (F) the addition of TTX (1 µg/ml). The APs were abolished and excitability was not restored by strong field stimulation in F. Thus, the hearts became progressively more responsive to TTX during development, i.e., the effect of TTX increased with increasing embryonic age. The upper traces give dV/dt; this trace has been shifted relative to the V-t trace to prevent obscuring dV/dt. The horizontal broken line in each panel represents zero potential. dV/dt calibration (in E) and voltage and time calibrations (in F) pertain to all panels. (Modified from Reference 12.)

membrane. By day 8, depolarization to less than -50 mV abolishes excitability. This indicates that the AP-generating Na^+ channels now consist predominantly of fast Na^+ channels. The density of fast Na^+ channels continues to increase until about day 18, when the adult maximal rate of rise of about 150 v/s is achieved. This conclusion of an increase in number of fast Na^+ channels has been strengthened by confirmatory observations reported by Iijima and Pappano[41] and Marcus and Fozzard.[42] During development, a large fraction of the slow Na^+ channels appears to have been lost functionally, and insufficient numbers remain to support regenerative excitation. Addition of some positive inotropic agents rapidly increases the number of slow Ca^{2+} channels available in the membrane and leads to the regaining of excitability in cells whose fast Na^+ channels have been voltage inactivated or blocked.[43-45]

TTX, a specific blocker of fast Na^+ channels, has no effect or only little effect on the AP rate of rise or overshoot of the young (2- to 3-d-old) hearts. In addition, TTX had no effect on the inward slow Na^+ current in voltage-clamp experiments. During the intermediate stage (days 5 to 7), TTX causes a reduction in $+\dot{V}_{max}$ to about 10 to 20 V/s, but the APs and accompanying contractions persist. After day 8, the APs are completely abolished by TTX despite increased stimulation intensity.

The Ca^{2+}-antagonist drugs, verapamil and D-600, block the APs of the young embryonic hearts.[46,47] In addition, it was found that nifedipine also blocked, whereas diltiazem, bepridil, and mesudipine did not.[48] Elevation of $[Ca]_o$ did not antagonize this block by the drugs, whereas elevation of $[Na]_o$ (starting from 50% of normal $[Na]_o$) did antagonize, consistent with the channel being a slow Na^+ rather than a slow Ca^{2+} channel. This indicates that these agents block slow Na^+ channels as well as slow Ca^{2+} channels, and so are not specific for blockade of Ca^{2+} current. In contrast, Mn^{2+} at 1 mM does not depress the APs of young hearts (Figure 3A), although it does block the contractions, indicating a greater specificity for slow Ca^{2+} channels and not for the slow Na^+ channels.[12]

Galper and Catterall[49] reported that the early embryonic chick heart was insensitive to TTX (with regard to contractions), but was sensitive to D-600. During subsequent development, the sensitivity to TTX increased and the sensitivity to D-600 decreased in a reciprocal manner. Kasuya et al.[50] also reported that the slowly rising APs of 3-d-old embryonic chick hearts involved cation channels that were pharmacologically different from those of old embryonic hearts. Nathan and De Haan[51] found that TTX-sensitive fast Na^+ conductance channels were absent or nonfunctional in cultured cell reaggregates derived from 3-d-old embryonic chick hearts. Ishima[52] reported that the contractions of 3- to 5-d-old embryonic chick hearts were not affected by TTX.

Consistent with the findings reported from several laboratories, from electrophysiologic experiments, that there is a large increase in number of fast Na^+ channels during development, it was demonstrated by Renaud et al.[53] that the number of specific TTX-binding sites (using ^{3}H-ethylenediamine-TTX) is very low in 3-d-old embryonic chick hearts, and increases fourfold to fivefold during development. They also confirmed that the APs in the 2- to 3-d hearts were insensitive to TTX. The K_D for TTX binding did not change. Iijima and Pappano[41] and Marcus and Fozzard[42] reported that there are some fast Na^+ channels detectable on day 3 and that the sensitivity of the fast Na^+ channels to blockade by TTX does not change during development. Renaud et al.[53] found that there was degradation of the TTX receptors with a half-time of 9 h in monolayer cultures that were reverted; in reaggregate cultures containing highly differentiated cells, the TTX receptors were stable 24 h after inhibition of protein synthesis.

Fujii et al.[54] reported results of their voltage clamp studies on isolated single ventricular cells from 2- to 7-d-old embryonic chick hearts. They found that the fast Na^+ current density increased about 17-fold between day 2 and day 7. The 7-d cells had a maximal fast Na^+ current of about 300 $\mu A/cm^2$ (equivalent to about 5 channels per μm^2), or about 4000 channels per cell. They concluded that the small fast Na^+ current normally makes no contribution to the AP in the 2-d-old hearts, because of their inactivation at the low resting potential or maximum diastolic potential at the stage of development. Other than the magnitude of the current, there were no differences in kinetics, voltage dependence, or TTX sensitivity of the fast Na^+ current in 2-d vs. 7-d hearts.

ORGAN-CULTURED YOUNG EMBRYONIC HEARTS

Cultivation of embryonic hearts *in vitro* aids in the analysis of the changes that occur during normal development *in situ*. When young (3-d-old) embryonic chick hearts, which have not yet become innervated by either cholinergic or adrenergic fibers, are placed into organ culture, they fail to gain TTX-sensitive fast Na^+ channels.[12,55] Instead, the APs continue to be generated by TTX-insensitive slow Na^+ channels, the rate of rise remain slow, and pacemaker potentials precede the APs. These APs resemble those recorded from

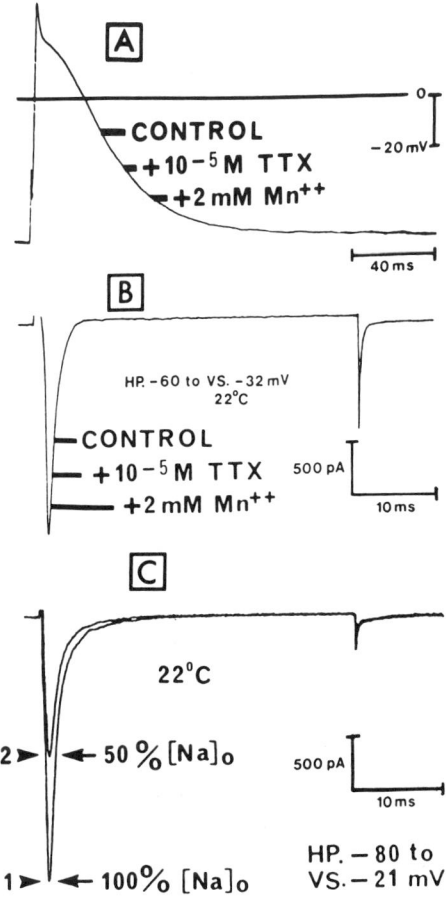

FIGURE 3. Evidence for a slow Na$^+$ current. The effects of TTX, Mn^{2+}, and extracellular Na$^+$ ([Na]$_o$) were determined on the amplitude of the inward slow current (I$_{si}$) of single, 3-d-old embryonic chick ventricular cells in culture. (A and B) Action potentials (A) and voltage step (VS) currents (B) recorded from the same cell by switching between current clamp and voltage clamp modes. (A) The slow action potential was insensitive to 2 mM Mn^{2+} and 10^{-5} M TTX. Three sweeps superimposed. (B) The inward slow current amplitude from the same cell was also insensitive to 10^{-5} M TTX and Mn^{2+}. Holding potential (HP) was -60 mV, and the VS was to -32 mV. Three sweeps superimposed. (C) The slow inward current (stepped from a HP of -80 to -21 mV) was first recorded in a medium containing 130 mM [Na]$_o$ (and 10^{-5} M TTX) (100% [Na]$_o$); decreasing [Na]$_o$ to 65 mM (50% [Na]$_o$) reduced the amplitude of I$_{si}$ by about 50%. Reperfusion with 100% [Na]$_o$ returned the current amplitude close to its control value (not shown). This experiment demonstrates that the slow inward current is carried by Na$^+$ ions. All experiments were done at 22°C; the frequency of stimulation was 0.02 Hz. (From Bkaily, G. and Sperelakis, N., unpublished observations.)

3-d-old hearts *in situ*. Similar findings were obtained when young hearts were grafted onto the chorioallantoic membrane of host chicks for blood perfusion.[56] Thus, organ-cultured young hearts do not differentiate much further *in vitro*, but appear to be arrested in the young embryonic state with respect to the number of fast Na$^+$ channels.

When young hearts which have been arrested in the early developmental state are treated with RNA-enriched fractions obtained from adult chicken hearts, they gain fast Na$^+$ channels and become completely sensitive to TTX;[57,58] i.e., young hearts *in vitro* can be induced to undergo further membrane differentiation.

However, slow Ca^{2+} channels were found to develop in young embryonic hearts during organ culture.[59] Embryonic chick hearts (3 d old) having slowly rising spontaneous APs dependent on TTX-insensitive slow Na$^+$ channels were placed into organ culture for 5 to 11 d. AP duration was markedly increased, and a notch appeared in some hearts between the initial spike phase and the plateau phase. Spike amplitude was mainly dependent on [Na]$_o$, whereas the plateau amplitude was dependent on [Ca]$_o$. The spike phase and the plateau phase of the slow APs are mainly dependent on currents through slow Na$^+$ channels and through slow Ca^{2+} channels, respectively. Mn^{2+} (0.5 mM) (a specific blocker of slow Ca^{2+} channels) and verapamil (5 μM) (a blocker of both slow Na$^+$ channels and slow Ca^{2+} channels) depressed the plateau duration and overshoot. High concentrations (10 to 30 μM) of verapamil depressed the AP amplitude and $\dot{V}_{max}$ and abolished automaticity. These results indicate that slow Ca^{2+} channels appear *de novo* during organ culture of young embryonic hearts, while the slow Na$^+$ channels are retained.

Thus, when young embryonic hearts are placed into organ culture, they gain slow Ca^{2+} channels, but fail to gain a large number of fast Na$^+$ channels. These findings suggest that differentiation of slow Ca^{2+} channels and of fast Na$^+$ channels during embryonic development *in situ* is controlled by different mechanisms. Thus, organ-cultured young embryonic hearts provide a useful model for studying the regulation of membrane electrical differentiation and appearance of ion channels.

CYCLIC NUCLEOTIDE LEVELS

Pronounced changes in cAMP content occur during embryonic development of the chick heart.[60-62] The cAMP level is highest in young hearts, and it decreases during development. McLean et al.,[60] found a cAMP level of 116 pmol/mg protein on day 4, and this decreased sharply by day 5 to about 41; there was a gradual further decline to 9.4 pmol/mg protein, which is about the adult level, by day 16. Renaud et al.[61] reported a cAMP level of 33.6 pmol/mg protein on day 4 and 11.7 on day 16. Thakkar and Sperelakis[62] reported that the basal cAMP content was 87.7 ± 1.3 pmol/mg of protein in 3-d-old embryonic chick hearts and decreased to 9.6 ± 0.6 in 9- to 19-d hearts (Figure 4). Changes in the cAMP levels in cultured heart cells and organ-cultured hearts are more variable.[60,61]

The relationship between changes in membrane properties and changes in cAMP levels during development of heart remains to be clarified. However, in cultured skeletal muscle, the cAMP level decreases sharply as the myoblasts fuse into myotubes, i.e., when the cells further differentiate.[63] In addition, increase in cAMP level is associated with increase in the number of available Ca^{2+} slow channels. Therefore, it is tempting to speculate that the decrease in the number of available slow channels during development of chick heart, as described above, results from the concomitant decreases in cAMP level. This would allow positive inotropic agents that increase the cAMP level to increase the number of available slow channels transiently back toward the density present in young embryonic hearts. In other words, the developmental change in the steady-state level of cAMP allows the fraction of slow channels that are available for activation to be modulated by inotropic agents.

Isoproterenol was capable of markedly elevating the cAMP level in all cases — young or old, cultured or noncultured — thus demonstrating the presence of functional β-adrenergic receptors. For example, 10^{-6} M isoproterenol raised the cAMP level of 7- to 8-d hearts from a control level of 35 to 80 pmol/mg of protein (at 3 min).[60] Thakkar and Sperelakis[62] found that 50 μm isoproterenol elevated the cAMP content from the basal level of 87.7 ± 1.3 to 212 ± 2.4 pmol/mg of protein in 3-d-old embryonic chick hearts. ACh (50 μM) and adenosine (Ado) (50 μM) were able to partly reverse the large elevation of cAMP produced by isoproterenol. These results clearly show that young hearts prior to innervation have functional β-adrenergic, muscarinic, and purinergic (P$_1$) receptors. The organ cultures and cell cultures of 4- and 16-d-old hearts also responded markedly to isoproterenol.[61]

The cAMP level of rat heart and skeletal muscle was reported to increase between birth

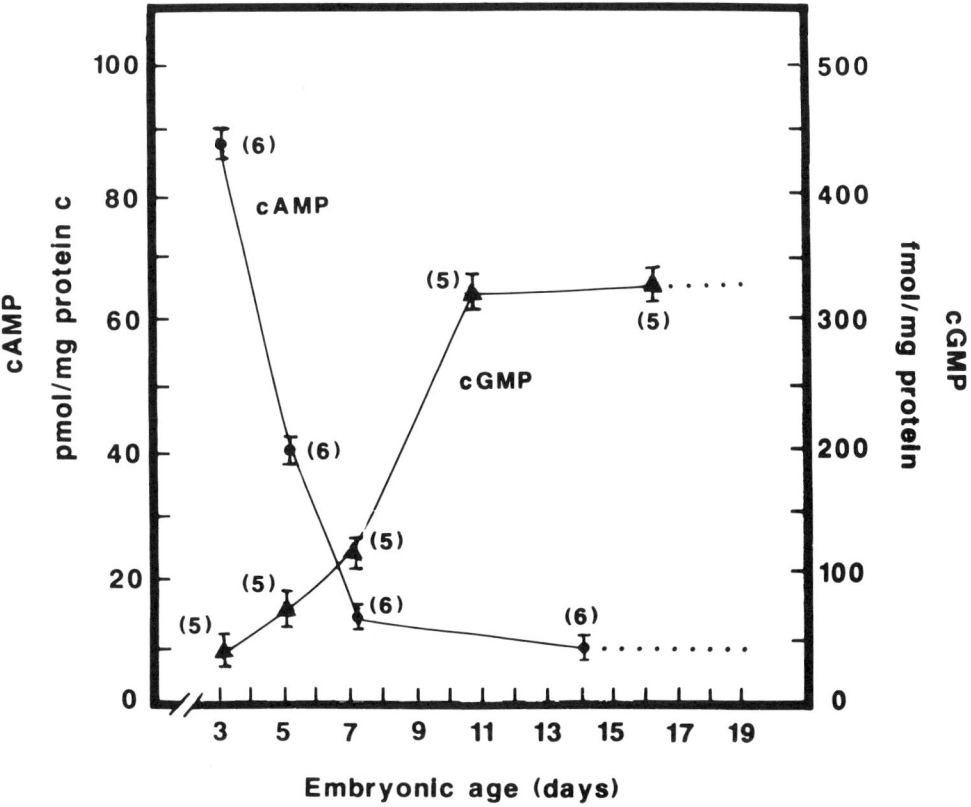

FIGURE 4. Changes in the basal levels of cAMP and cGMP as a function of embryonic age. Hearts from chick embryos were removed, incubated in MEM, and cyclic nucleotide levels determined as described in the methods. cAMP (●-●), cGMP (▲-▲). The data point plotted at 10.5 d represents the average of cGMP levels from days 9 to 12, and the data point plotted at 16.5 d represents the average from days 14 to 19. The average of cAMP levels from days 9 to 19 is represented as a data point plotted at day 14. (From Thakkar, J. K. and Sperelakis, N., *J. Dev. Physiol.*, 9, 497, 1987. With permisison.)

and postnatal day 10.[64] Likewise, in embryonic chick skeletal muscle, there was an increase in cAMP level between the 7th and 15th days.[65]

In contrast to cAMP levels, the cGMP level was reported by Thakkar and Sperelakis[62] to be very low in young embryonic chick hearts and to increase during development (Figure 4). The cGMP content was 45.5 ± 2.3 fmol/mg of protein in 3-d-old hearts and 338 ± 15.0 in 14- to 19-d hearts. Therefore, there was a reciprocal relationship between the basal cAMP and cGMP levels. Nitroprusside ($5 \times 10^{-5} M$) and hydrogen peroxide (0.1%), agents known to stimulate guanylate cyclase and elevate cGMP, increased cGMP level (in 3 min) in embryonic chick hearts 5 d old or older, but not in 3-d-old hearts. In young 3-d-old embryonic chick hearts, ACh (50 μM) and Ado (50 μM) increased cGMP concentration only slightly, whereas these substances produced a large increase (e.g., about twofold) in cGMP level in hearts 5 d old and older.[62] Therefore, it appears that the guanylate cyclase activity is very low in young 3-d-old hearts.

REGULATION OF Ca^{2+} SLOW CHANNELS BY CYCLIC NUCLEOTIDES AND PHOSPHORYLATION

A number of positive inotropic agents exert an effect on the myocardial cell membrane to increase the number of available slow channels, and this action may at least partly explain their effect on increasing cardiac contractility. It is through the slow channels that Ca^{2+} influx occurs during the cardiac AP, and the amount of Ca^{2+} ion entering the myocardial cell controls the force of contraction. Positive inotropic agents which have been shown to affect the number of available slow channels include catecholamines (β-adrenergic receptor agonists), histamine (H_2 receptor), and methylxanthines (phosphodiesterase inhibitor). The action of these agents is very rapid, the peak effect often occurring within 1 to 3 min. Histamine and β-adrenergic agonists lead to rapid stimulation of adenylate cyclase, with resultant elevation of cAMP levels. The methylxanthines enter into the myocardial cells and inhibit phosphodiesterase, thus causing an elevation of cAMP.

One method of detecting the effect of positive inotropic agents on the myocardial Ca^{2+} slow channels is to first block the fast Na^+ channels by TTX or to voltage inactivate them by partially depolarizing the cells (e.g., to -40 mV) in elevated $[K]_o$ (e.g., 25 mM). Then, addition of agents, such as beta agonists, which rapidly increase the number of slow channels available for activation upon stimulation, causes the appearance of slowly rising overshooting APs accompanied by contractions. The slow APs are blocked by agents which block inward slow Ca^{2+} current.[47,66]

cAMP is somehow involved in the functioning of the slow channels.[45,67] Consistent with this, cAMP and its dibutyryl and 8-bromo derivatives also induce the slow APs, but only after a long lag period (peak effect is 15 to 30 min), as expected from slow penetration through the membrane.[45,66] The addition of a GTP analog, GPP(NH)P (10^{-5} to 10^{-3} M), to directly activate adenylate cyclase, induced slow APs in cultured reaggregates of chick heart cells within 10 to 20 min.[68] Treatment with cholera toxin for 1 to 3 h increased adenylate cyclase activity, cAMP content, and contractility in embryonic chick ventricles.[69] Li and Sperelakis[70] showed that intracellular injection of GPP(NH)P and cholera toxin produced rapid and prolonged stimulation of the inward slow current. These results support the hypothesis that the intracellular level of cAMP controls the availability of the Ca^{2+} slow channels.

When cAMP was microinjected intracellularly (electrophoretically) into Purkinje fibers and ventricular muscle cells, slow APs were induced in the injected cell for a transient period of about 1 min; the amplitude and duration of the induced slow AP was a function of the amount of cAMP injected.[71] cAMP injections also potentiated for a transient period the rate of rise and amplitude of on-going slow APs induced by theophylline. Similar results were obtained when cAMP was injected intracellularly by pressure injection.[70] The responses were evident within 1 to 2 s after the injection was stopped, and the responses disappeared within about 30 s. These studies indicate that the average life span of the phosphorylated Ca^{2+} slow channels must be less than 30 s, and most likely is of the order of a few seconds at most. It is possible that the Ca^{2+} slow channels are phosphorylated and dephosphorylated during each cardiac cycle. cAMP injection into cultured heart cells by the phosphatidylcholine liposome method confirmed the results obtained by iontophoretic and pressure injections.[72]

Cyclic GMP was shown to have opposite effects of those of cAMP. Bath application of 8-Br-cGMP produced a relatively slow inhibition of ongoing slow APs, and intracellular pressure injection of cGMP produced a relatively rapid, transient depression of Ca^{2+}-dependent slow APs.[73] Similar results were obtained by liposome injection of cGMP into cultured chick heart cells.[72] In addition, 8-Br-cGMP was shown to block the inward slow current in voltage clamp experiments on single ventricular cells from old embryonic chick hearts (Figure 5).

The slow APs are blocked by hypoxia, ischemia, and metabolic poisons within 5 to 15 min, accompanied by a lowering of the cellular ATP level.[44] This suggests that interference with metabolism leads to blockade of the Ca^{2+} slow channels. In contrast, the fast APs are nearly unaffected at this time, thus indicating that the fast Na^+ channels are unaffected. Thus, there is a dependence of the functioning of the Ca^{2+} slow channels on metabolic

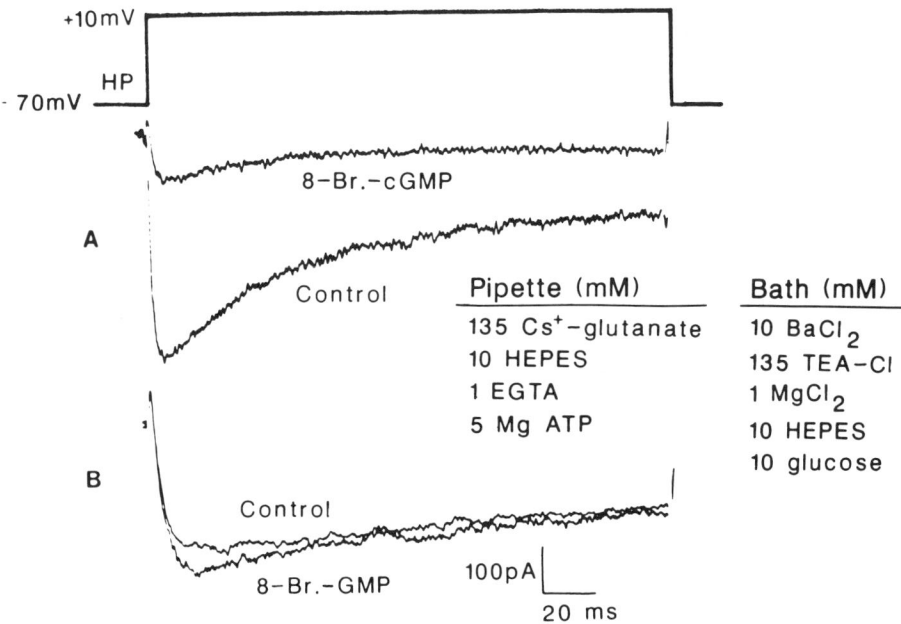

FIGURE 5. cGMP inhibition of the Ca^{2+} slow current in a cultured single ventricular cell from a 17-d-old embryonic chick heart. The cell was voltage stepped from a holding potential of -70 to $+10$ mV, and the currents were recorded. (A) Superimposed records of the control inward current and after addition of 1 mM 8-Br-cGMP to the bath. Note the marked inhibition of the inward current. (B) Lack of effect of the noncyclic 8-Br-GMP on the inward slow current in another cell. The composition of the pipette solution and bath solution is given in the figure. Ba^{2+} was used to carry the slow current, and Cs^+ and TEA^+ were used to block the outward K^+ currents. (From Wahler, G., Rusch, N., and Sperelakis, N., unpublished observations.)

energy. The contractions accompanying the fast APs are depressed or abolished, indicating that contraction is uncoupled from excitation, as expected if the slow channels were blocked. This ATP dependence of the Ca^{2+} slow channels is consistent with the phosphorylation hypothesis.

The myocardial slow channels are selectively blocked by acidosis. For example, the slow APs are depressed in rate of rise, amplitude, and duration as the pH of the perfusing solution is lowered below 7.0.[74] The slow APs are completely abolished at about pH 6.1, and 50% inhibition occurs at about pH 6.6. The contractions are also depressed by acidosis. Acidosis has little or no effect on the fast APs. Acidification of the cell membrane could change the surface charge of the membrane and/or change the conformation of the protein components of the slow channels. The effects of ischemia and hypoxia on the slow channels are partly mediated by the accompanying acidosis.[75]

Because of the relationship between cAMP and the number of available slow channels, and because of the dependence of the functioning of the slow channels on metabolic energy, it was postulated that a membrane protein must be phosphorylated in order for the slow channels to become available for voltage activation.[44,45,76-79] cAMP activates a cAMP-dependent protein kinase, and this enzyme, in the presence of ATP, phosphorylates a variety of proteins, including several membrane proteins.[80-82] Bkaily and Sperelakis[83] demonstrated that intracellular injection (by the liposome method) of a protein inhibitor of cyclic AMP-dependent protein kinase into cultured chick heart cells blocked ongoing Ca^{2+}-dependent slow APs, and this block could be reversed by injection of the catalytic subunit of the cAMP-protein kinase.

The protein that is phosphorylated might be a constituent of the Ca^{2+} slow channel itself

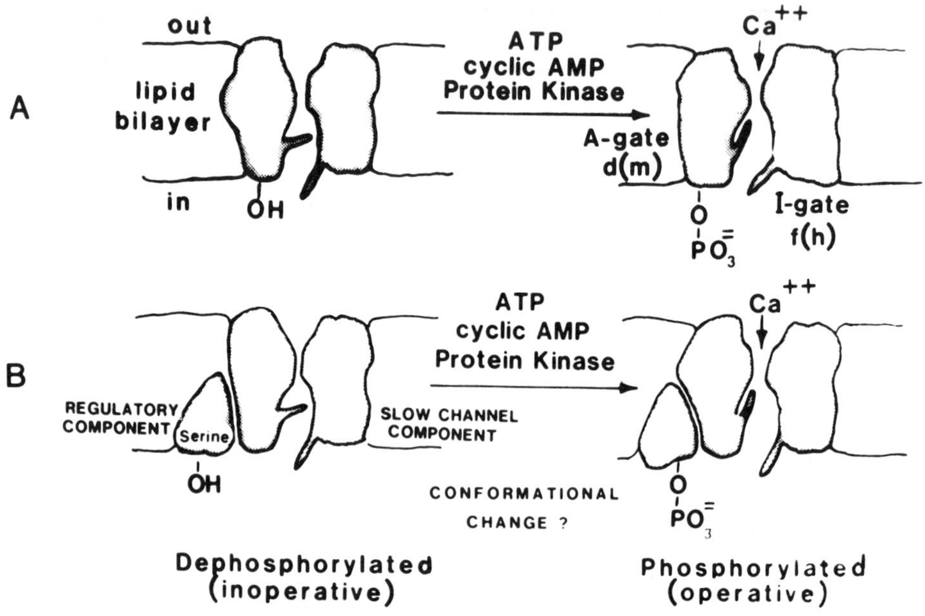

FIGURE 6. Cartoon model for a slow channel in myocardial cell membrane in two hypothetical forms: dephosphorylated form (left diagrams) and phosphorylated form (right diagrams). The phosphorylation hypothesis assumes that a protein constituent of the slow channel itself (A) or a stimulatory regulatory protein associated with the slow channel (B) must be phosphorylated in order for the channel to be in a functional state available for voltage activation. Phosphorylation occurs by activation of a cAMP-dependent protein kinase. A phosphoprotein phosphatase dephosphorylates the slow channel. Panel C depicts the possibility that cGMP inhibits the Ca^{2+} slow channel by phosphorylation of an inhibitory regulatory protein via a cGMP-dependent protein kinase.

or an associated regulatory protein (Figure 6). Phosphorylation could make the slow channel available for activation by causing a conformational change that either allowed the activation gate to be opened upon depolarization or increased the diameter of the water-filled central pore so that Ca^{2+} could pass through. The phosphorylated form of the slow channel would be the active (operational) form, and the dephosphorylated form would be the inactive (inoperative) form, i.e., only the phosphorylated form would be available for activation upon depolarization to threshold, the dephosphorylated channels being electrically silent. Thus, agents that act to elevate the cAMP level would increase the fraction of the slow channels that are in the phosphorylated form, and hence available for voltage activation. Positive inotropic agents that do not elevate cAMP, but yet potentiate the Ca^{2+} slow current, may act by inhibiting the phosphatase which dephosphorylates the slow channels.[84]

Ca^{2+} influx into the myocardial cell can be controlled by extrinsic factors, such as by circulating hormones or neurotransmitter release. This extrinsic control of the Ca^{2+} influx is enabled by the special properties of the slow channels as, for example, the requirement for phosphorylation. These actions are mediated, at least in part, by changes in the levels of the cyclic nucleotides. In addition, the myocardial cell itself can exercise control over its Ca^{2+} influx, i.e., there is intrinsic control. For example, under conditions of transient regional ischemia, many of the slow channels through which Ca^{2+} passes become unavailable (or silent). This effect may be mediated by lowering the ATP level of the affected cells and by the accompanying acidosis. In addition, a rise in intracellular Ca^{2+} can also inactivate Ca^{2+} channels.[85] Thus, the myocardial cell can partially or completely suppress its Ca^{2+} influx under adverse conditions. This causes the affected cells to contract weakly or not at all (uncoupling contraction from excitation), and since most of the work done by the cell is

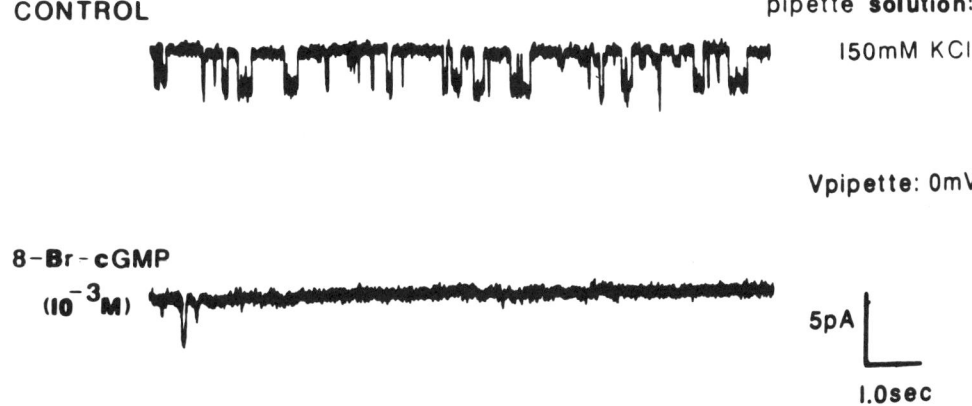

FIGURE 7. cGMP inhibition of single steady-state K$^+$ channel currents in a single 17-d-old embryonic chick heart cell (cultured for 4 d). At a V$_{pip}$ of 0 mV (i.e., at the resting potential) under control conditions, there was considerable channel activity. Following addition of 10^{-3} M 8-Br-cGMP, the K$^+$ channel activity was nearly abolished. (From Wahler, G. and Sperelakis, N., unpublished observations.)

mechanical, this serves to conserve ATP. Such a mechanism may serve to protect the myocardial cells under adverse conditions such as transient regional ischemia. Cyclic nucleotides may also regulate the functioning of other ionic channels, such as one of the K$^+$ channels (Figure 7).

REGULATION OF SLOW CHANNELS BY CYCLIC NUCLEOTIDES AND PHOSPHORYLATION IN YOUNG HEARTS

The effects of cyclic nucleotides were tested in isolated ventricles from 3-d-old embryonic chicks.[85a] Addition of 8-Br-cAMP to the bath (1 mM) increased $\dot{V}_{max}$, APD (action potential duration), AP amplitude, and AP overshoot. Forskolin (2 μM), a direct activator of adenylate cyclase, exerted a similar potentiation of these AP parameters, and increased the maximum diastolic potential (MDP) (i.e., hyperpolarized). Isoproterenol (0.5, 5 μM), an indirect stimulator of adenylate cyclase via the beta-adrenoceptor and G$_s$ coupling protein, had effects identical to those of forskolin and 8-Br-cAMP. These results indicate that cAMP potentiates the inward slow Na$^+$ current responsible for the upstroke of the AP in young embryonic chick hearts.

Addition of 8-Br-cGMP to the bath (1 mM) depressed the spontaneous APs of 3-d-old embryonic chick hearts under basal conditions.[85a] $\dot{V}_{max}$, APD, AP amplitude, AP overshoot, and MDP were all decreased within 30 to 60 min. APs potentiated by prior addition of 8-Br-cAMP, forskolin, or isoproterenol were also inhibited by 8-Br-cGMP. 8-Br-cGMP also inhibited the inward slow current that was potentiated by Bay-K-8644, a calcium slow channel agonist (direct activator) and slow Na$^+$ channel agonist.[86-89] These results indicate that cGMP inhibits the inward slow Na$^+$ current responsible for the upstroke of the AP, in a manner antagonistic to that of cAMP. Therefore, the slow Na$^+$ channels present in young hearts are regulated by cAMP and cGMP in a manner similar to that for the slow Ca^{2+} channels present in old embryonic and adult hearts.

Intracellular injection of calcineurin, a Ca^{2+}-calmodulin-dependent phosphatase, by the phosphotidylcholine liposome method, into 3-d-old embryonic chick hearts produced reversible inhibition of the inward current responsible for the AP upstroke.[85a] $\dot{V}_{max}$ of the basal APs was depressed by injection of calcineurin plus calmodulin, indicating that calcineurin inactivated the voltage-operated slow channels. Therefore, it appears that dephosphorylation of the slow Na^+ channel protein, or of an associated regulatory protein, inactivates the channel, consistent with the phosphorylation hypothesis for functioning of slow channels.

Diacetyl monoxime (DAM), a dephosphorylating agent, has a negative inotropic action on the heart, including a decrease in the Ca^{2+} sensitivity of the contractile proteins.[90] In embryonic chick hearts, DAM (10 to 20 mM) rapidly and reversibly depressed and blocked the Ca^{2+}-dependent slow APs in old (13- to 18-d-old) embryonic hearts and the Na^+-dependent slow APs in young (3-d-old) hearts.[86,87] Based on the fact that DAM suppressed the slow Na^+ current in young hearts as well as the slow Ca^{2+} current in old embryonic hearts, it was suggested that the slow Na^+ channels may need to be phosphorylated like the slow Ca^{2+} channels.

In addition, the pharmacology of these two types of slow channels, i.e., the slow Na^+ channel and the slow Ca^{2+} channel, appears to be similar. For example, the calcium antagonistic drugs have similar effects on both types.[47,48] In addition, the local anesthetics (lidocaine, procaine, procainamide, and quinidine) blocked the slow Na^+ channels of young embryonic chick hearts at concentrations similar to those for blockade of fast Na^+ channels in adult hearts.[91] It was shown that the local anesthetics also block the Ca^{2+} slow channels in concentrations similar for blockade of fast Na^+ channels.[92,93]

ACKNOWLEDGMENTS

The work of the author summarized and reviewed in this chapter was supported by a grant from the National Institutes of Health (HL-31942). The author is grateful to Rhonda S. Hentz for excellent assistance in preparation of the manuscript and to Paul Goosman for preparation of the figures.

REFERENCES

1. **Sperelakis, N.**, Developmental changes in membrane electrical properties of the heart, in *Physiology and Pathophysiology of the Heart,* Sperelakis, N., Ed., Martinus Nijhoff, Boston, 1984, 543.
2. **Sperelakis, N.**, Electrical field model for electric interactions between myocardial cells, in *Activation, Metabolism, and Perfusion of the Heart — Simulation and Experimental Models,* Sideman, S. and Bevar, R., Eds., Martinus Nijhoff, The Hague, 1987, 77.
3. **Sperelakis, N.**, Regulation of calcium slow channels in myocardial cells by cyclic nucleotides and phosphorylation, in *Protein Phosphorylation in Heart Muscle,* Solaro, R. J., Ed., CRC Press, Boca Raton, FL, 1986, 55.
4. **Sperelakis, N. and Bkaily, G.**, Cultured cell models for studying problems in cardiac toxicology, in *In Vitro Methods in Toxicology,* Atterwill, C. K. and Steele, C. E., Eds., Cambridge University Press, London, 1987, 77.
5. **Sperelakis, N. and Pappano, A.**, Physiology and pharmacology of developing heart cells, in *International Encyclopedia of Pharmacology and Therapeutics,* Papp, J. Gy., Ed., Pergamon Press, Oxford, 1983; *Pharmacol. Ther.,* 2, 1, 1983.
6. **Sperelakis, N. and Pappano, A.**, Physiology and pharmacology of developing heart cells, in *International Encyclopedia of Pharmacology and Therapeutics,* Papp, J. Gy., Ed., Pergamon Press, Oxford, in press.
7. **Bernard, C.**, Establishment of ionic permeabilities of the myocardial membrane during embryonic development of the rat, in *Developmental and Physiological Correlates of Cardiac Muscle,* Lieberman, M. and Sano, T., Eds., Raven Press, New York, 1976, 169.
8. **Boethius, J. and Knutsson, E.**, Resting membrane potential in chick muscle cells during ontogeny, *J. Exp. Zool.,* 174, 281, 1970.
9. **Couch, J. R., West, T. C., and Hoff, H. E.**, Development of the action potential of the prenatal rat heart, *Circ. Res.,* 24, 19, 1969.
10. **McDonald, T. F. and De Haan, R. L.**, Ion levels and membrane potential in chick heart tissue and cultured cells, *J. Gen. Physiol.,* 61, 89, 1973.
11. **Shimizu, Y. and Tasaki, K.**, Electrical excitability of developing cardiac muscle in chick embryos, *Tohoku J. Exp. Med.,* 88, 49, 1966.

12. **Sperelakis, N. and Shigenobu, K.**, Changes in membrane properties of chick embryonic hearts during development, *J. Gen. Physiol.*, 60, 430, 1972.
13. **Yeh, B. K. and Hoffman, B. F.**, The ionic basis of electrical activity in embryonic cardiac muscle, *J. Gen. Physiol.*, 52, 666, 1967.
14. **LeDouarain, G., Obrecht, G., and Coraboeuf, E.**, Déterminations régionales dans l'aire cardiaque presomptive mises en évidence chez l'embryon de poulet par la méthode microelectrophysiologique, *J. Embryol. Exp. Morphol.*, 15, 153, 1966.
15. **Sperelakis, N., Shigenobu, K., and McLean, M. J.**, Membrane cation channels: changes in developing hearts, in cell culture, and in organ culture, in *Developmental and Physiological Correlates of Cardiac Cells*, Lieberman, M. and Sano, T., Eds., Raven Press, New York, 1976, 209.
16. **Pappano, A. J.**, Sodium-dependent depolarization of non-innervated embryonic chick heart by acetylcholine, *J. Pharmacol. Exp. Ther.*, 180, 340, 1972.
17. **Carmeliet, E. E., Horres, C. R., Lieberman, M., and Vereecke, J. S.**, Developmental aspects of potassium flux and permeability of the embryonic chick heart, *J. Physiol. (London)*, 254, 673, 1976.
18. **De Haan, R. L.**, The potassium-sensitivity of isolated embryonic heart cells increases with development, *Dev. Biol.*, 23, 226, 1970.
19. **Löffelholz, K. and Pappano, A. J.**, Increased sensitivity of sinoatrial pacemaker to acetylcholine and to catecholamines at the onset of autonomic neuroeffector transmission in chick embryo heart, *J. Pharmacol. Exp. Ther.*, 191, 479, 1974.
20. **Pappano, A. J.**, Action potentials in chick atria: ontogenic changes in the dependence of tetrodotoxin-resistant action potentials on calcium, strontium, and barium, *Circ. Res.*, 39, 99, 1976.
21. **Sperelakis, N.** Electrical properties of embryonic heart cells, in *Electrical Phenomena in the Heart*, De Mello, W. C. Ed., Academic Press, New York, 1972, 1.
22. **Sperelakis, N.**, (Na^+-K^+)-ATPase activity of embryonic chick heart and skeletal muscles as a function of age, *Biochim. Biophys. Acta.*, 266, 230, 1972.
23. **Harsch, M. and Green, J. W.**, Electrolyte analyses of chick embryonic fluids and heart tissue, *J. Cell. Comp. Physiol.*, 62, 319, 1963.
24. **Klein, R. L.**, Ontogenesis of K and Na fluxes in embryonic chick heart, *Am. J. Physiol.*, 199, 613, 1970.
25. **Thureson-Klein, A. and Klein, R. L.**, Cation distribution and cardiac jelly in early embryonic hearts: a histochemical and electron microscopic study, *J. Mol. Cell. Cardiol.*, 2, 31, 1971.
26. **Sperelakis, N.**, Origin of the cardiac resting potential, in *Handbook of Physiology: the Cardiovascular System*, Vol. 1, Berne, R. M. and Sperelakis, N., Eds., American Physiological Society, Bethesda, MD, 1979, 187.
27. **Glitsch, H. G.**, An effect of the electrogenic sodium pump on the membrane potential in beating guinea-pig atria, *Pfluegers Arch.*, 344, 169, 1973.
28. **Isenberg, G. and Trautwein, W.**, The effect of dihydroouabain and lithium ions on the outward current in cardiac Purkinje fibers: evidence for electrogenicity of active transport, *Pfluegers Arch.*, 350, 41, 1974.
29. **Noma, A. and Irisawa, H.**, Contribution of an electrogenic sodium pump to the membrane potential in rabbit sinoatrial node cells, *Pfluegers Arch.*, 358, 289, 1975.
30. **Page, E. and Storm, S. R.**, Cat heart muscle *in vitro*. VIII. Active transport of sodium in papillary muscles, *J. Gen. Physiol.*, 48, 957, 1965.
31. **Vassalle, M.**, Electrogenic suppression of automaticity in sheep and dog Purkinje fibers, *Circ. Res.*, 27, 361, 1970.
32. **Lieberman, M., Horres, C. R., Aiton, J. F., and Johnson, E. A.**, Active transport and electrogenicity of cardiac muscle in tissue culture, in *27th Proc. Int. Congr. of Physiological Sciences*, Vol. 13, Paris, 1977, 446.
33. **Pelleg, A., Vogel, S., Belardinelli, L., and Sperelakis, N.**, Overdrive suppression of automaticity in cultured chick myocardial cells, *Am. J. Physiol.*, 238, H24, 1980.
34. **Jongsma, H. J., Masson-Pevet, M., and De Bruyne, J.**, Synchronization of the beating frequency of cultured rat heart cells, in *Developmental and Physiological Correlates of Cardiac Muscle*, Lieberman, M. and Sano, T., Eds., Raven Press, New York, 1976, 185.
35. **McLean, M. J. and Sperelakis, N.**, Retention of fully differentiated electrophysiological properties of chick embryonic heart cells in culture, *Dev. Biol.*, 50, 134, 1976.
36. **Sperelakis, N.**, Electrophysiology of cultured chick heart cells, in *Electrophysiology and Ultrastructure of the Heart*, Sano, T., Mizuhira, V., and Matsuda, K., Eds., Bunkodopp, Tokyo, 1967, 81.
37. **Sperelakis, N. and Lehmkuhl, D.**, Effect of current on transmembrane potentials in cultured chick heart cells, *J. Gen. Physiol.*, 47, 895, 1964.
38. **Sperelakis, N. and Lehmkuhl, D.**, Ionic interconversion of pacemaker and nonpacemaker cultured chick heart cells, *J. Gen. Physiol.*, 49, 867, 1966.
39. **Pappano, A. J. and Sperelakis, N.**, Low K^+ conductance and low resting potentials of isolated single cultured heart cells, *Am. J. Physiol.*, 217, 1076, 1969.
40. **Shigenobu, K. and Sperelakis, N.**, Development of sensitivity to tetrodotoxin of chick embryonic hearts with age, *J. Mol. Cell. Cardiol.*, 3, 271, 1971.

41. **Iijima, T. and Pappano, A. J.**, Ontogenetic increase of the maximal rate of rise of the chick embryonic heart action potential: relationship to voltage, time and tetrodotoxin, *Circ. Res.*, 44, 358, 1979.
42. **Marcus, N. C. and Fozzard, H.**, Tetrodotoxin sensitivity in the developing and adult chick heart, *J. Mol. Cell. Cardiol.*, 13, 335, 1981.
43. **Pappano, A. J.**, Calcium-dependent action potentials produced by catecholamines in guinea pig atrial muscle fibers depolarized by potassium, *Circ. Res.*, 27, 379, 1970.
44. **Schneider, J. A. and Sperelakis, N.**, The demonstration of energy dependence of the isoproterenol-induced transcellular Ca^{2+} current in isolated perfused guinea pig hearts — an explanation for mechanical failure of ischemic myocardium, *J. Surg. Res.*, 16, 389, 1974.
45. **Shigenobu, K. and Sperelakis, N.**, Ca^{2+} current channels induced by catecholamines in chick embryonic hearts whose fast Na^+ channels are blocked by tetrodotoxin or elevated K^+, *Circ. Res.*, 31, 932, 1972.
46. **McLean, M. J., Shigenobu, K., and Sperelakis, N.**, Two pharmacological types of slow Na^+ channels as distinguished by verapamil blockade, *Eur. J. Pharmacol.*, 26, 379, 1974.
47. **Shigenobu, K., Schneider, J. A., and Sperelakis, N.**, Blockade of slow Na^+ and Ca^{2+} currents in myocardial cells by verapamil, *J. Pharmacol. Exp. Ther.*, 190, 280, 1974.
48. **Kojima, M. and Sperelakis, N.**, Calcium antagonistic drugs differ in ability to block the slow Na^+ channels of young embryonic chick hearts, *Eur. J. Pharmacol.*, 94, 9, 1983.
49. **Galper, J. B. and Catterall, W. A.**, Developmental changes in the sensitivity of embryonic heart cells to tetrodotoxin and D 600, *Dev. Biol.*, 65, 216, 1978.
50. **Kasuya, Y., Matsuki, N., and Shigenobu, K.**, Changes in sensitivity to anoxia of the cardiac action potential plateau during chick embryonic development, *Dev. Biol.*, 58, 124, 1977.
51. **Nathan, R. D. and De Haan, R. L.**, *In vitro* differentiation of a fast Na^+ conductance in embryonic heart cell aggregates, *Proc. Natl. Acad. Sci. U.S.A.*, 75, 2776, 1978.
52. **Ishima, Y.**, The effect of tetrodotoxin and sodium substitution on the action potential in the course of development of the embryonic chicken heart, *Proc. Jpn. Acad.*, 44, 170, 1978.
53. **Renaud, J. F., Romey, G., Lombet, A., and Lazdunski, M.**, Differentiation of the fast Na^+ channel in embryonic heart cells: interaction of the channel with neurotoxins, *Proc. Natl. Acad. Sci. U.S.A.*, 78, 5248, 1981.
54. **Fujii, S., Ayer, R. K., Jr., and De Haan, R. L.**, Differentiation of transmembrane ionic currents in the early embryonic chick heart, *Prog. Dev., Biol. (Part A)*, 353, 1986.
55. **Sperelakis, N., Forbes, M. S., Shigenobu, K., and Coburn, S.**, Organ-cultured chick embryonic hearts of various ages. II. Ultrastructure, *J. Mol. Cell. Cardiol.*, 6, 473, 1974.
56. **Renaud, J.-F. and Sperelakis, N.**, Electrophysiological properties of chick embryonic heart grafted and organ cultured *in vitro*, *J. Mol. Cell. Cardiol.*, 8, 889, 1976.
57. **McLean, M. J., Renaud, J.-F., Sperelakis, N., and Niu, M. C.**, mRNA induction of fast Na^+ channels in cultured cardiac myoblasts, *Science*, 191, 297, 1976.
58. **Sperelakis, N., McLean, M. J., Renaud, J.-F., and Niu, M. C.**, Membrane differentiation of cardiac myoblasts induced *in vitro* by an RNA-enriched fraction from adult heart, in *The Role of RNA in Development and Reproductions*, 2nd Int. Symp., Niu, M. C. and Chuang, H. H., Eds., Van Nostrand Reinhold, New York, 1981, 730.
59. **Kojima, M. and Sperelakis, N.**, Development of slow Ca^{2+}-Na^+ channels during organ culture of young embryonic chick hearts, *J. Dev. Physiol.*, 7, 355, 1985.
60. **McLean, M. J., Lapsley, R. A., Shigenobu, K., Murad, R., and Sperelakis, N.**, High cyclic AMP levels in young embryonic chick hearts, *Dev. Biol.*, 42, 196, 1975.
61. **Renaud, J.-F., Sperelakis, N., and LeDouarin, G.**, Increase of cyclic AMP levels induced by isoproterenol in cultured and non-cultured chick embryonic hearts, *J. Mol. Cell. Cardiol.*, 10, 281, 1978.
62. **Thakkar, J. K. and Sperelakis, N.**, Changes in cyclic nucleotide levels during embryonic development of chick hearts, *J. Dev. Physiol.*, 9, 497, 1987.
63. **Reporter, M.**, An ATP pool with rapid turnover, within the cell membrane, *Biochem. Biophys. Res. Commun.*, 48, 598, 1972.
64. **Novak, E., Drummond, G. I., Skala, J., and Hahn, P.**, Development changes in cyclic AMP, protein kinase, phosphorylase kinase, phosphorylase in liver, heart and skeletal muscle of the rat, *Arch. Biochem. Biophys.*, 150, 511, 1972.
65. **Zalin, R. J. and Montague, W.**, Changes in cyclic AMP, adrenylate cyclase and protein kinase levels during the development of embryonic chick skeletal muscle, *Exp. Cell Res.*, 93, 55, 1975.
66. **Schneider, J. A. and Sperelakis, N.**, Slow Ca^{2+} and Na^+ current channels induced by isoproterenol and methylxanthines in isolated perfused guinea pig hearts whose fast Na^+ channels are inactivated in elevated K^+, *J. Mol. Cell. Cardiol.*, 7, 249, 1975.
67. **Watanabe, A. M. and Besch, H. R., Jr.**, Cyclic adenosine monophosphate modulation of slow calcium influx channels in guinea pig hearts, *Circ. Res.*, 35, 316, 1974.
68. **Josephson, I. and Sperelakis, N.**, 5'-Guanylimidodiphosphate stimulation of slow Ca^{2+} current in myocardial cells, *J. Mol. Cell. Cardiol.*, 19, 1157, 1978.

69. **Pappano, A. J., Hartigan, P. M., and Coutu, M. D.,** Acetylcholine inhibits the positive inotropic effect of cholera toxin in ventricular muscle, *Am. J. Physiol./Heart Circ. Physiol.*, 12, H343, 1982.
70. **Li, T. and Sperelakis, N.,** Stimulation of slow action potentials in guinea pig papillary muscle cells by intracellular injection of cyclic AMP, Gpp(NH)p, and cholera toxin, *Circ. Res.*, 52, 111, 1983.
71. **Vogel, S. and Sperelakis, N.,** Induction of slow action potentials by microiontophoresis of cyclic AMP into heart cells, *J. Mol. Cell. Cardiol.*, 13, 51, 1981.
72. **Bkaily, G. and Sperelakis, N.,** Injection of guanosine 5'-cyclic monophosphate into heart cells blocks calcium slow channels, *Am. J. Physiol./Heart Circ. Physiol.*, 17, H745, 1985.
73. **Wahler, G. M. and Sperelakis, N.,** Intracellular injection of cyclic GMP depresses cardiac slow action potentials, *J. Cyclic Nuleotide Protein Phosphorylation Res.*, 10, 83, 1985.
74. **Vogel, S. and Sperelakis, N.,** Blockade of myocardial slow inward current at low pH, *Am. J. Physiol.*, 233, C99, 1977.
75. **Belardinelli, L., Vogel, S. M., Sperelakis, N., Rubio, R., and Berne, R. M.,** Restoration of inward slow current in hypoxic heart muscle by alkaline pH, *J. Mol. Cell. Cardiol.*, 11, 877, 1979.
76. **Reuter, H. and Scholz, H.,** The regulation of the calcium conductance of cardiac muscle by adrenaline, *J. Physiol. (London)*, 264, 49, 1977.
77. **Sperelakis, N., Belardinelli, L., and Vogel, S. M.,** Electrophysiological aspects during myocardial ischemia, in *Proc. VIII World Congr. of Cardiology,* Excerpta Medica, Amsterdam, 1979, 229.
78. **Sperelakis, N. and Schneider, J. A.,** A metabolic control mechanism for calcium ion influx that may protect the ventricular myocardial cell, *Am. J. Cardiol.*, 37, 1079, 1976.
79. **Tsien, R. W., Giles, W., and Greengard, P.,** Cyclic AMP mediates the effects of adrenaline on cardiac Purkinje fibers, *Nature (London), New Biol.*, 240, 181, 1972.
80. **Greengard, P.,** in *Cyclic Nucleotides, Phosphorylated Proteins and Neuronal Function,* Raven Press, New York, 1978.
81. **Rinaldi, M. L., Capony, J.-P., and Demaille, J. C.,** The cyclic AMP-dependent modulation of cardiac sarcolemmal slow calcium channels, *J. Mol. Cell. Cardiol.*, 14, 279, 1982.
82. **Watanabe, A. M., Lindemann, J. P., Jones, L. R., Besch, H. R., Jr., and Bailey, J. C.,** Biochemical mechanisms mediating neural control of the heart, in *Disturbances in Neurogenic Control of the Circulation,* Abbound, F. M., Fozzard, H. A., Gilmore, J. P., and Reis, D. P., Eds., American Physiological Society, Bethesda, MD, 1981, 189.
82a. **Thakkar, J. K. and Sperelakis, N.,** unpublished observations.
83. **Bkaily, G. and Sperelakis, N.,** Injection of protein kinase inhibitor into cultured heart cells blocks calcium slow channels, *Am. J. Physiol./Heart Circ. Physiol.*, 246, H630, 1984.
84. **Vogel, S., Sperelakis, N., Josephson, I., and Brooker, G.,** Fluoride stimulation of slow Ca^{2+} current in cardiac muscle, *J. Mol. Cell. Cardiol.*, 9, 461, 1977.
85. **Tsien, R. W.,** Calcium channels in excitable cell membranes, *Annu. Rev. Physiol.*, 45, 341, 1983.
85a. **Tripathi, O. and Sperelakis, N.,** unpublished observations.
86. **Sada, H., Sada, S., and Sperelakis, N.,** Actions of the slow channel activator, Bay-K-8644, on the electrical activity of 3-day-old embryonic chick hearts, *Clin. Exp. Pharmacol. Physiol.*, 12, 521, 1985.
87. **Sada, H., Sada, S., and Sperelakis, N.,** Effects of diacetyl monoxime (DAM) on slow and fast action potentials of young and old embryonic chick hearts and rabbit hearts, *Eur. J. Pharmacol.*, 112, 145, 1985.
88. **Sada, H., Sada, S., and Sperelakis, N.,** Reactivation processes of three inward current systems involved in the rising phase of the action potentials in embryonic chick hearts, *Can. J. Physiol. Pharmacol.*, 64, 125, 1986.
89. **Sada, H., Sada, S., and Sperelakis, N.,** Recovery of the slow action potential is hastened by the calcium slow channel agonist, Bay-K-8644, *Eur. J. Pharmacol.*, 120, 17, 1986.
90. **Li, T., Sperelakis, N., TenEick, R. E., and Solaro, R. J.,** Effects of diacetyl monoxime on cardiac excitation-contraction coupling, *J. Pharmacol. Exp. Ther.*, 232 (3), 688, 1985.
91. **Riccioppo-Neto, F. and Sperelakis, N.,** Effects of lidocaine, procaine, procainamide and quinidine on electrophysiologic properties of cultured embryonic chick hearts, *Br. J. Pharmacol.*, 86, 817, 1986.
92. **Coyle, D. E. and Sperelakis, N.,** Bupivacaine and lidocaine blockade of calcium mediated action potentials in guinea pig ventricular muscle, *J. Pharmacol. Exp. Ther.*, 242, 1001, 1987.
93. **Josephson, I. and Sperelakis, N.,** Local anesthetic blockade of Ca^{2+}-mediated action potentials in cardiac muscle, *Eur. J. Pharmacol.*, 40, 201, 1976.

FUNCTIONAL DEVELOPMENT OF THE HEART: HEMODYNAMICS

Roger N. Ruckman, Susan A. O'Brien, and Donna J. Messersmith

INTRODUCTION

After the embryonic heart has developed from a straight tube to a looped series of chambers, and after initiation of the heart beat, certain hemodynamic characteristics of the heart become evident. This chapter reviews these hemodynamic effects and examines the relationship of form and function in normal cardiogenesis.

CELLULAR CHANGES

Migration, Torsion, Flexion

Determination of regulatory mechanisms is essential for an appreciation of normal cardiogenesis and equally important in understanding the etiology of congenital heart malformations. Some influential factors of normal cardiogenesis are listed in Table 1.

Although little is certain concerning the biochemical and morphogenetic processes taking place in cardiogenesis, influences of shape and function have been the subject of much investigation. Cell migration is evident in the chick embryo during tubular heart formation, as early as Hamburger-Hamilton (H-H) stage 9.[1] Continued growth and deformation play a major role in heart shaping and eventual looping.[2] At H-H stage 10 elongation of the embryonic chick heart tube is due to cell multiplication *in situ* as well as progressive addition of differentiated tissue from the caudal portion of the tube.[3,4] In addition, deformations dominate rapid morphogenetic events, whereas differential growth dominates slower morphogenetic events.[2]

In studies involving the embryonic chick heart, torsion has been observed to exert its influence, initially at H-H stage 10,[5,6] on the ultimate morphology of the heart. Torsion is an inherent property, dependent upon the intrinsic difference in potential between the two fused primordia of the cardiac tubes.[7-9] Because the cranial and caudal ends of the cardiac tube are fixed, differential growth and deformations result in torsion of the heart within this restricted region. Formation of the bulboventricular loop and later displacement of its convexity toward the caudal end of the embryo eventually results in a cephalic atrial region with respect to the caudally positioned ventricles.[1]

Toward the end of stage 10 (38 to 40 h) flexion becomes influential in regulation of morphogenesis, directing the entire embryo to adopt a "C" configuration. Cranial flexion is responsible for displacement of the head in a caudal direction until it contacts the cardiac tube, thus contributing to caudal movement of the tube itself. As a result, the primitve ventricle maintains its left-sided position and becomes oriented ventral and caudal to the atrium and dorsal to the bulbus cordis. Thus both processes, torsion and flexion, are influential in determining the eventual position of the heart.[1]

Cell Shape Change, Death, and Proliferation
Cell Shape Change

The transition from a straight cardiac tube to a four-chambered heart requires not only partitioning, but also shifts and absorption of tissue components. Implicit in this transition is change in cell shape. Endothelial cells have been of particular interest since they impinge on the lumen of the developing heart and are potentially susceptible to changes in morphology with alteration in hemodynamic function. Cells in the endocardium contribute to the makeup of cushion tissue, which, in turn, participates in intracardiac and outflow tract septation.

Table 1
DEVELOPMENTAL INFLUENCE ON MORPHOLOGY AND FUNCTION DURING ACTIVE CARDIOGENESIS

Source of influence	Effect	Stage of initial influence	
		Chick[a,b]	Human[b-e]
1. Initial cell migration, torsion, flexion	Establishment of primitive pump with regular heart rhythm	H-H 9 (29 h)	X (22 ± 1 d)
2. Cell shape change, death, and proliferation	Further morphologic development	H-H 9 (29 h)	X (22 ± 1 d)
3. Contractile activity	1. Establishment of forward flow	H-H 10 (33—38 h)	X (22 ± 1 d)
	2. Influence on flow-dependent morphology		
	3. Cardiac output		
4. Blood volume	1. Increased stroke volume	H-H 12 (45 h)	XI (24 ± 1 d)
	2. Increased blood pressure		
	3. Influences on flow-dependent morphology		
5. Cardiac jelly	1. Enhanced contractile force transmission	H-H 12 (45 h)	XI (24 ± 1 d)
	2. Valve action enhancement		
	3. Preservation of ventricular shape		
6. Heaped up endocardium	1. Enhanced contractile force	H-H 18 (65 h)	XI (24 ± 1 d)
	2. Uniform blood flow		
	3. Initiation of septation		
7. Innervation	1. Regulation of cardiac output	H-H 22 (84 h)	XI (24 ± 1 d)
8. Circulating catecholamines	1. Increased heart rate	Undetermined	Undetermined
	2. Influence on vascular tone		

[a] Hamburger V. and Hamilton, H. L., *J. Morphol.*, 88, 49, 1951.
[b] Sissman, N. J., *Am. J. Cardiol.*, 25, 141, 1970.
[c] Streeter, G. L., *Contrib. Embryol.*, 30, 211, 1942.
[d] Streeter, G. L., *Contrib. Embryol.*, 31, 27, 1945.
[e] Streeter, G. L., *Contrib. Embryol.*, 32, 133, 1948.

Early studies[10] suggested that endocardial thickenings of the atrioventricular (AV) canal and bulbus cordis might be in response to deforming forces from blood flow through the heart.

During looping, myocardial cells of the greater ventricular curvature change shape.[11] As the cells of greater curvature increase their apical surface area, the cells of lesser curvature do not, thus altering the shape of the entire organ. Whether these cells cause looping or act in response to bending of the heart shape is not clear. Hydration of glycosaminoglycans (GAG) within the cardiac jelly may provide enough internal force to promote looping. Thus cardiac jelly may be more actively influential in looping, whereas cellular shape change may play a passive role.[12-15]

Recent work looking at cell shape change with scanning electron microscopy has suggested that resorption of cardiac jelly under endocardial cells, possibly through the action of hyaluronidase, may allow portions of the endocardium to project into the lumen while other portions recede as infoldings.[16] Others have noted in the tubular heart that hyaluronidase injections into the cardiac jelly cause the tubular heart to shrink and become flabby,[17] i.e., the hydrostatic support system for the myocardium is lost. The ability to maintain an intact heart wall, therefore, depends in part on reinforcing fibers in the wall. Evidence exists that myofibrils align according to mechanical stresses generated within the heart.[18] The resultant arrangement of myofibrils helps to stabilize the deforming forces on cells that result from pumping action of the heart.

Experimentation concerning the effect of alcohol on the embryonic chick addresses the importance of myofibrillar integrity. Analysis of electron micrographs shows that myofibrils are among the last cellular components to break down in response to alcohol exposure and may be responsible for maintaining integrity of myocardial function in the stressed embryo.[19-22]

Cell Death

In addition to cellular growth and deformation, shape change occurs in response to physiologic cell death. A large body of data has been developed showing the importance of cell death in both normal and abnormal cardiac development.[23] Hughes[24] and Menkes et al.[25] reported the occurrence of cell death in the chick embryo during remodeling of the dorsal aorta, involution of the ductus arteriosus, formation of the aorticopulmonary septum, and absorption of the cardiac bulb. Pexieder[23] and Pexieder and Paschoud[26] demonstrated cell death foci in the AV cushions, zone of fusion of the AV cushions, conotruncal (bulbus) cushions, zone of fusion of the conotruncal (bulbus) cushions, walls of the aorta and the pulmonary artery, and semilunar valves of the aorta and pulmonary artery. Some of the observed cell death has been shown to be interrelated with hemodynamics of the heart. Menkes et al.[27] pointed out that cell death in abnormal areas may lead to abnormal cardiac development. Certain experimental interventions that change hemodynamics will increase cell death in key developmental areas. For example, manipulation of the aortic arches causes appearance of a new cell death focus in the sensitive conotruncal region. The result is incomplete ventricular septation and the persistence of ventricular septal defects (Figure 1).[28,29]

Cell Proliferation

Cell proliferation is also important in cardiogenesis, and regional variations of growth have been shown in the heart.[30] Autoradiography, using tritiated thymidine[31] and mitotic indices in different portions of the heart, has been most commonly used.[32-34] However, the interrelationship, if any, with hemodynamic function remains unclear.

HEMODYNAMIC EFFECTS

Experimental investigations in embryonic hearts have provided information regarding

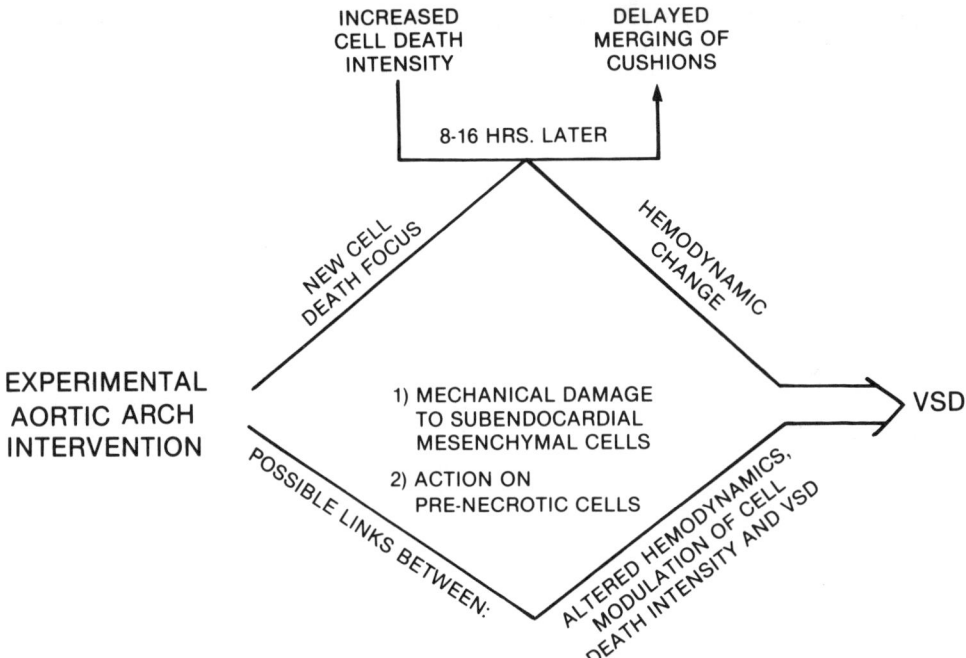

FIGURE 1. Diagram illustrating the effects of aortic arch intervention on cell death and merging of cushions (above) with proposed mechanisms of action (below).[29] Rychter and Lemez noted a 92% rate of ventricular septal defects using this hemodynamic intervention.[28]

early functional changes in heart rate, blood volume, stroke volume, cardiac output, and blood pressure (Figure 2).

Heart Rate

By direct inspection, the earliest contractile activity, usually occurring along the right margin of the developing ventricular component of the cardiac tube, is irregular and localized. This is noted at ten somites in the chick.[35,36] Within an hour, the opposite ventricular wall shows contractions, gradually becoming synchronous such that the opposing walls of the primitive ventricle move rhythmically toward each other. Initial heart rates may be as low as eight beats per minute with frequent pauses. By 12 to 13 somites, the atrium also shows contractile activity. Consequently, the sequence of atrial, then ventricular contraction is established, thus permitting blood flow (16 to 17 somites). The initial progression of blood is jerky with both forward and backward movement of red blood cells. Development of uniform forward flow from the initial peristaltoid contractions becomes evident with the appearance of heaped-up endocardium serving as a "valve" in the AV canal.[37] In addition, measurement of electrograms and intracellular action potentials confirms a cephalocaudal rate gradient in the heart tube with evidence that the pacemaker is in the sinoatrial portion of the heart from the beginning of contraction.[38] For the first 3 d of development, heart rate is independent of direct nervous control and the influence of circulating catecholamines.[39] Heart rate varies directly with metabolic rate and temperature and inversely with body weight.[40] In addition, rate will increase with rising blood pressures.[41,42] The most rapid rise in rate occurs in the 48 h following the start of contractions, with more gradual rise to a peak and subsequent plateau level reached by the eighth day of incubation, all such changes occurring during active cardiogenesis.[40]

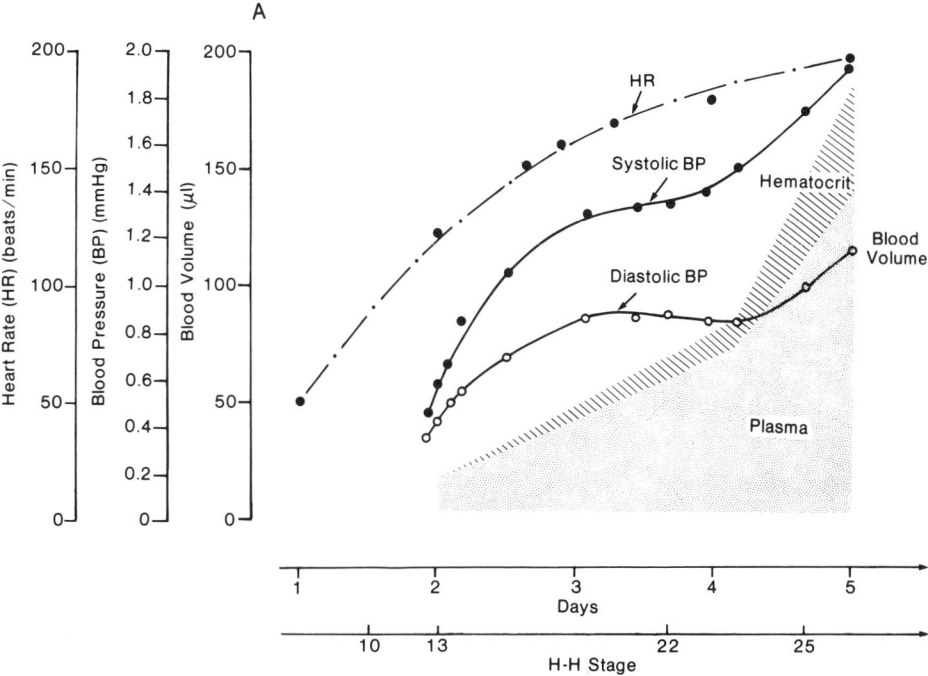

FIGURE 2. Early morphologic and functional changes in (chick) heart development. (A) Illustrates average heart rate, systolic and diastolic blood pressures, and total blood volume with hematocrit and plasma components. Morphologic heart changes at approximate corresponding Hamburger-Hamilton stages and days of incubation are included at center.[1,40,43,52] (B) Illustrates average dorsal aortic blood velocity, stroke volume, cardiac output, and shortening fraction (contractility). Heart rate is included for comparison.[48,53]

Blood Volume

Dye dilution studies, using 1% Geigy blue solution, have shown that circulating blood volume increases exponentially after day 2 in the chick embryo heart. The increase in blood volume, as confirmed by serial determinations of erythrocyte count, hematocrit, and hemoglobin assay, is due primarily to changes in plasma volume. When total blood volume is divided by embryo weight, there is a continuous decrease in the ratio after day 5.[43] However, hemoglobin per gram of embryo remains relatively constant after day 8 in the chick.

Stroke Volume

Along with an increase in circulating plasma volume, there is an associated change in stroke volume of the primitive ventricle. By inspection, changes in ventricular size have been noted.[44] We have made measurements of these changes in dimension at both end diastole and end systole, using cinephotography.[45] Such measurements, in turn, allow calculation of end diastolic volume, end systolic volume, stroke volume, and cardiac output. In addition, indices of function, shortening fraction, and ejection fraction can be determined. Our findings confirm the observations, from inspection and from micromethod determination of blood volume,[43] that stroke volume increases with time in the primitive ventricle.

Frank-Starling Relationship

Investigations in fetal lambs, examining cardiac function in late gestation, have demonstrated that changes in resting myocardial fiber length help determine stroke volume and cardiac output. The response to changes in filling pressure has been noted for both left[46]

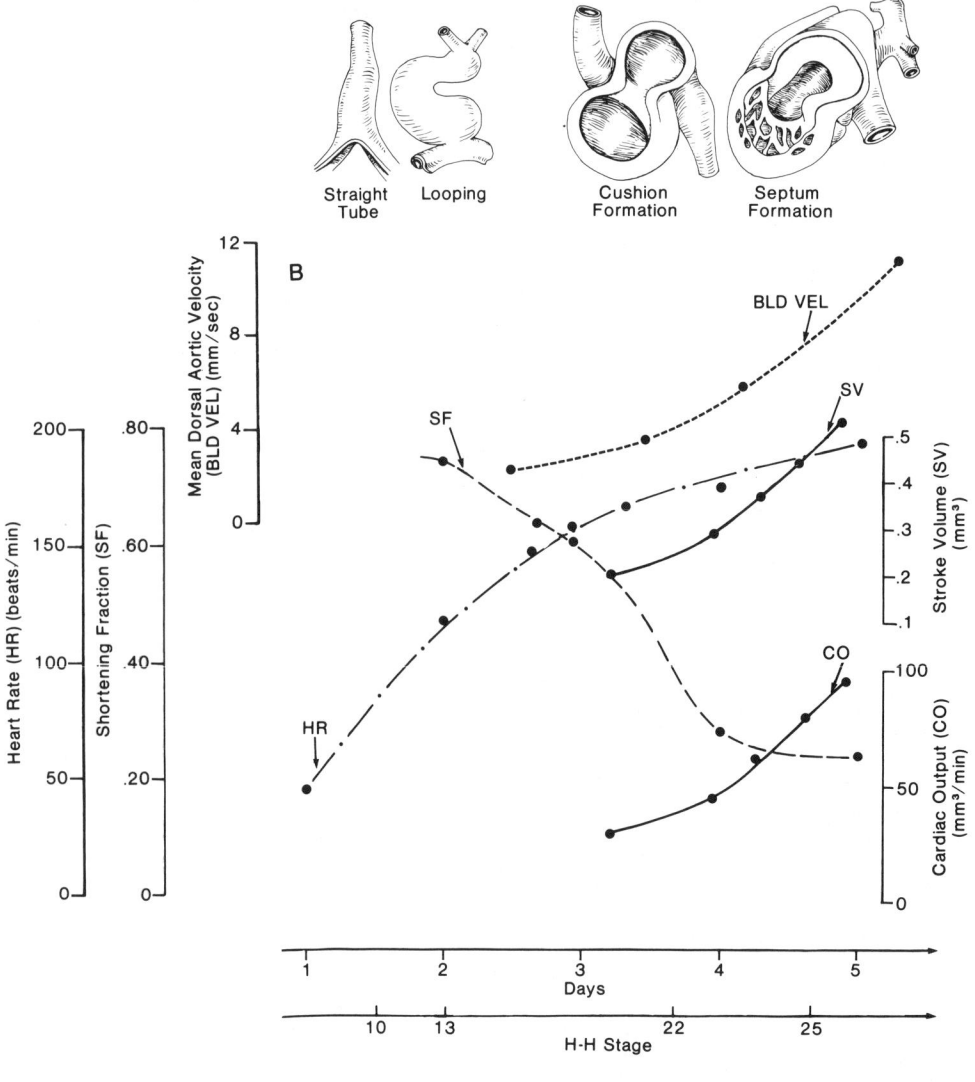

FIGURE 2B.

and right[47] ventricles. Study of the early embryonic chick has suggested a similar stretch response that affects both stroke volume[48] and heart rate.[42]

Blood Pressure

With the advent of micropuncture techniques, it became possible to cannulate various points of the embryonic circulation. Most studies have used glass capillaries drawn out to tip diameters from 5 μm to 0.8 mm depending on the point of cannulation. The capillary tubes are in turn connected to a manometer either directly[42,49-51] or with electronic signal processing.[52,53] In the chick, systolic pressures in the range of 0.5 mmHg have been noted after 2 d incubation with gradual rise over the next 10 d to over 10 mmHg.[51,52]

Closer examination of specific systolic and diastolic pressures shows a gradual, but not linear, rise in each pressure with close correlation to embryonic weight.[53] Regional differences in blood pressure in the aortic arches have been described and are felt to be related to different flow patterns.[54]

Of particular interest are experiments involving direct ventricular puncture. By stage

26, the pressure tracings show that reflux of blood into the ventricle during diastole is prevented by an arterial "valve" mechanism produced by masses of endocardial tissue in the outflow portion of the primitive ventricle.[36,38] Other ventricular puncture experiments in the 3-d chick embryo have demonstrated a relationship between blood pressure and heart rate.[42] Introduction of fluid into the ventricle with associated increase in ventricular pressure causes a rise in heart rate, probably through a stretch mechanism.[42] Nonphysiologic increases of circulating fluids due to hypoxia[55] or trypan blue[56] will cause both increases in blood pressure and distension of vessels[56] or the ventricle itself.[57] Endothelial integrity can thus be lost with hemorrhage and associated hematoma formation.

Flow Patterns

After the heart develops from a straight tube through the looping stage, distinct streams of blood are noted through the heart and into the arch vessels. Early experimental data suggested spiral streaming of blood, which could participate in shaping the heart during cardiogenesis.[58] Many papers have been subsequently published, some asserting that streams of blood flow are major determinants of heart shape[59-67] and others suggesting that the form of the heart develops independently of any stream effects.[68] A recent study, using microangiography, helps put the issue of flow-molding theory in pespective.[69] Methylene blue, injected into vitelline veins proximal to the sinus venosus, divides into two streams (A and B) which run respectively dorsal and ventral in the sinus venosus, cranial and caudal in the atrium and AV canal, ventral and dorsal in the primitive ventricle and conus cordis, and finally separate either into the left or right branchial arches. It is noteworthy that the streams run parallel rather than spiralling in the conus cordis. In addition, this longitudinal separation of streams was maintained in embryos with slower heart rates (from cooling), and the stream pattern in the chick reverses itself from stage 18 to stage 22. Since streaming is neither spiral in spatial pattern nor fixed in temporal development, questions are raised concerning the role of streaming in initiating septation or in subsequent molding of the heart (Figure 3).

However, recent work does suggest that hemodynamics can affect valve development. Examination of abnormal flow states and their consequences on morphogenesis can help our understanding of the normal. Such examination has been done in 3.5-d chick embryos, subjected to insertion of a fine wire behind the trunco-conal portion of the heart. This technique, modified from a similar procedure to look at abnormal septation,[70,71] allows examination of downstream valve development, independent of any direct tissue injury from the wire insertion.[72] Abnormalities of semilunar valve leaflets are noted in the experimental embryos, but not in controls. Overall, it appears that blood flow helps to determine shape and orientation of endothelial cells.[73,74] Microscopic studies have shown that the endothelial cells of the semilunar valves are normally arranged in conformation with the direction of flow, but are arranged randomly when flow is disturbed.[72] Others have suggested that altered local flow first acts on the mesenchymal core of the valve cusps with alteration of extracellular materials and then changes in endothelial cell arrangement.[75,76]

Cardiac Jelly

The extracellular matrix (ECM), termed cardiac jelly, is considered to be the largest component of the early developing embryonic heart.[77] As a transmitter of contractile force[44,45] and a source of stability and shape change,[78] ECM is a key influence in normal development.[2] However, elaboration of the cardiac jelly is one of the most important morphogenetic events taking place in the early stages of the embryonic chick heart.[79] Its alteration or disturbance can lead to experimentally induced cardiac malformations which confirm ECM as a potential influence in abnormal development.[78,80]

The early vertebrate heart (H-H stages 10 to 13) consists of an outer myocardium and

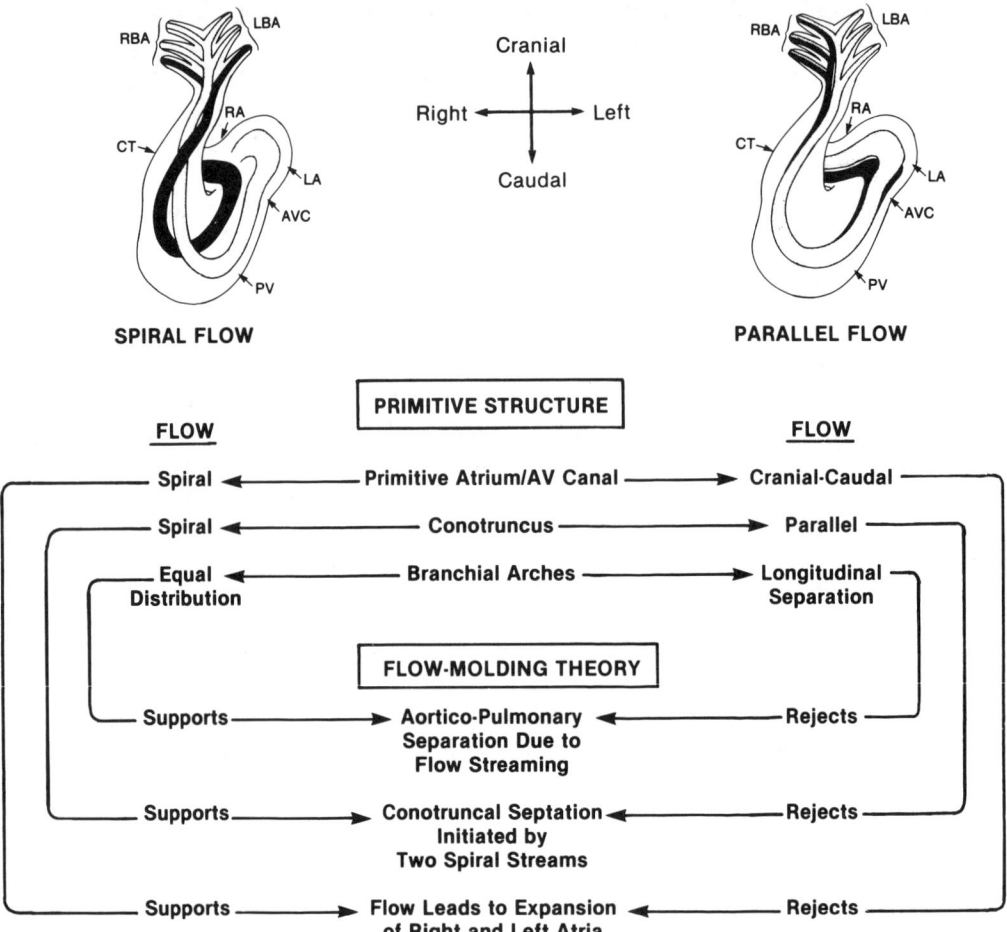

FIGURE 3. Parallel flow, supported by Yoshida et al., argues against the traditional flow-molding theory: RBA, right branchial arches; LBA, left branchial arches; RA, right side of primitive atrium; LA, left side of primitive atrium; CT, conotruncus; AVC, atrioventricular canal; RV, primitive ventricle. (From Yoshida, H., Manasek, F., and Arcilla, R. A., *Circ. Res.*, 53, 363, 1983. With permission.)

an inner endocardium separated by cardiac jelly. The matrix is produced largely by the myocardium, and most of its constituent macromolecules include type I collagenous molecules,[81] GAG,[82-85] and glycoproteins.[86] The origin of these macromolecules in the early chick heart appears to be myoblasts and myocytes.[87] Later in development, the endocardium and endocardial cushion cells appear to contribute to the matrical components of the conotruncus and AV canal.[82,88,89] Endocardial cushion cells may be directed by the ECM in their centrifugal migration through the cardiac jelly to the myocardium (see Figure 5). Manasek[87] suggested that such proliferation in restricted regions may take place as a result of "regional composition differences" in the ECM.

Cardiac jelly appears to facilitate the transmission of the force of myocardial contraction during systole through its elastic properties.[44,45] By stage 18, cardiac jelly enhances the "valve" action of the endocardium heaped up in the AV canal and also helps to preserve functional shape of the developing ventricle due to its inherent stability. Studies of isolated cardiac jelly have shown that the presence of GAG with associated hydration as well as the interaction of macromolecules with a filamentous network in the jelly are responsible for this stability.[12] It has been further shown that the filaments are oriented radially along the

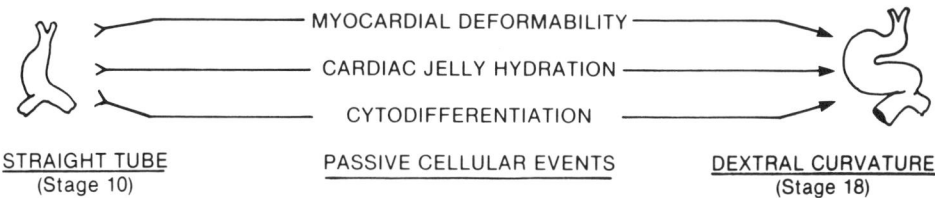

FIGURE 4. Three key factors central to normal cardiac looping. (From Manasek, F. J., *Fed. Proc., Fed. Am. Soc. Exp. Biol.,* 40, (7), 2011, 1981. With permission.)

lines of principal tensile stress, and cardiac jelly thickness increases with each systole.[17] Therefore, these characteristics also increase the efficiency of pumping action.

The interaction of distending forces and fibrillogenesis in the cardiac jelly may also be involved in morphogenesis of the myocardial wall and the looping cardiac tube.[13] The optimal radial distribution[17] of the filaments permits looping of the cardiac tube where a random network of fibers would inhibit bending and require additional force to complete looping. In a likewise manner, longitudinal fibers would restrict bending and, if fixed, would require breaking for bending to occur.

Recent investigation suggests that myocardial cells may play a passive role in looping, whereas hydration of the cardiac jelly and reversible swelling of the GAG could provide enough internal force to deform the heart.[13] In addition, according to this model, changes in myocardial elasticity or a decrease in force would terminate bending. Cytodifferentiation may be related to such changes.[90,91] In fact, myofibrillar accumulation[92] within the myocardium may reduce the myocardial deformability of the constituent cells and consequently stop bending.[13] Thus, interplay of three key elements appears central to the resulting heart shape: differential myocardial deformability, increase in internal pressure, and cytodifferentiation[2] (Figure 4).

Endocardial Cushion Formation

As the paired primordial heart tubes fuse along the midline, the mesenchyme around them thickens to form the myoepicardial mantle. This layer, which gives rise to the myocardium and epicardium, is separated from the inner endocardium by an acellular matrix called cardiac jelly. Investigation has shown that cells from the inner endocardium migrate through the cardiac jelly to the myocardium, where they form the endocardial cushions[93,37] (Figure 5).

Endocardial cushions appear in the region of the future AV canal and outflow tract (bulbus cordis) shortly after the onset of cardiac function and after the heart establishes asymmetry (D-loop). The other two regions of the primitive heart (atrium and ventricle) remain acellular. Appearing as swellings and functioning as primitive valves,[37] the cushions constitute the primordia of the adult valves and membranous septa.[94] The cushions are composed of endothelium, cushion cells, and cardiac jelly.

Determination of cellular origin of cushion mesenchyme is essential for an understanding of possible mechanisms leading to anomalous cushion development and resulting valvular defects. Results of an investigation by Patten[37] suggest that mesenchymal cells are derived from endothelium. In support of this finding, tunicates, which possess no such endothelium, have no cushion tissue.[95,96] The lack of organized myofibrils in the earliest cushion tissue is an indication that the origin of cushion cells is not the myocardium. However, Viragh and Challice[97] presented evidence suggesting the splanchnic mesoderm to be the contributor of mesenchymal cells to the cardiac jelly of the distal truncus in the 9-d mouse heart.

Markwald et al.[93] conducted an extensive investigation using ultrastructural, cytochemical, histologic, and autoradiographic techniques to determine cushion tissue origin and mechanisms underlying cellular transformation and eventual migration of presumptive cush-

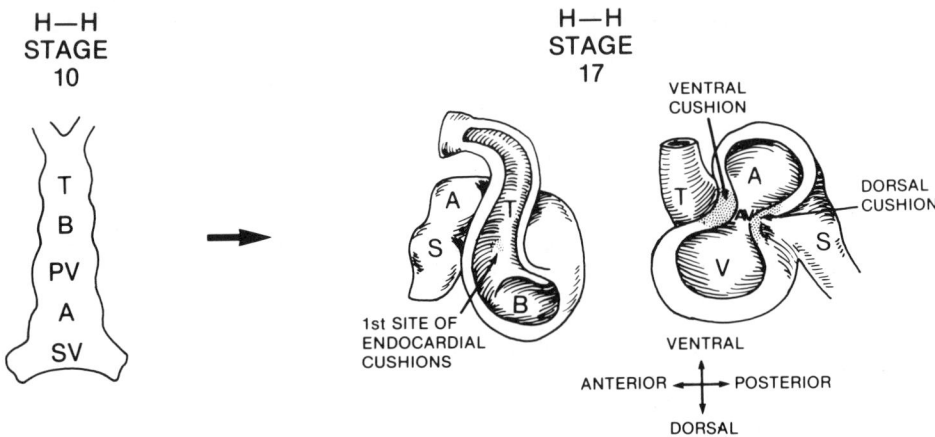

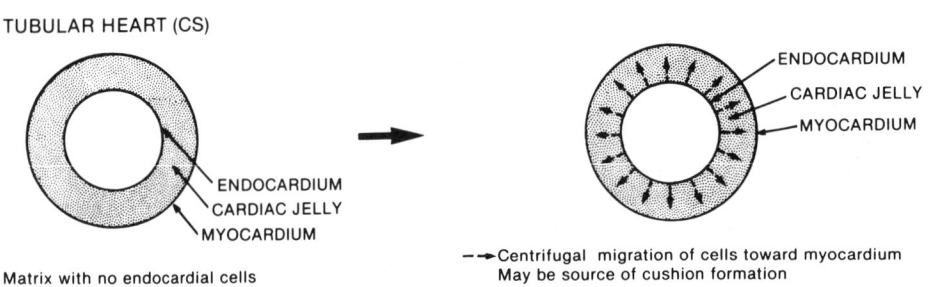

FIGURE 5. A schematic representation of endocardial cushion cell migration and subsequent accumulation at the conotruncus and atrioventricular canal. Cellular events depicted on the bottom underlie gross morphologic changes illustrated above: T, truncus; B, bulbus; PV (V), primitive ventricle; A, atrium; SV (S), sinus venosus; AV, atrioventricular canal; CS, cross section. (Upper left, from Keith, A., in *Studies in Pathology*, Bullock, E. H., Ed., University of Aberdeen, Scotland, 1906, 55; Keith, A., The Hunterian lectures on malformations of the heart, *Lancet*, 2, 433, 1909; Keith, A., Fate of the bulbus cordis in the human heart, *Lancet*, 2, 1267, 1924.) (Upper right, from de la Cruz, M. V., Muñoz-Armas, S., and Muñoz-Castellanos, L., *Development of the Chick Heart*, The Johns Hopkins University Press, Baltimore, 1972. With permission.)

ion cells to the myocardium. The endocardium is surrounded by a layer of simple squamous epithelium. Following activation, the endothelium undergoes an "epithelial-mesenchymal" transition, whereby the cardiac jelly is seeded with mesenchymal cells. Collagenous myofibrils present in the cardiac jelly serve as a scaffold through which mesenchymal cells migrate to the myocardium where they undergo stratification and differentiation.

The endothelial layer is absent of basal lamina which might otherwise restrict cellular movement. In addition, appendages (filopodia, pseudopodia) have been observed immediately prior to the appearance of the first cushion cells. Fitzharris[16] noted these projections, or "flutes", in chicks from H-H stage 13 (immediately prior to the onset of cushion formation), through H-H stage 25 (completion of cushion tissue formation).

Histochemical studies have cited hyaluronate (produced by the myocardium) as a major constituent of cardiac jelly.[82] Rapid hyaluronate synthesis precedes cell migration and results in expansion of extracellular space.[98-100] Toole[101] proposed that hyaluronate promotes cell locomotion through activation of cell surface receptors and demonstrated that hyaluronate must be eliminated to allow immobilization or interaction with other matrical components.

The critical role potentially played by GAG has been a subject of intensive investigation.[93] Sulfated GAG synthesis occurs in migrating cushion cells. Radioactively labeled ^{35}S-GAG is taken up by the matrix and incorporated into chondroitin sulfate, possibly enabling the matrix to play a directional role during cushion formation.[93] Exposure of 2- and 3-d-old chick embryos to N-acetylsalicylic acid may inhibit GAG synthesis, such as hyaluronic acid. Surviving embryos show defects in the formation of cushion cells as well as in the matrix. Such cardiac alterations may be implicated in the developmental malformation of the AV canal and outflow tract.[79] Additional experimentation involving teratogenic exposure has demonstrated decreased incorporation of labeled sulfate into GAG[102] and inhibition of GAG synthesis,[78] thus inhibiting cardiac matrix formation. Observed defects include outflow obstruction, valvular atresia, and septal defects. In addition, a defect in one or more of the three component processes leading to cushion tissue formation may cause maldevelopment of valves as well as septal defects with resultant altered hemodynamics.[87,88,103]

Innervation and Catecholamine Actions

The developing chick embryo is free of receptors for either nervous or biochemical stimulation during the first 3 d of incubation. However, by stage 22 ($3^1/_2$ d), histologically definable nervous tissue is present in the heart.[104-106] Sympathomimetic receptors are noted in earliest form by day 3,[107] in more defined form on the fourth day,[39] and vagal control is present by day 5.[108] Consequently, powerful modulators of hemodynamic function are appearing during active morphogenesis of the heart. Experiments in which catecholamines have been given to 5-d chick embryos have shown a profound decrease in stroke volume and associated cardiac output.[109] Such a change in flow may have impact on normal development of the heart and aortic arches. A study in 3- to 4-d chick embryos showed that propranolol, a blocker of beta-adrenergic effect, causes slowing of the heart rate. Embryos in the study were subsequently observed to show a variety of cardiac anomalies, presumably related to the change in hemodynamics.[110] Similar changes in cardiac anatomy were noted after prolonged hypothermia, which causes a similar reduction in heart rate.[111] By contrast, the vagus nerve appears to have little influence on heart rate at this stage in the chick.[112]

Catecholamines may affect vascular tone related to the presence or absence of receptors. Administration of epinephrine to 2- to 2.5-d chicks causes a fall in both systolic and diastolic pressure, whereas no change is noted at 3 to 3.5 d. After 3.5 to 4 d, there is a prominent increase in systolic, diastolic, and pulse pressure. The effect of epinephrine on heart rate was found to be inconsistent in these studies.[113] Another study confirmed chronotropic sensitivity to catecholamines at 4 to 5 d and sensitivity of blood pressure at 6 d. Repetition of catecholamine dose decreases the response. Both alpha and beta vascular receptor activity is demonstrable using blocking agents at 7 d.[114] In addition, it is suspected that the change in blood pressure caused by catecholamines is mediated primarily through the extra embryonic vasculature.[113]

CONCLUSION

The relationship between form and function has repeatedly manifested itself in the regulatory mechanisms underlying normal cardiogenesis. Moreover, the interrelation of form and function has been shown to be a potential source of cardiac maldevelopment. Thus, as hemodynamic effects are central to our understanding of normal cardiogenesis, they are equally essential for a proper understanding of cardiac maldevelopment.

REFERENCES

1. **de la Cruz, M. V., Muñoz-Armas, and Muñoz-Castellanos, L.**, *Development of the Chick Heart*, The Johns Hopkins University Press, Baltimore, MD, 1972.
2. **Manasek, F. J.**, Determinants of heart shape in early embryos, *Fed. Proc., Fed. Am. Soc. Exp. Biol.*, 40(7), 2011, 1981.
3. **Lopori, N. G.**, Research on heart development in chick embryo under normal and experimental conditions, *Monit. Zool. Ital.*, 1, 159, 1967.
4. **Stalsberg, H. and De Haan, R. L.**, The precardiac areas and formation of the tubular heart in the chick embryo, *Dev. Biol.*, 19, 128, 1969.
5. **Hamburger, V. and Hamilton, H. L.**, A series of normal stages in the development of the chick embryo, *J. Morphol.*, 88, 49, 1951.
6. **Hamilton, H. L.**, *Lillie's Development of the Chick*, 3rd ed., Holt, Rinehart, & Winston, New York, 1952.
7. **De Haan, R. L.**, Cell interactions and oriented movements during development, *J. Exp. Zool.*, 157, 127, 1964.
8. **Orts Llorca, F. and Gil, D. R.**, A casual analysis of the heart curvatures in the chick embryo, *Wilhelm Roux Arch. Entwicklungsmech. Org.*, 158, 52, 1967.
9. **Stalsberg, H.**, Mechanism of dextral looping of the embryonic heart, *Am. J. Cardiol.*, 25, 265, 1970.
10. **Chang, C.**, On the reaction of the endocardium to the blood stream in the embryonic heart, with special reference to the endocardial thickenings in the atrioventricular canal and the bulbus cordis, *Anat. Rec.*, 51, 253, 1932.
11. **Manasek, F. J., Burnside, M. B., and Waterman, R. E.**, Myocardial cell shape change as a mechanism of embryonic heart looping, *Dev. Biol.*, 29, 349, 1972.
12. **Nakamura, A. and Manasek, F. J.**, Experimental studies of the shape and structure of isolated cardiac jelly, *J. Embryol. Exp. Morphol.*, 43, 167, 1978.
13. **Nakamura, A., Kulikowski, R. R., Lacktis, J. W., and Manasek, F. J.**, Heart looping: a regulated response to deforming forces, in *Etiology and Morphogenesis of Congenital Heart Disease*, Van Praagh, R. and Takao, A., Eds., Futura, New York, 1980.
14. **Wainwright, S. A., Biggs, W. D., Currey, J. D., and Gosline, J. M.**, *Mechanical Design in Organisms*, John Wiley & Sons, New York, 1976, 293.
15. **Lacktis, J. W. and Manasek, F. J.**, An analysis of deformation during a normal morphogenic event, in *Morphogenesis and Malformation of the Cardiovascular System*, Rosenquist, G. C. and Bergsma, D., Eds., Alan R. Liss, New York, 1978, 205.
16. **Fitzharris, T. P.**, Endocardial shape change in the truncus during cushion tissue formation, in *Perspectives in Cardiovascular Research*, Vol. 5, Pexieder, T., Ed., Raven Press, New York, 1981, 227.
17. **Nakamura, A. and Manasek, F. J.**, Cardiac jelly fibrils: their distribution and organization, in *Morphogenesis and Malformation of the Cardiovascular System*, Rosenquist, G. C. and Bergsma, D., Eds., Alan R. Liss, New York, 1978, 229.
18. **Thornell, L. E., Sjorstrom, M., and Andersson, K. E.**, The relationship between mechanical stress and myofibrillar organization in heart Purkinje fibers, *J. Mol. Cell Cardiol.*, 8, 689, 1976.
19. **Ruckman, R. N., Messersmith, D. J., O'Brien, S. A., and Morse, D. E.**, Alcohol sensitivity in the embryonic heart, *Pediatr. Cardiol.*, 5, 254, 1984.
20. **Ruckman, R. N., O'Brien, S. A., Messersmith, D. J., Boeckx, R. L., and Morse, D. E.**, Changes in myocardial structure and function in the embryonic chick exposed to ethanol, *Teratology*, 31, 31A, 1985.
21. **Ruckman, R. N., O'Brien, S. A., Messersmith, D. J., Boeckx, R. L., and Morse, D. E.**, Alterations of myocardial function and structure in a model of alcohol embryopathy, *Am. Heart J.*, 110, 705, 1985.
22. **Ruckman, R. N., Messersmith, D. J., O'Brien, S. A., Getson, P. R., Boeckx, R. L., and Morse, D. E.**, Chronic ethanol exposure in the embryonic chick heart: effect on myocardial function and structure, *Teratology*, 37, 317, 1988.
23. **Pexieder, T.**, Cell death in the morphogenesis and teratogenesis of the heart, *Adv. Anat. Embryol. Cell Biol.*, 51/3, 1, 1975.
24. **Hughes, A. F. W.**, The histogenesis of arteries in the chick embryo, *J. Anat.*, 77, 266, 1948.
25. **Menkes, B., Alexandru, C., Pavkov, A., and Mircova, O.**, Researches on the formation and the elastic structure of the aorticopulmonary septum in the chick embryo, *Rev. Roum. Embryol. Cytol. Serv. Embryol.*, 2, 79, 1965.
26. **Pexieder, T. and Paschoud, N.**, La stabilité phylogénétique des zones de la mort cellulaire physiologique dans l'organogenèse du coeur, *Acta Anat. (Basel)*, 86, 321, 1973.
27. **Menkes, B., Sandor, S., and Ilies, A.**, Cell death in teratogenesis, *Adv. Teratol.*, 4, 170, 1970.
28. **Rychter, Z. and Lemez, L.**, The vascular system of the chick embryo. III. On the problem of septation of the heart bulb and trunk in chick embryos, *Cesk. Morphol.*, 7, 21, 1959.

29. **Pexieder, T.**, Über die Wirkung der Haemodynamik auf den Zelluntergang in den Herzbulbuswülsten des Hühnerembryos, *Acta Anat. (Basel)*, 82, 459, 1972.
30. **DeVries, P. A. and Saunders, C. M.**, The development of ventricles and spiral outflow tract in the human heart: a contribution to the development of the human heart from age group IX to age group XV, *Carnegie Inst. Washington Contrib. Embryol.*, 37, 87, 1962.
31. **Sissman, N. J.**, Cell multiplication rates during development of the primitive cardiac tube in the chick embryo, *Nature (London)*, 210, 504, 1966.
32. **Stalsberg, H.**, Regional mitotic activity in the precardiac mesoderm and differentiating heart tube in the chick embryo, *Dev. Biol.*, 20, 18, 1969.
33. **Thompson, R. P. and Fitzharris, T. P.**, Morphogenesis of the truncus arteriosus of the chick embryo heart: the formation and migration of mesenchymal tissue, *Am. J. Anat.*, 154, 545, 1979.
34. **Rosenquist, G. C.**, Endoderm/mesoderm multiplication rates in stage 5—12 chick embryos, *Anat. Rec.*, 202, 95, 1982.
35. **Sabin, F. R.**, Studies on the origin of blood-vessels and of red blood-corpuscles as seen in the living blastoderm of chicks during the 2nd day of incubation, *Carnegie Inst. Washington Contrib. Embryol.*, 9, 213, 1920.
36. **Patten, B. M.**, Initiation and early changes in the character of the heart beat in vertebrate embryos, *Physiol. Rev.*, 29, 31, 1949.
37. **Patten, B. M.**, Valvular action in the embryonic chick heart by localized apposition of endocardial masses, *Anat Rec.*, 102, 299, 1948.
38. **Van Mierop, L. H. S.**, Location of pacemaker in chick embryo heart at the time of initiation of heart beat, *Am. J. Physiol.*, 212, 407, 1967.
39. **Paff, G. H. and Glander, T. P.**, Observations on the development of the electrocardiogram, *Anat. Rec.*, 160, 405, 1968.
40. **Romanoff, A. L.**, *The Avian Embryo*, Macmillan, New York, 1960.
41. **Barry, A.**, The effect of exsanguination on the heart of the embryonic chick, *J. Exp. Zool.*, 88, 1, 1941.
42. **Rajala, G. M., Kalbfleisch, J. H., and Kaplan, S.**, Evidence that blood pressure controls heart rate in the chick embryo prior to neural control, *J. Embryol. Exp. Morphol.*, 36(3), 685, 1976.
43. **Rychter, Z., Kopecky, M., and Lemez, L.**, A micromethod for determination of the circulating blood volume in chick embryos, *Nature (London)*, 175, 1126, 1955.
44. **Barry, A.**, The functional significance of the cardiac jelly in the tubular heart of the chick embryo, *Anat. Rec.*, 102, 289, 1948.
45. **Ruckman, R. N., Cassling, R. J., Clark, E. B., and Rosenquist, G. C.**, Cardiac function in the embryonic chick, in *Perspectives in Cardiovascular Research*, Vol. 5, Pexieder, T., Ed., Raven Press, New York, 1981, 407.
46. **Kirkpatrick, S. E., Pitlick, P. T., Naliboff, J., and Friedman, W. F.**, Frank-Starling relationship as an important determinant of fetal cardiac output, *Am. J. Physiol.*, 231, 495, 1976.
47. **Thornburg, K. L. and Morton, M. J.**, Filling and arterial pressures as determinants of RV stroke volume in the sheep fetus, *Am. J. Physiol.*, 244, H656, 1983.
48. **Faber, J. J., Green, T. J., and Thornburg, K. L.**, Embryonic stroke volume and cardiac output in the chick, *Dev. Biol.*, 41, 14, 1974.
49. **Hughes, A. F. W.**, The blood pressure of the chick embryo during development, *J. Exp. Biol.*, 19, 232, 1942.
50. **Paff, G. H., Boucek, R. J., and Gutten, G. S.**, Ventricular blood pressures and competency of valves in the early embryonic chick heart, *Anat. Rec.*, 151, 119, 1965.
51. **Girard, H.**, Arterial pressure in the chick embryo, *Am. J. Physiol.*, 224(2), 454, 1973.
52. **Van Mierop, L. H. S. and Bertuch, C. J.**, Development of arterial blood pressure in the chick embryo, *Am. J. Physiol.*, 212, 43, 1967.
53. **Clark, E. B. and Hu, N.**, Developmental hemodynamic changes in the chick embryo from stage 18 to 27, *Circ. Res.*, 51, 810, 1982.
54. **Pexieder, T.**, Blood pressure in the 3rd and 4th aortic arch and morphogenetic influence of laminar blood streams in the development of the vascular system of the chick embryo, *Folia Morphol. (Prague)*, 17(3), 273, 1969.
55. **Grabowski, C. T. and Schroeder, R. E.**, A time lapse study of chick embryos exposed to teratogenic doses of hypoxia, *J. Embryol. Exp. Morphol.*, 19, 347, 1968.
56. **Rajala, G. M. and Kaplan, S.**, Abnormally elevated blood pressure in the trypan blue treated chick embryo during early morphogenesis, *Teratology*, 21, 247, 1980.
57. **Ruckman, R. N., Rosenquist, G. C., and Rademaker, D. A.**, The effect of graded hypoxia on the embryonic heart, *Am. J. Cardiol.*, 49(4), 939, 1982.
58. **Spitzer, A.**, Virchows, Ueber den Bauplan des normalen und missbildeten Herzens. Versuch einer phylogenetischen Theorie, *Arch. Pathol. Anat.*, 243, 81, 1923.
59. **Bremer, J. L.**, The presence and influence of two spiral streams in the heart of the chick embryo, *Am. J. Anat.*, 49, 409, 1932.

60. **Goerttler, K.,** Uber Blutstromwirkung als Gestaltungsfaktor fur die Entwicklung des Herzens, *Beitr. Pathol. Anat. Allg. Pathol.,* 115, 33, 1955.
61. **Barthel, H.,** Missbildungen des menschlichen Herzens, *Entwicklunggeschichte und Pathologie,* Georg Thieme Verlag, Stuttgart, 1960.
62. **Jaffee, O. C.,** Hemodynamics and cardiogenesis. I. The effects of altered vascular patterns on cardiac cell death, *J. Morphol.,* 110, 217, 1962.
63. **Jaffee, O. C.,** Bloodstreams and the formation of the interatrial septum in the anuran heart, *Anat. Rec.,* 147, 355, 1963.
64. **Jaffee, O. C.,** Hemodynamic factors in the development of the chick embryo heart, *Anat. Rec.,* 151, 69, 1965.
65. **Jaffee, O. C.,** Rheological aspects of the development of blood flow patterns in the chick embryo heart, *Biorheology,* 3, 59, 1966.
66. **Jaffee, O. C.,** Hemodynamic analysis of experimentally produced cardiac malformations, *Anat. Rec.,* 154, 509, 1966.
67. **Jaffee, O. C.,** The development of the arterial outflow tract in the chick embryo heart, *Anat. Rec.,* 158, 35, 1967.
68. **Manasek, F. J. and Monroe, R. G.,** Early cardiac morphogenesis is independent of function, *Dev. Biol.,* 27, 584, 1972.
69. **Yoshida, H., Manasek, F. J., and Arcilla, R. A.,** Intracardiac flow pattern in early embryonic life. A re-examination, *Cir. Res.,* 53, 363, 1983.
70. **Gessner, I. H.,** Spectrum of congenital cardiac anomalies produced in chick embryos by mechanical interference with cardiogenesis, *Cir. Res.,* 18, 625, 1966.
71. **Gessner, I. H. and Van Mierop, L. H. S.,** Experimental production of cardiac defects: the spectrum of dextroposition of the aorta, *Am. J. Cardiol.,* 25, 1970, 272.
72. **Colvee, E. and Hurle, J. M.,** Malformations of the semilunar valves produced in chick embryos by mechanical interference with cardiogenesis, *Anat. Embryol.,* 168, 59, 1983.
73. **Flaherty, J. T., Pierce, J. E., Ferrans, V. J., Dali, J. P., Tucker, W. K., and Fry, D. L.,** Endothelial nuclear patterns in the canine arterial tree with particular reference to the hemodynamic events, *Circ. Res.,* 30, 23, 1972.
74. **Reidy, M. A. and Langville, B. L.,** The effect of local blood flow patterns on endothelial cell morphology, *Exp. Mol. Pathol.,* 32, 276, 1980.
75. **Jackman, R. W.,** Persistence of axial orientation cues in regenerating intima of cultured aortic explants, *Nature (London),* 296, 80, 1982.
76. **Buck, R. C.,** The longitudinal orientation of structures in the subendothelial space of rat aorta, *Am. J. Anat.,* 156, 1, 1979.
77. **Manasek, F. J.,** Extracellular matrix: introduction, in *Perspectives in Cardiovascular Research,* Vol. 5, Pexieder, T., Ed., Raven Press, New York, 1980, 165.
78. **Gessner, I. H.,** Some biochemical and anatomic effects of sodium salicylate on the chick embryo heart, in *Pathophysiology of Congenital Heart Disease,* Adams, F. H., Swan, H. J. C., and Hall, V. E., Eds., University of California Press, Berkeley, 1970, 17.
79. **Lopez, C. A. and Martinez, M. S.,** The importance of extracellular matrix components in development of the embryonic chick heart, in *Perspectives in Cardiovascular Research,* Vol. 5, Pexieder, T., Ed., Raven Press, New York, 1981, 167.
80. **Gessner, I. H. and Bostrom, H.,** "In vitro" studies on S-sulfate incorporation into the acid mucopolysaccharides of chick embryo cardiac jelly, *J. Exp. Zool.,* 160, 283, 1965.
81. **Johnson, R. C., Manasek, F. J., Vinson, W. C., and Seyer, J. M.,** The biochemical and ultrastructural demonstration of collagen during early heart development, *Dev. Biol.,* 36, 252, 1974.
82. **Markwald, R. R. and Adams Smith, W. N.,** Distribution of mucosubstances in the developing rat heart, *J. Histochem. Cytochem.,* 20, 896, 1972.
83. **Manasek, F. J., Reid, M., Vinson, W., Seyer, T., and Johnson, R.,** Glycosaminoglycan synthesis by the early embryonic chick heart, *Dev. Biol.,* 35, 332, 1973.
84. **Gessner, I. H., Lorincz, A. E., and Bostrom, H.,** Acid mucopolysaccharide content of the cardiac jelly of the chick embryo, *J. Exp. Zool.,* 160, 291, 1965.
85. **Ortiz, C. E.,** Estudio histoquimico de la gelatina cardiaca en el embrion de pollo, *Arch. Inst. Cardiol. Mex.,* 28, 244, 1958.
86. **Manasek, F. J.,** Glycoprotein synthesis and tissue interaction during establishment of the functional embryonic chick heart, *J. Mol. Cell. Cardiol.,* 8, 389, 1976.
87. **Manasek, F. J.,** The extracellular matrix of the early embryonic heart, in *Developmental and Physical Correlates of Cardiac Muscle,* Lieberman, M. and Sano, T., Eds., Raven Press, New York, 1976, 1.
88. **Markwald, R. R., Fitzharris, T. P., and Adam Smith, W. N.,** Structural analysis of endocardial cytodifferentiation, *Dev. Biol.,* 42, 160, 1975.
89. **Markwald, R. R., Fitzharris, T. P., Bank, H., and Bernanke, D. H.,** Structural analysis on the matrical organization of glycosoaminoglycans in developing endocardial cushions, *Dev. Biol.,* 62, 292, 1978.

90. **Manasek, F. J.**, Heart development: interactions involved in cardiac morphogenesis, in *The Cell Surface in Animal Embryogenesis and Development,* Poste, G. and Nicolson, G., Eds., North-Holland, Amsterdam, 1976, 545.
91. **Manasek, F. J., Kulikowski, R. R., and Fitzpatrick, L.,** Cytodifferentiation: a casual antecedent of looping?, in *Morphogenesis and Malformation of the Cardiovascular System,* Rosenquist, G. and Bergsma, D., Eds., Alan R. Liss, New York, 1978, 161.
92. **Orts Llorca, F. and González-Santander, R.,** Estudio electromicroscopica de la primera aparicion y desairollo de los miofilamentos cardiacos (miofibrillogenesis) en el embrion de pollo, *Rev. Esp. Cardiol.,* 4, 537, 1969.
93. **Markwald, R. R., Fitzharris, T. P., and Manasek, F. J.,** Structural development of endocardial cushions, *Am. J. Anat.,* 148, 85, 1977.
94. **Van Mierop, L. H. S., Alley, R. D., Kausel, H. W., and Stranahan, A.,** The anatomy and embryology of endocardial cushion defects, *J. Thorac. Cardiovasc. Surg.,* 43, 71, 1963.
95. **Kalk, M.,** The organization of a tunicate heart, *Tissue Cell,* 2, 99, 1970.
96. **Oliphant, L. W. and Cloney, R. A.,** The ascidian myocardium's sarcoplasmic reticulum and excitation-contraction coupling, *Z. Zellforsch. Mikrosk. Anat.,* 129, 395, 1972.
97. **Viragh, S. and Challice, C. E.,** Origin and differentiation of cardiac muscle cells in the mouse, *J. Ultrastruct. Res.,* 42, 1, 1973.
98. **Pratt, R. M., Larsen, M. A., and Johnston, M. C.,** Migration of cranial neural crest cells in the cell-free hyaluronate-rich matrix, *Dev. Biol.,* 44, 298, 1975.
99. **Meier, S. and Hay, E. D.,** Control of corneal differentiation by extracellular materials. Collagen as a promoter and stabilizer of epithelial stroma production, *Dev. Biol.,* 38, 249, 1974.
100. **Toole, B. P. and Gross, J.,** The extracellular matrix of the regenerating next limb: synthesis and removal of hyaluronate prior to differentiation, *Dev. Biol.,* 25, 57, 1971.
101. **Toole, B. P.,** Hyaluronate and hyaluronidase in morphogenesis and differentiation, *Am. Zool.,* 13, 1061, 1973.
102. **Overman, D. O. and Beaudoin, A. R.,** Early biochemical changes in the embryonic rat heart after teratogenic treatment, *Teratology,* 3, 183, 1971.
103. **Manasek, F. J.,** The extracellular matrix: a dynamic component of the developing embryo, in *Current Topics in Developmental Biology,* Vol. 10, Moscona, A. A. and Monroy, A., Eds., Academic Press, New York, 1975, 35.
104. **Szepsenwol, J. and Bron, A.,** Le premier contact du système nerveux vagosympathique avec l'appareil cardio-vasculaire chez les embryons d'oiseaux (canard et poulet), *C. R. Soc. Biol.,* 118, 946, 1935.
105. **Szepsenwol, J. and Bron, A.,** L'origine des cellules nerveuses sympathiques dans le coeur des oiseaux (embryons de canard et de poulet), *C. R. Soc. Biol.,* 118, 1030, 1935.
106. **Sissman, N. J.,** Developmental landmarks in cardiac morphogenesis: comparative chronology, *Am. J. Cardiol.,* 25, 141, 1970.
107. **St. Petery, L. B. and Van Mierop, L. H. S.,** Evidence for the presence of adrenergic receptors in 3-day-old chick embryo, *Am. J. Cardiol.,* 232, H250, 1977.
108. **Legrande, M. C., Paff, G. H., and Boucek, R. J.,** Initiation of vagal control of heart rate in the embryonic chick, *Anat. Rec.,* 155, 163, 1966.
109. **Cheung, M. D., Gilbert, E. F., Bruyere, H. J., Ishikawa, S., and Hodach, R. J.,** Chronotropism and blood flow patterns following teratogenic doses of catecholamines in 5-day-old chick embryos, *Teratology,* 16, 337, 1977.
110. **Giliani, S. H. and Silvestri, A.,** The effect of propranolol upon chick embryo cardiogenesis, *Exp. Cell Biol.,* 45, 158, 1977.
111. **de la Cruz, M. V., Campillo-Sainz, C., and Muñoz-Armas, S.,** Congenital heart defects in chick embryos subjected to temperature variations, *Circ. Res.,* 18, 257, 1966.
112. **Bogue, J. Y.,** Electrocardiogram of developing chick, *J. Exp. Biol.,* 10, 286, 1933.
113. **Hoffman, L. E. and Van Mierop, L. H. S.,** Effect of epinephrine on heart rate and arterial blood pressure of the developing chick embryo, *Pediatr. Res.,* 5, 472, 1971.
114. **Girard, H.,** Adrenergic sensitivity of circulation in the chick embryo, *Am. J. Physiol.,* 224(2), 461, 1973.

DEVELOPMENT OF CARDIOVASCULAR FUNCTION

George A. Brooks

INTRODUCTION

Cardiovascular function is relatively well developed at birth, and, as with other functions and systems, cardiovascular function expands with somatic development and maturation. At particular stages in the developmental process (e.g., at puberty) cardiovascular function may develop at an accelerated rate, but the general trend during development is for a consistent increase in various parameters of cardiovascular function (e.g., O_2 uptake, delivery, and utilization).[1-3] In absolute terms, cardiovascular function appears to increase up until the age of 25 years; thereafter, a slow, progressive decline occurs, especially if age-related changes are accompanied by disuse atrophy.[4]

Since the beginning of investigations into developmental changes in cardiovascular function,[5] it has been apparent that young individuals respond to exercise much like their older counterparts and that they possess large reserves of cardiovascular function. In fact, it is not uncommon to find in 8-year-old boys and girls that values for the relative measure of cardiorespiratory function (maximal O_2 consumption [$\dot{V}O_2$max] in milliliters per kilogram per minute) are comparable to those found in adult athletes.[6] This conclusion about the relative cardiovascular capacities of children is known by every parent who has taken their 4- to 6-year olds hiking or backpacking. All during the day the children will complain about being tired. They will carry nothing; they will require rests and chocolates every few minutes, and they will seemingly cry continuously. However, at the end of the day, when the parents are almost too tired to set up camp, the children will tirelessly chase fireflies or similarly interesting wildlife specimens.

In this chapter, the results of an extensive body of literature on exercise physiology and biochemistry will be highlighted to describe and understand developmental changes in cardiovascular function. Again, because in healthy individuals significant reserves exist in cardiovascular function, the performance characteristics and limits of cardiovascular function are best described by the capacity to respond to the external stress of exercise. The exercise testing protocols utilized to study children as well as adults can be classified into two general categories: these are progressive and continuous protocols. In a progressive testing protocol, the external power output (treadmill speed and gradient or cycle ergometer resistance during constant speed cycling) increases periodically, usually every 1 to 4 min. In a continuous exercise protocol, subjects are asked to maintain a constant, submaximal exercise power output. Most frequently, the power output selected is relative to the subject's maximal capacity, determined on a previous, progressive protocol. Then, at various points during exercise, as well as during rest and recovery, parameters such as cardiac output, heart rate, oxygen consumption, blood pressure, and pulmonary ventilation are determined.

Because exercise stress testing involves significant understanding, interest, and motivation on the part of the subjects as well as the achievement of significant degrees of muscle strength and coordination, for practical considerations most of the data reviewed herein are on children of age 5 years and older. Additionally, this represents the beginning of school age when groups of children become available to researchers. We shall also consider static measurements of cardiovascular dimensions to describe cardiovascular development between birth and the school years.

The development of ventilatory (respiratory) capacity is not covered here, but is covered in the accompanying chapter by T. D. Fahey. While the cardiovascular functions of normal healthy children are sufficient to cope with a variety of environmental stresses such as

Table 1
CARDIOVASCULAR CAPACITIES OF COLLEGE ATHLETES AFTER BEDREST AND TRAINING[9-11]

	Students			Olympic athletes
	Control	After bedrest	After training	
$\dot{V}O_2$max (l/min)	3.30	2.43	3.91	5.38
Arterial O_2 content (vol%)	21.9	20.5	20.8	22.8
Maximal cardiac output (l/min)	20.0	14.8	22.8	30.4
Stroke volume (ml/beat)	104	74	120	167
Maximal heart rate (beats/min)	192	197	190	182
Systemic (a-v)O_2 (vol%)	16.2	16.5	17.1	18.0
Resting blood pressure (mmHg)	120/70	—	—	120/70
Exercise blood pressure (mmHg)	180/70	—	—	180/70

Note: Values are compared to Olympic athletes.

exercise, the absence of ability to respond to exercise stress is highly suggestive of a pathology. Therefore, the use of exercise as a strategy in pediatric medicine is becoming accepted.[7,8] Although it is not a purpose of this review to describe the use of exercise testing in pediatrics, it should be noted that exercise testing is used in the diagnosis of conditions such as congenital heart defects, asthma, and growth hormone deficiency. Further, exercise training of children is considered useful in the management of selected pathologies such as muscle dystrophies, diabetes, obesity, cystic fibrosis, cerebral palsy, and asthma. The use of exercise in pediatrics has recently been reviewed by Bar-Or.[8]

MAXIMAL OXYGEN CONSUMPTION IN THE ADULT

Functional capacity of the cardiovascular system can be described in two general ways: these are in terms of oxygen flow and utilization (i.e., the Fick principle), and in terms of pressure-resistance interactions. In this section, relationships among components of the Fick principle are discussed. Discussion of pressure-resistance relationships follows.

Among the various functions of the cardiovascular system, those of O_2 delivery and CO_2 removal can be viewed as most important. And, although the Fick equation can be written for both O_2 and CO_2, for our purposes in terms of describing limitations to function, the Fick relationship is best described in terms of O_2:

$$\dot{V}O_2 = Q(a-v)O_2 \qquad (1)$$

where: $\dot{V}O_2$ is the rate of O_2 consumption (in liters per minute), Q is the cardiac output (in liters per minute), and (a-v)O_2 is the arterial-venous O_2 difference (in liters of O_2 per liter of blood).

According to the Fick relationship, the maximal rate of O_2 consumption ($\dot{V}O_2$max) is the product of two quantities, each of which is determined in whole, or in large part, by the cardiovascular system. As seen in Table 1,[9-11] in untrained adults maximal aerobic power ($\dot{V}O_2$max) can increase approximately ten-fold. After intense physical training, $\dot{V}O_2$max can increase an additional 20%. Although larger increases in $\dot{V}O_2$max due to training have been reported in particular individuals,[12] this value of approximately 20% improvement in $\dot{V}O_2$max appears to be the upper limit of the environmentally induced (training) response in most individuals.

In the transition from rest to exercise up to maximal intensity exercise, (a-v)O_2 increases.

Because arterial O_2 content is well maintained even during maximal work, increases in the (a-v)O_2 during exercise are attributable to decreases in venous O_2 content (CvO_2). Additionally, because the venous blood draining muscles of both trained and untrained individuals during maximal exercise contains little O_2, the ability to increase (a-v)O_2 by decreasing CvO_2 through endurance training is limited. This aspect of cardiorespiratory adaptation will be discussed further.

Depending on an individual's resting heart rate, the relative ability to increase cardiac frequency during exercise varies. However, maximal heart rate is largely genetically predetermined, and is infrequently much different from 200 bpm in young adults.

Because cardiac output is the product of heart rate and stroke volume:

$$Q = (fH)(vS) \qquad (2)$$

and because maximal heart rate does not increase in response to training or other forms of environmental stress, increments in cardiac output, if any, are due to increases in stroke volume (vS) (Table 1).

Stroke volume is affected by several myocardial and hemodynamic factors. Myocardial contractility increases from rest to exercise so that end systolic volume is decreased compared to rest, and the chamber volume is ejected within a considerably shorter time. In addition to the increase in myocardial contractility ($+dp/dt$), the rate of myocardial relaxation ($-dp/dt$) also increases in the transition from rest to exercise. As a result, end diastolic volume during exercise is approximately the same as during rest. As opposed to exercise in the recumbent or supine positions which favor venous return (e.g., swimming), exercise in the upright position (e.g., cycling, walking, running) imposes additional stress on the cardiovascular system especially when heart rate is elevated and ventricular filling time is decreased due to tachycardia. Having a high rate of venous return enhances diastolic ventricular filling and systolic stroke volume through the Frank-Starling mechanism.[13,14]

The maintenance or decrease in end systolic volume during exercise and the absence of distension in capacitance vessels during exercise implies a close correspondence of cardiac output with venous return. In fact, these observations not only support the importance of the Frank-Starling mechanism in determining cardiac output during exercise, they suggest that cardiac output is determined by venous return.

Given that $\dot{V}O_2$max is limited by maximum cardiac output, which is in turn determined by maximum stroke volume, it is important to recognize that maximal stroke volume is determined by ventricular volume. Thus, as heart size increases during growth and development, the absolute limits of cardiovascular capacity will be determined.

As illustrated in Table 1, maximal cardiac output decreases approximately 10% with bed rest and increases approximately 20% with endurance training. However, compared to Olympic athletes, the cardiovascular capacities of the students studied by Saltin et al.[10] after training was significantly lower. This result is interpreted to mean that absolute ability to increase cardiac output is largely genetically determined by the size of the heart.

The influence of endurance training on cardiac dimensions has been the subject of several investigations employing both cross-sectional and longitudinal designs.[1-3,15,16] Although the results of these investigations are somewhat variable, it is apparent that the ability to alter ventricular end diastolic diameter through training is limited to increments of only a few percent.[14] Therefore, despite the fact that a change in ventricular volume is related to cubic function of the chamber radius, the ability to improve cardiac output by training is constrained, by anatomy of the heart, to a stroke volume change of 10% or less. Experimentally, with laboratory animals surgical removal of the pericardial sac surrounding the heart has been observed to allow ventricular expansion and increases in Q, vS, and $\dot{V}O_2$max.[17]

Increases in O_2 delivery during exercise are also mediated, in part by an expansion in blood volume. Effects of training on hematocrit and hemoglobin concentrations are variable and probably not significant. However, training can result in 10% expansion in plasma, total blood, and erythrocyte volumes.[9,22] The major effect of these adaptations is to increase preload on the heart, and thus stroke volume. These adaptations, however, are transient and are lost after several months of inactivity.

The form of exercise engaged in can result in dramatically different myocardial adaptations. In contrast to the volume overload (preload)-induced adaptations to endurance exercise training, heavy resistance exercises such as weight lifting result in increased afterload stresses and ventricular hypertrophy, but not increase in resting end diastolic volume.[14,18,19] This hypertrophy probably results from the tremendous intrathoracic and abdominal pressures which result during the isometric contraction phases of weight lifting exercises. In heavy weight lifters, systolic blood pressures exceeding 400 mmHg have been observed during exercise in normotensive subjects.[20,21]

Blood Flow Dynamics

During exercise in healthy individuals, the ventricular ejection fraction is high and the end diastolic volume is low regardless of the state of training.[9,13,14] Despite all of the positive inotropic effects on the myocardium during exercise and the superimposed effects of endurance training, it is apparent that the large increases in cardiac output during exercise could not be accomplished without peripheral vascular regulation and adaptation.

Seen from the perspective of the basic relationship between blood flow, arterial blood pressure, and total peripheral resistance:

$$Q = P/TPR \qquad (3)$$

the importance of arteriolar vasomotor activity during exercise becomes apparent.

In a normotensive individual, systolic blood pressure may increase from a value of 120 mmHg to a maximal value which is probably less than 200 mmHg during maximal sustained exercise. In contrast, diastolic blood pressure may change little, or may even decrease, especially if the exercise is prolonged. Therefore, mean blood pressure rises only moderately during sustained exercise.[22]

In the transition from rest to exercise, cardiac output increases approximately fourfold in normal, untrained adults (Table 1). By comparison, cardiac output may increase sixfold to eightfold in Olympic athletes. However, because the blood pressure responses in normal and highly trained individuals are essentially the same, in trained individuals there must be significant changes in the regulation of arteriolar blood flow. During exercise, especially during exercise in trained individuals, a release of vasoconstrictor tone is prerequisite to the large increases in cardiac output observed.[9,13]

Regional blood flow in muscle is controlled by both local and autonomic factors.[22] Although it is apparent that autonomic influences on regional blood flow during exercise are very important, it is unclear whether or not trained and untrained individuals differ in their autonomic responses during maximal exercise.[9,13] Sympathetic nervous system activity appears to be reduced in trained as compared to untrained individuals during rest and given levels of submaximal exercise, but during maximal exercise trained and untrained individuals both experience fivefold increases in circulating levels of catecholamines.[22]

In summary of this section, it is apparent that the capacity to increase cardiac output during exercise and environmentally induced changes in cardiac output are well matched to peripheral adaptations in the regulation of blood flow. Training-induced adaptations in the regulation of peripheral blood flow have, unfortunately, not been studied in detail.

The Peripheral Use of Oxygen

From the perspective of cardiovascular function during rest and exercise, it is the respiratory capacities of mitochondria in peripheral tissue beds which are served. Respiratory (mitochondrial) capacities of muscle vary widely, depending on fiber type.[23] Respiratory capacities of highly oxidative (red) fibers exceed those of fast twitch-glycoglytic (white) fibers by threefold to fivefold.[24] Endurance capacity may double the mitochondrial content of muscle.[24,25] Therefore, it is possible to observe that the muscle respiratory capacities of leg muscles in marathon runners exceed those of sedentary controls by tenfold.[26]

Compared to the ability for cardiovascular oxygen transport:

$$TO_2 = (CaO_2)Q \qquad (4)$$

the ability of the peripheral (muscle) mitochondria to utilize O_2 is much greater. Physical activities requiring a critical muscle mass of approximately 50% of the musculature will elicit $\dot{V}O_2$max.[22] As already mentioned, the ability to increase $\dot{V}O_2$max through training by expanding the (a-v)O_2 is small because of the low content of O_2 in venous blood draining contracting muscles during exercise. Therefore, the large increase in muscle mitochondrial capacity seen with endurance training increases $\dot{V}O_2$max slightly (by expanding the [a-v]O_2), but largely the adaptation probably serves other purposes. These likely include increasing the capacity to utilize fat (and thereby sparing limited glycogen reserves), increasing the ability to clear lactate, and minimizing the effects of oxidative damage during prolonged exercise.

These conclusions regarding the interrelationships among $\dot{V}O_2$max muscle respiratory capacity and exercise endurance are supported by the following observations. The correlation between $\dot{V}O_2$max and exercise endurance ranges from 0.6 to 0.8.[25,27,28] However, the correlation between muscle respiratory capacity and exercise endurance in adults is significantly higher, 0.8 to 0.9.[25,29,30,32] Therefore, it appears that while cardiovascular function determines the upper limit of $\dot{V}O_2$max, a high muscle mitochondrial capacity is also required for high endurance capacity.

GENETIC LIMITATIONS IN CARDIOVASCULAR ADAPTATION

Bouchard and Lortie[32] have recently reviewed the relationship between heredity and endurance performance. According to them, endurance performance is seen as a multifactorial phenotype influenced by both genetic and nongenetic factors. According to them,

$$Vp = Vg + Ve + Vge + E \qquad (5)$$

where Vp is the total variation in endurance performance; Vg is the variation attributable to genotype; Ve represents the nongenetic, environmental effects (such as training and nutrition); Vge is the interaction effect due to the interactions between genotype and environment (trainability); and E is the random error component.

The genetic effects on cardiovascular capacity, muscle fiber type, and, hence, endurance, have been described on studies of identical and fraternal twins. Figures 1 and 2 are based on a series of investigations by Klissouvas and co-workers.[33] The high correlation between the $\dot{V}O_2$max of fraternal twin siblings, and the even higher correlation in identical twins, leads to the conclusion that $\dot{V}O_2$max is predominantly determined by heredity.[34]

Komi and Karlsson[35] investigated the interrelationships among a large number of physiological and performance variables in mono- and dizygous twin boys and girls. These investigators confirmed the strong genetic influence on muscle fiber type. Figure 3 illustrates the strong correlation between percent slow twitch (high oxidative) fiber types in twins. Komi and Karlsson calculated that the heritability of variation (Hest) in muscle fiber type was 99.5%.

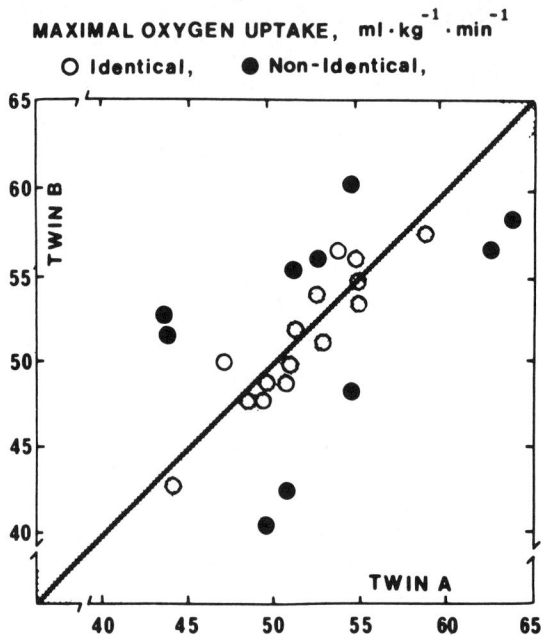

FIGURE 1. The correlation between maximal oxygen uptake ($\dot{V}O_2$) values in identical (monozygous) and nonidentical (dizygous) twins. The correlation is clearly higher in identical than in nonidentical twins. (From Klissouras, V., *Can. J. Sports Sci.*, 1, 195, 1976. With permission.)

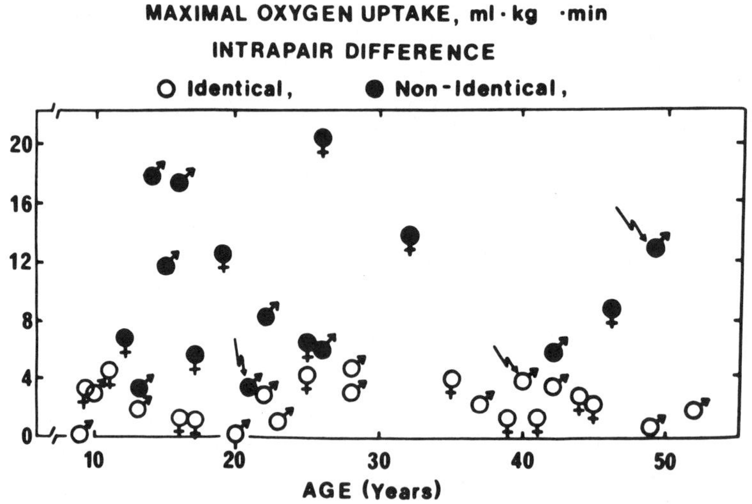

FIGURE 2. Intrapair differences in $\dot{V}O_2$max between identical and nonidentical twins across age. Identical twins are clearly more similar (less different) than are nonidentical twins, regardless of the age of testing. (From cross-sectional data of Klissouras, V., *Can. J. Sports Sci.*, 1, 195, 1976. With permission.)

In the investigation of Komi and Karlsson,[35] strong genetic links in $\dot{V}O_2$max and muscle fiber type were demonstrated. However, while there was a moderate though significant correlation between these variables in monozygous twins (r = 0.47), the correlation in dizygous twins was insignificant.

In their recent review, Bouchard and Lortie[32] examined the effects of a number of

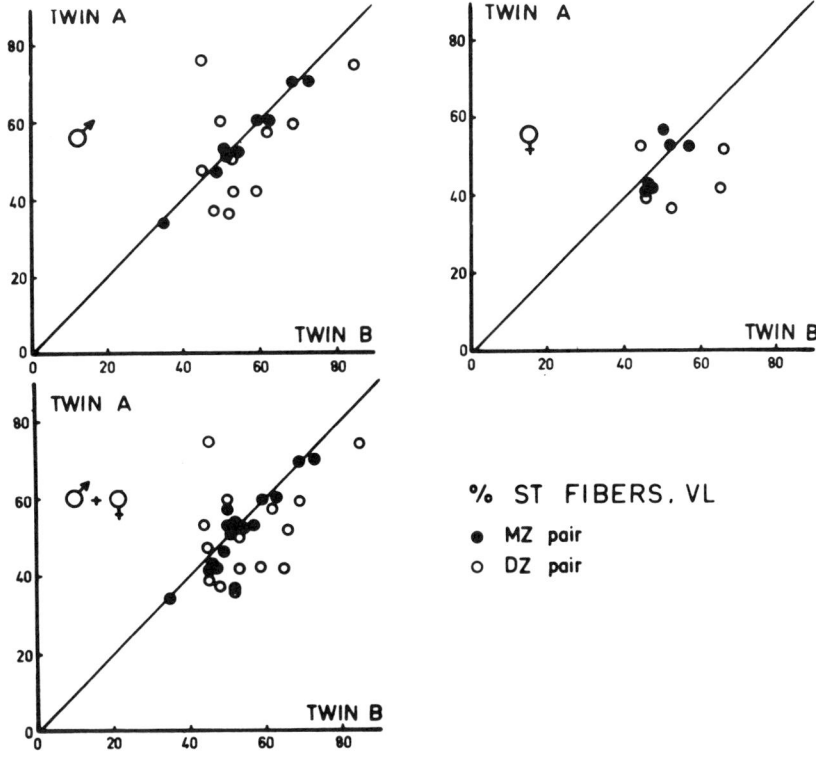

FIGURE 3. Intrapair comparisons of slow twitch fiber type distribution in medial vastus lateralis in mono- and dizygous twins. (From Komi, P. V. and Karlsson, J., *Acta Physiol. Scand. Suppl.*, 462, 1979. With permission.)

heredity and environment related factors on endurance performance. Their opinions, based on considerable experience as well as reevaluation of published data, suggest that while the effects of heredity are important in determining the cardiovascular and other factors determining exercise performance, environmental factors, such as training, are important also. The effects of training are manifest in two ways. First, the amount and quality of training, and, second, the interaction between genetics and environment (training). This latter factor is sometimes referred to as "trainability".

In their review, Bouchard and Lortie[32] presented evidence to suggest that there are high and low responders to habitual activity. They concluded that the differential sensitivity to training is genetically determined (hence the Vge component, above).

In Table 1, the training effect on $\dot{V}O_2$max was less than 20%. However, as noted above, Hickson et al.[12] have observed $\dot{V}O_2$max to improve approximately 49% due to training. Most recently, Bouchard and Lortie[32] trained 24 previously sedentary subjects for 20 weeks. The average improvement in $\dot{V}O_2$max was 33%, but the range in improvements was 5 to 88%. These results indicate vast relative differences in the population for improvements in aerobic power.

In making these types of comparisons, it must be acknowledged that the degree of improvement to be expected in response to training is negatively correlated ($r = -0.6$) to the initial degree of fitness (i.e., it is relatively easier for an unfit subject to improve than a fit one).[32] For this reason comparisons on identical twins again become informative.

In studies on ten pairs of monozygous twins trained for 20 weeks, Prud'homme et al.[36] observed $\dot{V}O_2$max to increase approximately 14%. However, the range of improvement was 0 to 41%. Interclass correlation analysis indicated that 74% of the intertwin response was

was genotype dependent. Therefore, it is possible to conclude that both the level of aerobic power (a cardiovascular parameter) as well as the ability to improve aerobic power are largely genetically determined.

DEVELOPMENT OF CARDIORESPIRATORY FUNCTION

The development of cardiorespiratory parameters in children, and the effects of exercise training thereon, have been studied utilizing cross-sectional, longitudinal, and combination designs. Of particular note are the studies on Swedish girl swimmers who were initially followed over the course of a year of training,[15] and then intermittently up to 10 years later.[3,16,37] In comparison to their nonathletic counterparts, girl swimmers (mean age 12.9 years) were characterized by greater heart and lung volumes, greater total hemoglobin, and $\dot{V}O_2$max. Over the course of the year of training, $\dot{V}O_2$max increased 12% (in liters per minute), but this increase was proportional to the increase in body mass. Therefore, the initial $\dot{V}O_2$max of girl swimmers (51 ml/kg/min) was essentially unchanged over the course of a year of training and rapid development. All of the women had stopped training 10 years later. During that time, heart volume declined 16%, but remained larger than expected in the general population. Similarly, vital capacity was well maintained in detrained swimmers, but ventilatory residual volume increased. Maximal O_2 consumption (measured in absolute terms of liters per minute) decreased 22% over the 10-year interval of measurement. The resulting mean $\dot{V}O_2$max values at age 23.9 (2.18 l/min; 36.6 ml/kg/min) were not particularly impressive when compared to the normal population. Therefore, the authors concluded that the functional capacity of the cardiovascular system declined more rapidly with sedentary living than its physical dimensions.[3]

Very informative of the effects of development and activity on cardiorespiratory capacity is that of Mirwald et al.[1] They reported on a 10-year longitudinal investigation of 14 physically active and 9 inactive boys in Saskatoon. Results are displayed in Figure 4. Over the years from age 7 to the pubertal growth spurt, $\dot{V}O_2$max increased continuously, but the rate of increase was essentially constant, or decreased slightly until the pubertal growth spurt. Results obtained on active and inactive boys changed in parallel during this period. During the pubertal spurt, $\dot{V}O_2$max increased rapidly, and the values obtained on active boys became significantly greater than on inactive boys. After the pubertal spurt, the rate of change in $\dot{V}O_2$max declined until adult values were approached at age 17.

As indicated earlier for girl swimmers, the pubertal spurt in lean mass is associated with the absolute increase in $\dot{V}O_2$max. Therefore, when the relative change in $\dot{V}O_2$max (i.e., milliliters per kilogram per minute) is calculated for children, the apparent increase in aerobic power associated with puberty is blunted. This effect is well illustrated in the data of Rutenfranz et al.,[2] who studied Norwegian and German children (Figure 5). In their study, relative $\dot{V}O_2$max was unaltered over the pubertal growth spurt.

CARDIOVASCULAR CAPACITY IN VERY YOUNG CHILDREN

The previous description of cardiovascular responses and development in children has been based upon the literature of exercise physiology. Because this sort of testing requires significant levels of intellectual and physical development, the data reviewed have focused on dynamic parameters determined on children 5 years and older. For describing the development of cardiovascular capacity in younger children, examination of static cardiovascular dimensions is useful.

Cardiac and body masses are known to increase proportionally from birth through maturation.[38] However, a cardiac mass of 20 g in a 3.0-kg infant[39] yields a body- to heart-mass ratio of 150. This is to be compared with a ratio of 233 in a 70-kg adult with a 300-

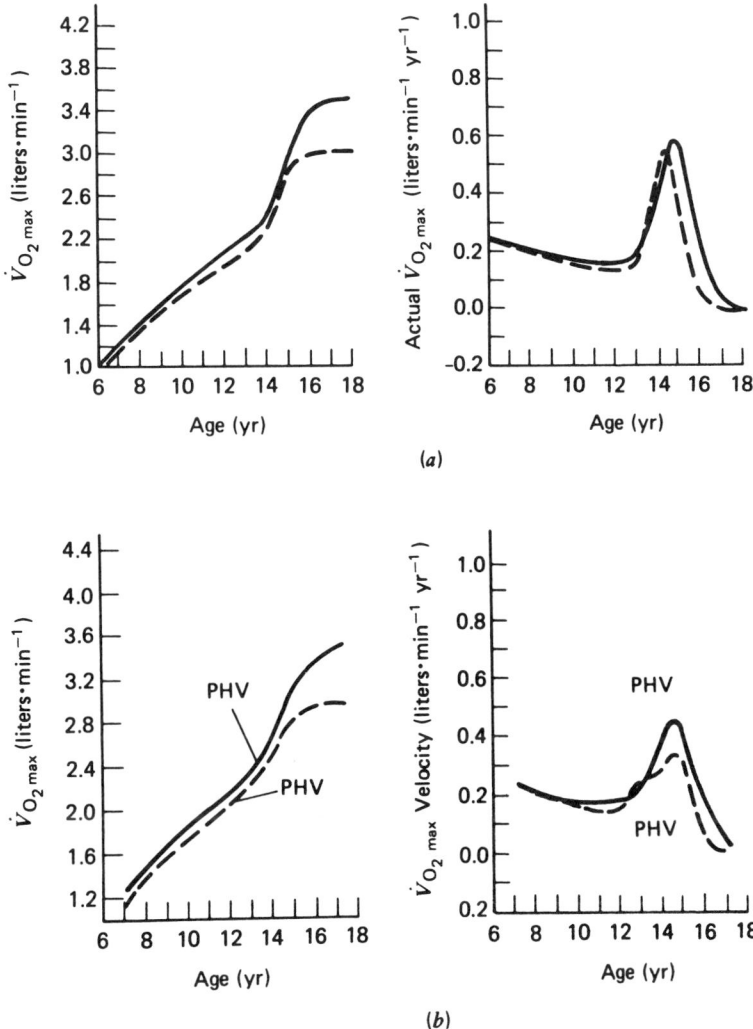

FIGURE 4. The absolute measure of cardiorespiratory capacity ($\dot{V}O_2$max, in liters per minute) increases continuously in childhood up to the age of puberty. During the pubertal growth spurt, $\dot{V}O_2$max increases at an accelerated rate. (a) The rate of increase in $\dot{V}O_2$max of active (solid lines) is slightly greater than in inactive children (broken lines). (b) In the bottom panels, curves are aligned on peak height velocity (PHV). (From Mirwald, R. L., Bailey, D. A., Cameron, N., and Rasmussen, R. L., *Ann. Hum. Biol.*, 8, 405, 1981. With permission.)

g heart.[40] The favorable ratio of cardiac to whole-body mass in the infant supports resting metabolic rates (7 ml of O_2 per kilogram per minute), cardiac outputs (140 ml/kg/min), and heart rates (140 bpm) which are approximately twice as high as in the adult. The high metabolic rate and levels of cardiovascular performance in infants as compared to adults reflects the high ratio of body surface area to body mass, and the need for relatively high heat production to offset loss in the infant.

Heart and internal chest diameters (as determined by X-ray) increase in rough proportion from birth to adulthood. However, the ratio of heart to internal chest diameter appears to be greatest in the newborn.[41,42] Though a very indirect measure of cardiovascular capacity, like the favorable body/heart weight ratio in infants, the heart/chest diameter ratio in infants supports the presence of relatively well-developed cardiovascular capacity in the infant.

Resting systolic blood pressures are relatively low in the newborn (60 mmHg), but the

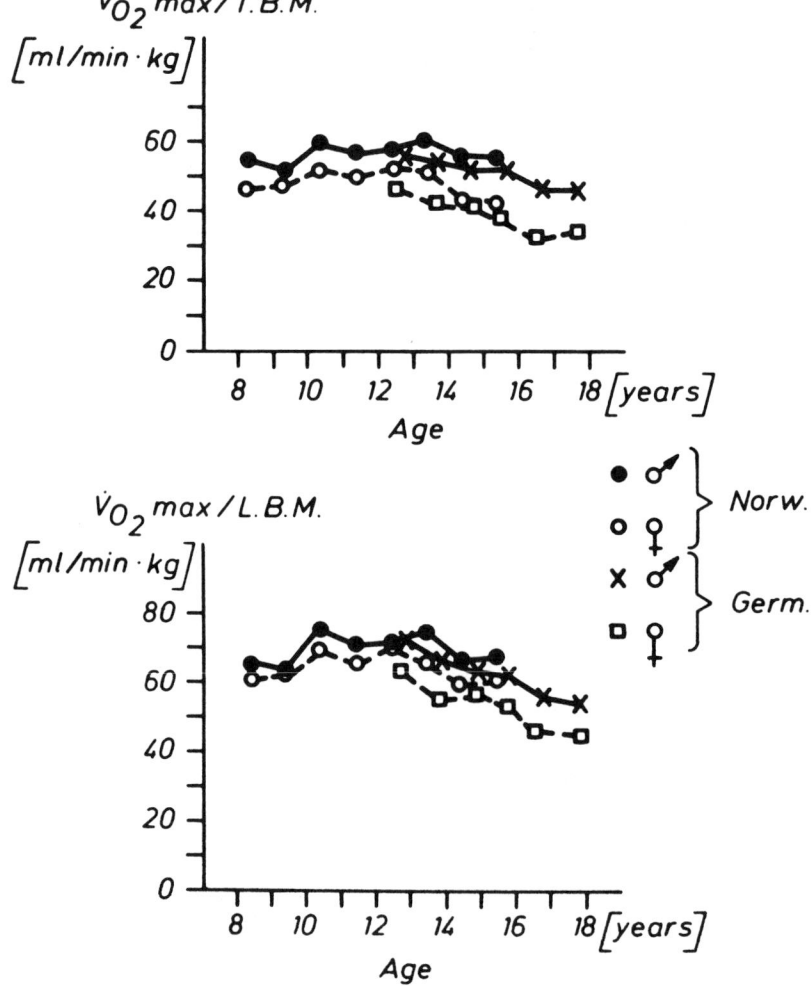

FIGURE 5. The relative measure of cardiorespiratory capacity ($\dot{V}O_2$max, milliliters per kilogram per minute) is essentially constant during the childhood years, with only a slight effect of the pubertal growth spurt apparent. This indicates that the expansion in cardiorespiratory capacity occurs in proportion to the general increase in total body mass (TBM) and lean body mass (LBM). (From Rutenfranz, J., Anderson, K. L., Seliger, V., Klimmer, F., Berndt, I., and Ruppel, M., *Eur. J. Pediatr.*, 136, 123, 1981. With permission.)

values increase progressively through childhood toward adult values (Table 2).[43] Because of the relatively high tissue perfusion levels in children (i.e., Q/Body Mass), the low blood pressures in children indicate low levels of total peripheral resistance (TPR). Although autoregulatory responses in children during exercise have not been studied in detail, the pressure-resistance relationships in children suggest that they rely more on increments in cardiac output to increase muscle perfusion during exercise than on the shunting mechanism. Thus, even though values of $\dot{V}O_2$max seen in children are relatively high, they are significantly less than in trained adults.

OPTIMIZING THE TRAINING EFFECTS; TRAINING DURING THE PUBERTAL GROWTH SPURT

The habitual exercise of children during the growth spurt has long been suspected as a

Table 2
NORMAL AVERAGE RESTING BLOOD PRESSURES AND HEART RATES AT VARIOUS AGES (mmHg AND bpm)

Age	Systolic	Diastolic	Heart rate
1 day	52[a]		140
1 month	80[a]		130
1 year	96	65	115
2 years	99	65	115
6 years	100	65	100
10 years	110	60	90
14 years	118	60	80
16 years	120	65	70

Note: Data are from Reference 43.

[a] Obtained by the "flush" method, represents a value between systolic and diastolic.

means to optimize the physiological, biochemical, and anatomical changes occurring during that time. However, this hypothesis remains controversial. As noted above, increased cardiovascular capacity resulting from endurance training can be observed during the growth spurt (Figure 4). However, as also noted above, the increases in cardiorespiratory capacity appear to be proportional to the increase in lean body mass (Figure 3). In a study of identical twins, in which half the twins trained, but the siblings did not,[44] the rate of increase in $\dot{V}O_2$max within twin pairs during the pubertal spurt was similar.

The enhancement of developmental processes by training during the pubertal growth spurt, if any, have been hypothesized to be due to endocrine influences. However, this hypothesis was investigated in detail by Fahey et al.,[45] who studied 22 boys from the ages of 5 to 18 before and after progressive intensity cycle ergometer exercise. Figure 6 clearly indicates that while exercise causes an increased level of circulating growth hormone, there does not occur an exaggerated growth hormone response to exercise at any pubertal stage. Further, Fahey et al. demonstrated what while serum testosterone levels clearly differ during various pubertal stages, progressive intensity exercise had no major effect on serum testosterone levels. Consequently, the authors could not confirm the hypotheses of an exaggerated anabolic endocrine responses to exercise in pubescent children.

"ANAEROBIC" WORK CAPACITY IN CHILDREN

As mentioned earlier, cardiovascular responses of children to exercise mimic those of adults. However, during exercise blood and muscle lactate levels in children are lower than in adults.[46-48] Muscle biopsies of children taken during rest and exercise indicate that muscle adenosine triphosphate (ATP) and creatine (CP) levels, as well as their responses during exercise, are similar to those in adults.[48] Therefore, the lower lactate levels in children during exercise appear to be due to lower resting glycogen levels,[48] and to lower levels of phosphofructokinase in children.[49] These differences indicate that the capacity for rapid glycogenolysis and glycolysis in immature skeletal muscle is not completely developed. Because the levels of muscle phosphagens are similar to those in adults, but because muscle glycogenolytic capacity in children is not yet fully developed, the propensity for children to perform short, sprint-type exercises of 5- to 10-s duration, but not long sprint exercises of 30- to 45-s duration, is understood.

In the recent past, there have been some attempts to evaluate the balance of aerobic and

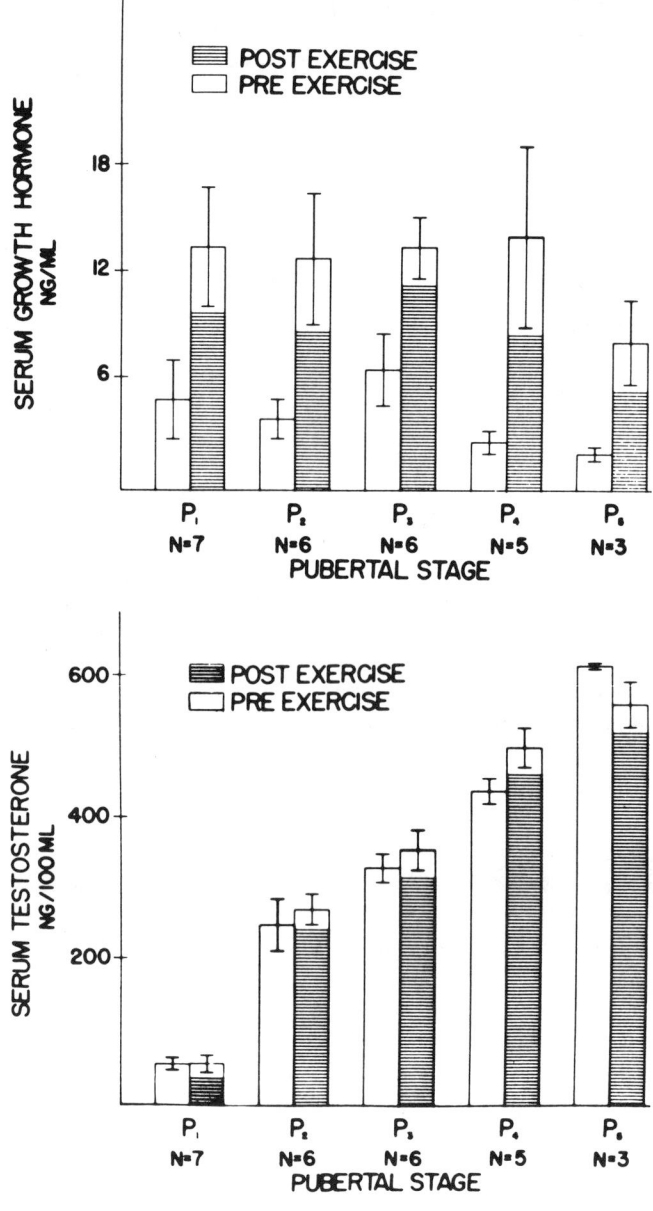

FIGURE 6. Pre- and postexercise levels in growth hormone and testosterone in boys at different pubertal stages. A clear growth hormone response to exercise is indicated, but the response is not exaggerated at any pubertal stage. Puberty affects testosterone levels, but again there is no exaggerated response to exercise. (From Fahey, T. D., Valle-Zuris, A. D., Oehlsen, G., Trieb, M., and Seymour, J., *J. Appl. Physiol.*, 46, 823, 1979. With permission.)

anaerobic metabolism in the muscles of children by measurements of ventilatory parameters during progressive exercise.[50] However, objections to this sort of procedure for children are essentially the same as those for adults.[51] Consequently, determination of the ventilatory, "anaerobic threshold" on children during progressive exercise does not significantly enhance our understanding of the balance of aerobic and anaerobic metabolism during exercise in children. The lower glycolytic and glycogenolytic capacities of children, as compared to adults, serve to illustrate the importance of cardiovascular function in compensating for external stresses such as exercise.[22]

SUMMARY

Cardiovascular function is relatively well developed at birth. Because cardiovascular capacity develops in proportion with muscle and whole-body development, significant reserves in cardiovascular function are well maintained even during the rapid growth spurt of puberty. In absolute terms, cardiovascular capacity reaches a peak at about age 25 years, and declines gradually thereafter, especially from disuse atrophy. The absolute upper limits of cardiovascular capacity appear to be largely genetically determined, but can be modified approximately 20% by endurance exercise training. In particular individuals, an interaction between genetics and environment is especially pronounced and the ability to increase cardiovascular capacity is much greater (approximately 40%). The improvement in cardiovascular capacity which results from normal development and training appears to be additive. Consequently, training during the pubertal growth spurt or other stages of life does not appear to result in extraordinary adaptations. Endurance exercise training in the adult years can be a valuable tool in managing cardiovascular disease risk factors.[8,52]

ACKNOWLEDGMENT

Research discussed in this chapter was supported by National Institutes of Health (NIH) Grant DK19577.

REFERENCES

1. **Mirwald, R. L., Bailey, D. A., Cameron, N., and Rasmussen, R. L.**, Longitudinal comparison of aerobic power in active and inactive boys ages 7.0 to 17.0 years, *Ann. Hum. Biol.*, 8, 405, 1981.
2. **Rutenfranz, J., Anderson, K. L., Seliger, V., Klimmer, F., Berndt, I., and Ruppel, M.**, Maximum aerobic power and body composition during the puberty growth period; similarities and differences between children of two European countries, *Eur. J. Pediatr.*, 136, 123, 1981.
3. **Eriksson, B. O., Engelstrom, I., Karlberb, P., Lunden, A., Saltin, B., and Thoren, G.**, Long-term effect of previous swimtraining in girls, a 10-year follow-up of the "girl swimmers", *Acta Paediatr. Scand.*, 67, 285, 1978.
4. **Åstrand, P.-O.**, Human physical fitness with special reference to sex and age, *Physiol. Rev.*, 36, 307, 1956.
5. **Robinson, S.**, Experimental studies of physical fitness in relation to age, *Arbeitsphysiologie*, 10, 251, 1938.
6. **Costill, D. L., Fink, W. J., and Pollock, M. L.**, Muscle fiber composition and enzyme activities of elite distance runners, *Med. Sci. Sports*, 8, 96, 1976.
7. **Thoren, C.**, Exercise testing in children, *Paediatrician*, 7, 100, 1978.
8. **Bar-Or, O.**, Clinical implications of paediatric exercise physiology, *Ann. Clin. Res.*, 34, 97, 1982.
9. **Bloomqvist, G. and Saltin, B.**, Cardiovascular adaptations to physical training, *Annu. Rev. Physiol.*, 45, 169, 1983.
10. **Saltin, B., Bloomqvist, G., Mitchell, J., Johnson, R. L., Wildenthal, K., and Chapman, C.**, Response to exercise after bed rest and after training, *Circulation*, 38(Suppl. 7), 1, 1968.
11. **Johnson, R. L.**, Oxygen transport, in *Clinical Cardiology*, Willerson, J. T. and Sanders, C. A., Eds., Grune & Stratton, New York, 1977, 74.
12. **Hickson, R. C., Bomze, H. A., and Holloszy, J. O.**, Linear increase in aerobic power induced by a strenuous exercise, *J. Appl. Physiol.*, 42, 372, 1977.
13. **Scheuer, J. and Tipton, C. M.**, Cardiovascular adaptions to physical training, *Annu. Rev. Physiol.*, 39, 221, 1977.
14. **Schaible, T. F. and Scheuer, J.**, Response of the heart to exercise training, in *Growth of the Heart in Health and Disease*, Zak, R., Ed., Raven Press, New York, 1984, 381.

15. Åstrand, P.-O., Engstrom, L., Eriksson, B. O., Karlberg, P., Saltin, B., and Thoren, C., Girl swimmers, *Acta Paediatr. Scand. Suppl.*, 147, 1963.
16. Engstrom, I., Eriksson, B. O., Karlberg, P., and Saltin, B., Preliminary report on the development of lung volumes in young girl swimmers, *Acta Paediatr. Scand.* 60(Suppl. 217), 73, 1971.
17. Stray-Gunderson, J., Musch, T. I., Haidet, G. C., Swain, D. P., Ordway, G. A., and Mitchell, J. H., The effect of pericardiectomy on maximal oxygen consumption and maximal cardiac output in untrained dogs, *Circ. Res.*, 58, 523, 1986.
18. Morganroth, J. B., Marioin, J. B., Henry, W. L., and Epstein, S. E., Comparative left ventricular dimensions in trained athletes, *Ann. Intern. Med.*, 82, 521, 1975.
19. Grossman, W., Jones, D., and McLaurin, L. P., Wall stress and patterns of hypertrophy in human left ventricle, *J. Clin. Invest.*, 56, 56, 1975.
20. MacDougall, J. D., Tuxen, D., Sale, D. G., Moroz, J. R., and Sutton, J. R., Arterial blood pressure response to heavy resistance exercise, *J. Appl. Physiol.*, 58, 785, 1985.
21. Keul, J., Dickhuth, H.-H., Simon, G., and Lehman, M., Effect of static dynamic exercise on heart volume, contractility and left ventricular dimension, *Cir. Res.*, 48(Suppl. I), 1162, 1981.
22. Brooks, G. A. and Fahey, T. D., *Exercise Physiology: Human Bioenergetics and Its Applications*, John Wiley & Sons, New York 1984.
23. Barnard, J., Edgerton, V. R., Furukawa, T., and Peter, J. B., Histochemical, biochemical, and contractile properties of red, white, and intermediate fibers, *Am. J. Physiol.*, 220, 410, 1971.
24. Holloszy, J. O., Effects of exercise on mitochondrial oxygen uptake and respiratory enzyme activity in skeletal muscle, *J. Biol. Chem.*, 242, 2278, 1967.
25. Davies, K. J. A., Packer, L., and Brooks, G. A., Biochemical adaptation of mitochondria, muscle, and whole-animal respiration to endurance training, *Arch. Biochem. Biophys.*, 209, 538, 1981.
26. Gohil, K., Jones, D. A., Corbucci, G. G., Krywawych, S., McPhail, G., Round, J. M., Montanari, G., and Edwards, R. H. T., Mitochondrial substrate oxidation, muscle composition, and plasma metabolite levels in marathon runners, in *Biochemistry of Exercise*, Knuttgen, H. G., Vogel, G. A., and Poortmans, J., Eds., Human Kinetics, Champaign, IL, 1982, 286.
27. Foster, C., Costill, D.-L., Costill, J. T., and Fink, J. T., Skeletal muscle enzyme activity, fiber composition, and $\dot{V}O_2$max in relation to distance running performance, *Eur. J. Appl. Physiol.*, 39, 73, 1978.
28. Fitts, R. H., Booth, F. W., Winder, W. W., and Holloszy, J. O., Skeletal muscle respiratory capacity, endurance, and glycogen utilization, *Am. J. Physiol.*, 228, 1029, 1975.
29. Booth, F. W. and Nahara, K. A., Vastus lateralis cytochrome oxidase activity and its relationship to oxygen consumption in man, *Pfluegers Arch.*, 349, 319, 1974.
30. Henrickson, J. and Reitman, J. S., Time course of changes in human skeletal muscle succinate dehydrogenase and cytochrome oxidase activities and maximal oxygen uptake with physical activity and inactivity, *Acta Physiol. Scand.*, 99, 91, 1977.
31. Ivy, J. L., Withers, R. T., Van Handel, P. J., Relger, D. H., and Costill, D. L., Muscle respiratory capacity and fiber type as determinants of lactate threshold, *J. Appl. Physiol.*, 48, 523, 1980.
32. Bouchard, C. and Lortie, G., Heredity and endurance performance, *Sports Med.*, 1, 38, 1984.
33. Klissouras, V., Prediction of athletic performance: genetic considerations, *Can. J. Sports Sci.*, 1, 195, 1976.
34. Klissouras, V., Pirnay, F., and Petit, J.-M., Adaptation to maximal effort: genetics and age, *J. Appl. Physiol.*, 36, 288, 1973.
35. Komi, P. V. and Karlsson, J., Physical performance, skeletal muscle enzyme activities, and fiber types in monozygous and dizygous twins of both sexes, *Acta Physiol. Scand. Suppl.*, 462, 1979.
36. Prud'homme, D., Bouchard, C., LeBlanc, C., Landry, F., and Fontaine, F., Sensitivity of maximal aerobic power to training is genotype dependent, *Med. Sci. Sports Exercise*, 16, 489, 1984.
37. Eriksson, B. O., Lundin, A., and Saltin, B., Cardiovascular function in former girl swimmers and the effects of physical training, *Scand. J. Clin. Lab. Invest.*, 35, 135, 1975.
38. Timiras, P. S. and Valcana, T., Body growth, in *Developmental Physiology and Aging*, Timiras, P. S., Ed., Macmillan, New York, 1972, 273.
39. Gruenwald, P. I. and Mink, H. N., Evaluation of body and organ weights in perinatal pathology. I. Normal standards derived from autopsies, *Am. J. Clin. Sci. Pathol.*, 34, 247, 1960.
40. Guyton, A. C., *Textbook of Medical Physiology*, W. B. Saunders, Philadelphia, 1976, 355.
41. McCammon, R. W., *Human Growth and Development*, Charles C. Thomas, Springfield, IL, 1970.
42. Malina, R. M. and Roche, A. F., *Manual of Physical Status and Performance in Childhood*, Vol. 2, Plenum Press, New York, 1983.
43. Watson, E. H. and Lowrey, G. H., *Growth and Development of Children*, Year Book Medical Publishers, Chicago, 1958, 164.
44. Weber, G., Kartodihardjo, Q., and Klissouras, V., Growth and physical training with reference to heredity, *J. Appl. Physiol.*, 40, 211, 1976.

45. **Fahey, T. D., Valle-Zuris, A. D., Oehlsen, G., Trieb, M., and Seymour, J.,** Pubertal stage differences in hormonal and hematological responses to maximal exercise in males, *J. Appl. Physiol.,* 46, 823, 1979.
46. **Macek, M., Vavra, J., and Novosadova, J.,** Prolonged exercise in prepubertal boys, cardiovascular and metabolic adjustment, *Eur. J. Appl. Physiol.,* 35, 291, 1976.
47. **Gadhoke, S. and Jones, N. L.,** The responses to exercise in boys aged 9—15 years, *Clin. Sci.,* 37, 789, 1969.
48. **Eriksson, O. and Saltin, B.,** Muscle metabolism during exercise in boys aged 11 to 16 years compared to adults, *Acta Paediatr. Belgica,* 28, 256, 1974.
49. **Eriksson, B. O., Gollnick, P. D., and Saltin, B.,** Muscle metabolism and enzyme activities after training in boys 11—13 years old, *Acta Physiol. Scand.,* 86, 485, 1973.
50. **Cooper, D. M. and Weiler-Ravel, D.,** Gas exchange response to exercise in children, *Am. Rev. Respir. Dis.,* Suppl. 129, S47, 1984.
51. **Brooks, G. A.,** Anaerobic threshold: review of concept and directions for future research, *Med. Sci. Sports Exercise,* 17, 22, 1985.
52. **Gilliam, T. D., Catch, V. L., Thorland, W., and Weltman, A.,** Prevalence of coronary heart disease risk factors in active children, 7—12 years of age, *Med. Sci. Sports,* 9, 21, 1977.

… Part B: Cardiovascular and Respiratory Development 101

DEVELOPMENT OF CARDIAC VASCULATURE

Karel Rakusan

PRENATAL DEVELOPMENT

Data concerning the chronology of anatomic changes in the developing heart in a group of seven frequently studied species are tabulated in a review by Sissman.[1] Important developmental landmarks in the human heart, such as first fusion of epimyocardial layers of bilateral heart together with the first appearance of myofibrils and myocardial contractions, are assigned to stage 10 (approximately 22 d past fertilization). The blood flow through the heart begins 2 d later. During this time, the earliest heart curvature is noticeable, and soon the S-shaped loop is formed.

Up to the final stages of loop formation the myocardium is a compact tissue layer composed exclusively of myocytes, and its nutrition is provided by simple diffusion. Subsequently, wide intercellular spaces appear between the myocardial cells in the inner part of the ventricular region. Thus, the myocardial wall is composed of two distinct zones—a compact outer layer and a spongy inner layer. Intercellular spaces of the inner layer are soon penetrated by endocardial outpockets. This development of endocardial diverticula guarantees that the individual myocytes are never far removed from the blood supply. The endocardial cells that penetrate the intercellular spaces are the first nonmuscle cells observed in the myocardial wall.

In contrast to the spongy inner layer of the heart, which contains sinusoids formed by indentation of endocardium into previously formed intercellular spaces, the development of the vasculature in the outer layer starts much later and is of different origin. The earliest vascular structures in this region are clusters of primitive erythroblasts surrounded by a thin layer of endothelium. Hirakow[2] calls them local vascular structures. He found them on a single occasion at stage 14 (approximately 29 d past fertilization), but by stage 16 (33 d) they were always present. Similarly, Heintzberger[3] describes the solitary endothelial plexus located in the subepicardial mesenchyme as the beginning of ventricular vascularization, and places this event around 30 d after fertilization. This plexus spreads out between the cells of the myocardium and finally connects with intertrabecular spaces. Subsequently, it also comes into contact with coronary veins and arteries. The major coronary veins initially develop as outgrowths of the endothelial lining from the left horn of the sinus venosus, the future coronary sinus. This process starts around day 35 and is soon followed by sprouting from the aortic sinuses of Valsalva, heralding the formation of coronary arteries. In this process, the formation of the left coronary artery precedes the formation of the right one. Around day 40, both coronary arteries are formed, becoming invested with a covering of one or two layers of mesenchymal cells and then connected to "primitive capillaries".

By stage 20 (approximately 41 d past fertilization), the basic pattern of cardiac blood supply as we know it from the adult heart is established. In the meantime, the sinusoidal system in the inner layer undergoes further modifications, during which at least part of these structures[5] is transformed into myocardial blood capillaries and Thebesian veins.

A brief survey of the literature on cardiac vessel formation in various mammalian hearts reveals three concepts of cardiac angiogenesis (for citations see Hirakow).[2] In the older literature, the prevailing notion is that terminal vessels are formed by sprouting and continuous extension of the coronary veins and arteries. A second group of studies, based mainly on the investigation of the rat heart, describes the formation of the coronary bed by sprouting of the sinusoid endothelium into intramyocardial gaps and clefts, which later connect to coronary veins and arteries. Finally, some authors, including those cited in the description

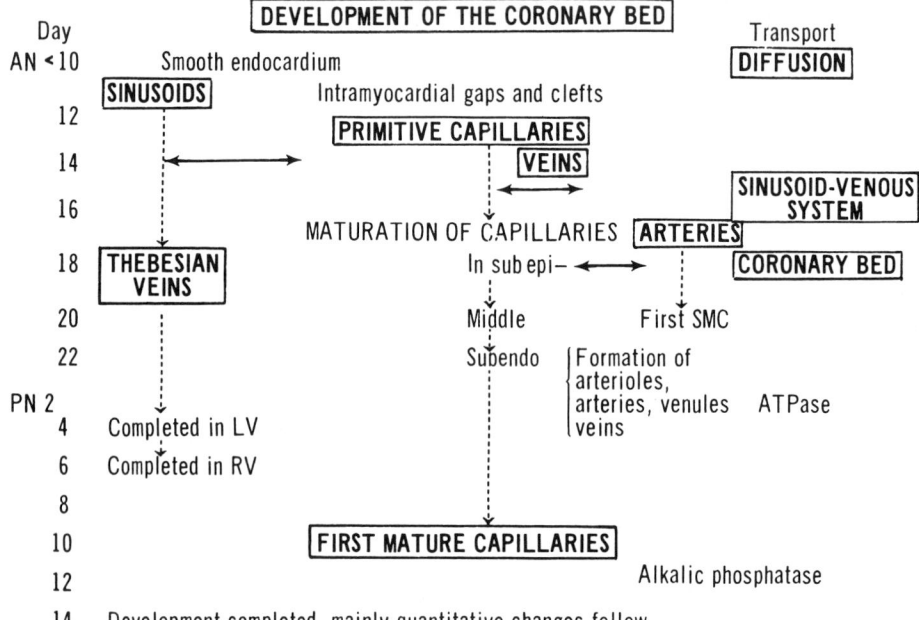

FIGURE 1. Development of the coronary bed in the rat heart. Flowchart is based mainly on the studies of Schiebler and co-workers.[4-7] For details see the text: LV, left ventricle; RV, right ventricle; SMC, smooth muscle cells; AN, antenatal; PN, postnatal.

of the characteristics of vascular development in the human heart, describe the independent appearance of endothelial channels under the epicardium, which subsequently connect to sinusoids of the inner layer on one side, and to coronary veins and arteries on the other side.

Development of the coronary bed has been studied extensively in the rodent heart.[4-8] Its most prominent features are summarized from various sources on a flowchart (Figure 1), which describes development in the rat heart. Up to day 10 past fertilization, the endocardium is smooth and the only mode of transport is diffusion. Afterward, one can see the formation of large intertrabecular sinusoids, lined by a very thin endothelium, but the basement membrane is absent. In addition, intramyocardial clefts appear which are originally without an endothelial lining. The outgrowth of endothelial cells into the clefts follows. However, the vascular wall is not always continuous and erythrocytes can apparently leave these structures without difficulty. Thus, the transport of oxygen and various substances from the vascular bed into the intercellular spaces is provided even to regions not yet reached by vascular proliferation. The connection with the cavities of the heart is present, but its functional importance is open to speculation. Viragh and Challice,[8] in their study on the murine heart, compare the circulation at this stage to the movement of water in and out of a sponge as it is alternatively squeezed and released.

The first rudiments of the embryonic capillaries are observed after day 13. These initially blind ending vessels, especially noticeable at the cardiac base and septum anastomose to form capillary nets which communicate with sinusoids. Maturation of primitive capillaries is a relatively long process. On the 18th day past fertilization, one can see the first capillaries with closed endothelial walls and a distinct basal membrane. The process is finished only by the end of the second postnatal week. The regional sequence of maturation is from the outer subepicardial region toward the endocardium. In the perinatal period following the establishment of blood flow from coronary arteries, some of the capillaries would transform into arterioles, venules, small veins, and terminal arteries.

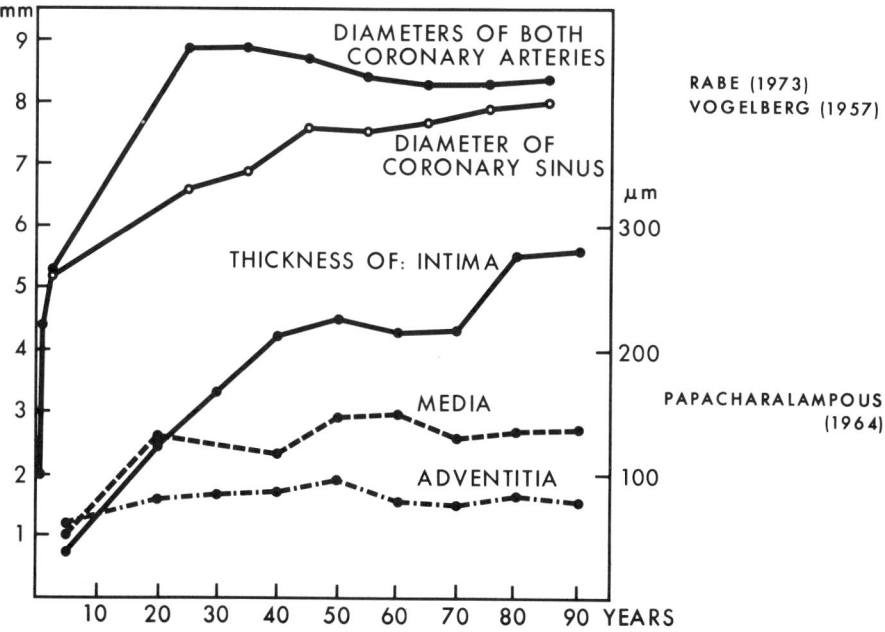

FIGURE 2. Age-related changes of coronary artery diameters and the diameter of the coronary sinus[10] (in millimeters) and the thickness of the individual layers of the coronary arteries[11] (in microns). (From Rakusan, K., in *Growth of the Heart in Health and Disease*, Zak, R., Ed., Raven Press, New York, 1984, 131. With permission.)

Coronary veins originate from the coronary sinus and fuse with the capillary and sinusoid network. First, the sinusoid-venous system is restricted to the septum and base (around day 14); later it spreads to the entire heart. Establishment of the coronary vascular system as we know it in the adult heart is associated with the appearance of coronary arteries. They originate as protuberances of the aortic sinus around day 16 to 17. Soon, they connect with the existing capillaries, and the adult pattern of coronary blood flow is established. The original sinusoids and intramyocardial clefts face three possible fates: some of these structures are obliterated, some are transformed into capillaries, and the remainder organize into Thebesian veins. Formation of the Thebesian venous system is completed only in the early postnatal period. The early postnatal period is also characterized by establishment of definite morphological and biochemical features of the vascular walls (for example, the ATPase reaction is uniform after the third postnatal day, and the reaction for alkalic phosphatase after day 11). The perinatal period is also characterized by the spatial arrangement of capillary orientation into longitudinal subepicardial and subendocardial layers which turn into the circular middle section. This process is finished by the end of the first postnatal week, and subsequent changes are mainly of a quantitative nature.

POSTNATAL DEVELOPMENT

My recent chapter on cardiac growth, maturation, and aging[9] includes most of the pertinent data on postnatal development of the coronary vasculature. As discussed in the previous section, the basic coronary vascular bed in the human heart is established well before birth. Apart from an increase in the number of small branches of the coronary arteries, the major change is the growth in size of the individual vessels. Age-related changes in the caliber of the human coronary arteries, based on the study of Rabe,[10] are depicted in Figure 2. According to this author, coronary artery diameters increase in size rapidly during the

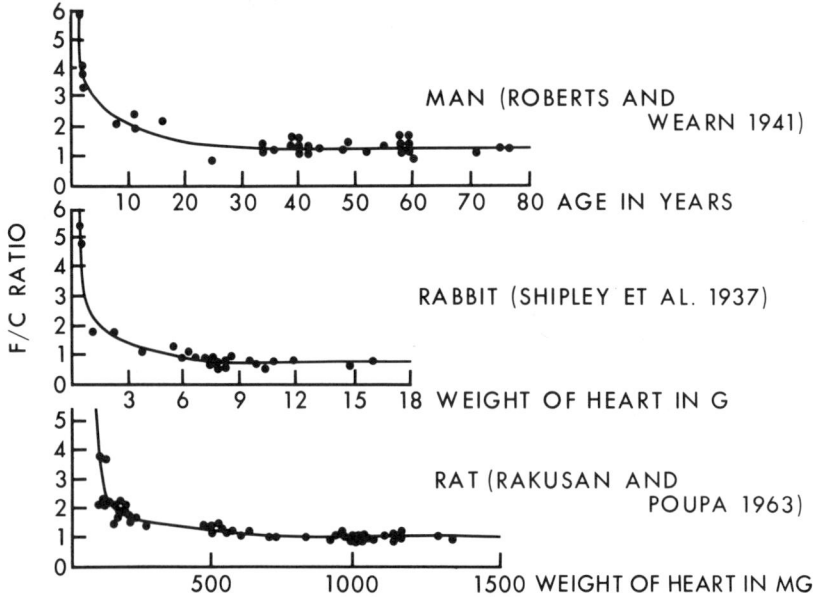

FIGURE 3. Postnatal development of the fiber/capillary (F/C) ratio in the hearts from three mammalian species: man,[12] rabbit,[13] and rat.[14] (From Rakusan, K., in *Growth of the Heart in Health and Disease*, Zak, R., Ed., Raven Press, New York, 1984, 131. With permission.)

first postnatal years and continue growing up to the age of 30 years. Beyond this age, the diameters of coronary arteries remain unchanged, irrespective of age and cardiac weights. In contrast, some other authors report an increase in the coronary lumen which is more or less proportional to the growth of the ventricular mass beyond 30 years of age.[9] Figure 2 also contains changes in the diameter of the coronary sinus, which increases moderately with age and changes in the thickness of the individual layers of the coronary arteries.[11] The thickness of the intima increases continuously with age in both coronary arteries, reaching values ten times higher in the oldest group (90 years) than in the youngest (first decade). The thickness of the media increases only moderately with age, while the adventitia remains more or less unchanged.

The only systematic description of postnatal changes in the capillary supply of the human heart is that by Roberts and Wearn.[12] These authors describe a rapid decrease of the fiber-to-capillary ratio from 6 in newborns to 1 in adults. This development is not peculiar to the human heart. As a matter of fact, almost identical changes were found in the rabbit and rat hearts, as displayed in Figure 3. Thus, in adult mammalian hearts, on the average, one capillary supplies a single muscle cell, while in newborn hearts several cells are supplied by a single capillary. The muscle cells in the newborn hearts, however, are of much smaller size. Therefore, the capillary density (number of capillaries per mm^2) decreases with age and growth of the cardiac mass. This starts around the weaning period and continues up to the senescence. Similar changes are found both in anatomic studies and by measurement of the intercapillary distance on the surface of the beating ventricles.[14,15]

Rapid decline in the fiber-to-capillary ratio in the early postnatal period is a reflection of the considerable proliferation of capillaries during this time. Recently, it was estimated that almost half of all the capillaries present in the normal adult rat heart is formed during the first 3 postnatal weeks.[16] Similarly, Olivetti and co-workers[17] reported that in rats during the first 10 postnatal days, capillaries grow two to three times more rapidly than the myocardial mass.

In the first section, maturation of capillaries was described as taking place in the perinatal period. The process is not interrupted by the birth itself. Postnatal changes are mainly characterized by the establishment of the definite orientation of the capillary net in various regions of the heart. This feature, together with the massive proliferation of capillaries after the birth, is probably responsible for the fact that the spacing of the capillary net in the young heart is quite irregular. Maturation of the heart is associated with increased homogeneity of capillary spacing.[16] The increasing homogeneity of the capillary spacing is indicative of improving morphological conditions for oxygen supply, and thus, at least in part, compensates for the decreasing capillary density, which has the opposite influence.

The mechanisms responsible for rapid capillary proliferation in the newborn heart and the reasons for subsequent slow-down and eventual cessation of capillary growth are not known. In a recent study, protamine exhibited inhibitory effects on capillary growth in rat hearts during the early postnatal period.[16] While one cannot exclude a direct effect of this compound on proliferating capillary endothelial cells, a more probable explanation lies in its blockage of the effect of heparin, locally released by mast cells. Heparin is only one of many putative substances leading to angiogenesis. Obviously, more research is necessary to elucidate this important biological phenomenon which has very practical implications for clinical medicine.

SUMMARY AND CONCLUSIONS

For practical reasons, this overview was divided into sections on prenatal and postnatal development, but it no doubt became obvious that the growth and development of cardiac vasculature is a continuous process uninterrupted by the birth itself. As a matter of fact, the logical dividing line probably should be placed into the early postnatal period. Up to this stage, qualitative development of the coronary vascular bed takes place, whereas subsequent changes are mainly of a quantitative nature. Also at this time, the final geometrical arrangement is being established. This is also a period characterized by a rapid growth of terminal vessels. The mechanisms regulating vascular growth in mammalian hearts are poorly understood, but it seems that vascular development and growth follows the same trend in all mammalian hearts, including the human heart.

REFERENCES

1. **Sissman, N. J.**, Developmental landmarks in cardiac morphogenesis: comparative chronology, *Am. J. Cardiol.*, 25, 141, 1970.
2. **Hirakow, R.**, Development of the cardiac blood vessels in staged human embryos, *Acta Anat.*, 115, 220, 1983.
3. **Heintzberger, C. F. M.**, The vascularization pattern in the ventricular wall in different species during development, in *The Coronary Sinus*, Mohl, W., Wolner, E., and Glogar, D., Eds., Steinkopff Verlag, Darmstadt, Germany, 1984, 47.
4. **Voboril, Z. and Schiebler, T. H.**, Uber die Entwicklung der Geffassversorgung des Rattenherzens, *Z. Anat. Entwicklungsgesch.*, 129, 24, 1969.
5. **Henningsen, B. and Schiebler, T. H.**, Zur Fruhentwicklung der herzeigenen Strombahn. Elektromikroskopische Untersuchung an der Ratte, *Z. Anat. Entwicklungsgesch.*, 130, 101, 1970.
6. **Ostadal, B. and Schiebler, T. H.**, Die Capillarentwicklung in Rattenherzen Elektromikroskopische Untersuchungen, *Z. Anat. Entwicklungsgesch.*, 133, 288, 1971.
7. **Blatt, H. L.**, Uber die Entwicklung der Coronararterien bei der Ratte. Licht- und elektronmikroskopische Untersuchungen, *Z. Anat. Entwicklungsgesch.*, 142, 53, 1973.
8. **Viragh, S. and Challice, C. E.**, The origin of the epicardium and the embryonic myocardial circulation in the mouse, *Anat. Rec.*, 201, 157, 1981.

9. **Rakusan, K.,** Cardiac growth, maturation and aging, in *Growth of the Heart in Health and Disease,* Zak, R., Ed., Raven Press, New York, 1984, 131.
10. **Rabe, D.,** Kalibermessungen an der Herzkranzarterien und dem Sinus coronarius, *Basic Res. Cardiol.,* 68, 356, 1973.
11. **Papacharalampous, N. X.,** Altersbedigte histologische und histochemische Veranderungen der Coronargeffasse, *Virchows Arch. Pathol. Anat. Physiol.,* 338, 187, 1964.
12. **Roberts, J. T. and Wearn, J. T.,** Quantitative changes in the capillary-muscle relationship in human hearts during normal growth and hypertrophy, *Am. Heart J.,* 21, 617, 1941.
13. **Shipley, R. A., Shipley, L. J., and Wearn, J. T.,** The capillary supply in normal and hypertrophied hearts of rabbits, *J. Exp. Med.,* 65, 29, 1937.
14. **Rakusan, K. and Poupa, O.,** Changes in the diffusion distance in the rat heart muscle during development, *Physiol. Bohemoslov.,* 12, 220, 1963.
15. **Henquell, L., Odoroff, C. L., and Honig, C. R.,** Coronary intercapillary distance during growth: relation to PCO_2 and aerobic capacity, *Am. J. Physiol.,* 231, 1852, 1976.
16. **Rakusan, K. and Turek, Z.,** Protamine inhibits capillary formation in growing rat hearts, *Circ. Res.,* 57, 393, 1985.
17. **Olivetti, G., Anversa, P., and Loud, A. V.,** Morphometric study of early postnatal development in the left and right ventricular myocardium of the rat. II. Tissue composition, capillary growth, and sarcoplasmic alterations, *Circ. Res.,* 46, 503, 1980.

MALFORMATIONS OF THE HEART AND BLOOD VESSELS

Roger N. Ruckman, Susan A. O'Brien, and Donna J. Messersmith

INTRODUCTION

This chapter will first describe the effects of environmental and genetic factors acting independently or through complex interactions to produce a wide spectrum of congenital heart defects. Secondly, proposed mechanisms of cardiac maldevelopment based on experimental animal models will be discussed. Finally, applications of experimental models to two cardiac lesions, ventricular septal defect (VSD) and transposition of the great arteries (TGA), will be reviewed.

Classification

Congenital heart disease (CHD) encompasses those cardiac defects present at birth and characterized by gross morphologic abnormalities which interfere with normal cardiac functioning. CHD comprises one third of all major birth defects.[1] Further, approximately 8 per 1000 of all live-born infants are afflicted with CHD.[2]

Classification of factors leading to CHD is complicated by the simultaneous interaction of multiple influences which obscure pathogenetic mechanisms. The majority of CHDs are believed to result from an interplay between genetic susceptibility and exposure to adverse environmental conditions, whereas few influences have been shown to adversely affect cardiogenesis independent of other factors (Figure 1).

ENVIRONMENTAL INFLUENCES

Classic Teratogens

Since their identification as teratogens, rubella and thalidomide have been regarded as classical models in teratology, and both have been intensively examined to determine mechanisms of action. Exposure to these agents at specific times in gestation is capable of producing a constellation of extracardiac and cardiovascular anomalies in 50% or more of all cases.[3] In addition, each of these environmental influences has demonstrated a link with CHD.

Rubella

Rubella virus, first recognized to have adverse effects by Gregg[4] in 1941, is the classical model for intrauterine infection of the human fetus. Rubella exhibits a gestational age dependency, a factor common to many teratogens. Therefore, frequency and specificity of resulting anomalies will vary dependent upon time of exposure. Table 1 shows the relationship between human gestational age of exposure to rubella and resulting percentage of abnormal infants.[5-7] Campbell[8] estimated the proportion of CHD due to maternal rubella during pregnancy to be 2 to 4%. The frequency of cardiovascular disease among infants who acquired the rubella syndrome during the 1964 to 1965 epidemic was reported to be 71%.[9] Rubella virus, like other teratogens, demonstrates an association between gestational age and frequency of cardiac anomalies. Table 2 shows stages of cardiac development with associated periods of vulnerability to teratogenic exposure.[9,10]

In addition to causing anomalies in the multiple extracardiac organ systems, rubella is associated with valvular heart lesions and myocarditis with myocardial necrosis.[11-13] Nora and Nora[9] found the most common cardiac malformations occurring among patients with rubella syndrome to be peripheral pulmonary artery stenosis (55%) and patent ductus arter-

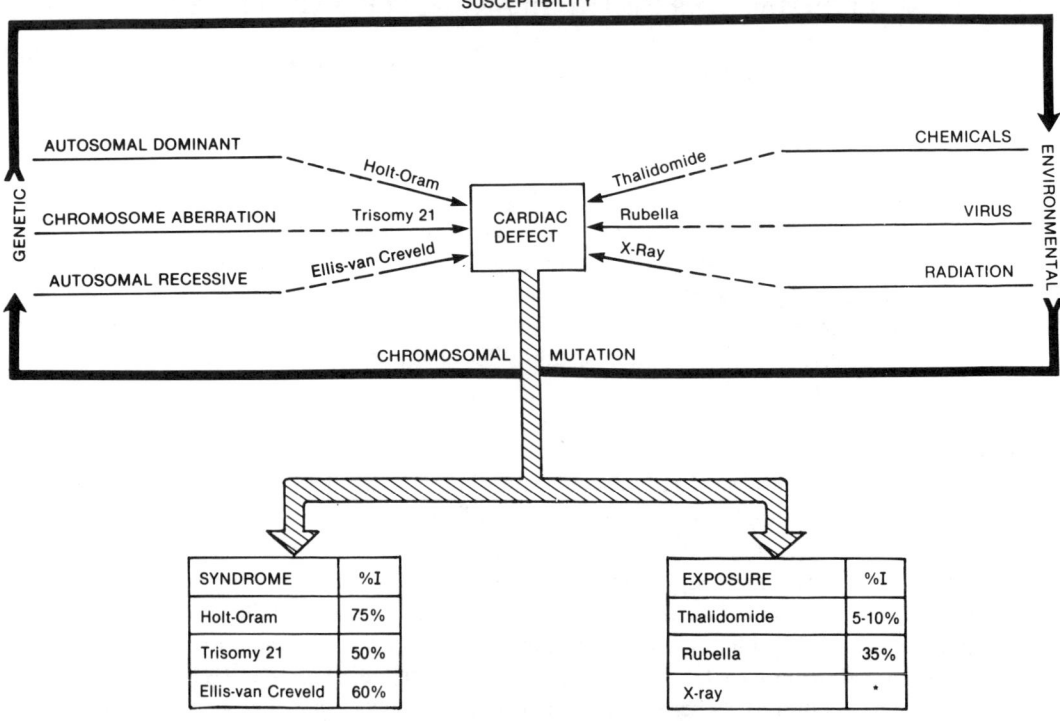

FIGURE 1. The majority of congenital heart disease (CHD) arises through a combined effect of genetic susceptibility and exposure to adverse environmental conditions, whereas few influences have been linked to CHD independent of this combined effect. Factors independently associated with CHD are illustrated below with corresponding incidence figures: %I, percent incidence of cardiovascular malformations; *, incidence figure unavailable. (Data from Smith, A. T., Sack, G. H., Jr., and Taylor, G. J., *J. Pediatr.*, 95, 538, 1979; Nora, J. J. and Nora, A. H., *Circulation*, 57, 205, 1978; McKusick, V. A., *Circulation*, 30, 326, 1964.)

Table 1
GESTATIONAL AGE DEPENDENCY
EXHIBITED BY THE RUBELLA VIRUS

Time of maternal exposure (months)	Abnormal infants (%)
1	50
2	22
3—5	6—8

iosus (43%). Furthermore, Campbell[8] noted patent ductus arteriosus to be the primary lesion associated with rubella, and 6% of those cases had associated VSD. Other cardiac lesions, atrial septal defect (ASD), tetralogy of Fallot, and pulmonary valvular stenosis, occurred in only 7% of all affected infants.

Thalidomide

Thalidomide, a sedative-hypnotic drug introduced to the German public in 1957, exhibits the same gestational age dependency that is observed with rubella. However, thalidomide exerts its effect within a much narrower range. In the human fetus, extreme sensitivity to thalidomide has been shown to exist during the 20th- to 35th-d postconception.[14] In a review by Nora and Nora,[9] the most common cardiac malformations associated with thalidomide

Table 2
POTENTIAL PERIODS OF VULNERABILITY TO TERATOGENIC EXPOSURE DURING CARDIOGENESIS

Abnormality/site of lesion	Vulnerable period (d)	Most sensitive period (d)
Truncoconal septation	14 to 34	18 to 29
Endocardial cushions	14 to 38	18 to 33
Ventricular septum	14 to ?	18 to 39
Atrial septum secundum	14 to ?	18 to 50
Semilunar valves	14 to ?	18 to 50
Ductus arteriosus	14 to ?	18 to 60
Coarctation of aorta	14 to ?	18 to 60

From Nora, J. J. and Nora, A. H., *Genetics and Counseling in Cardiovascular Diseases*, Charles C Thomas, Springfield, IL, 1978. With permission.

exposure were reported to be tetralogy of Fallot, VSD, ASD, and truncus arteriosus.[9] Among exposed infants, thalidomide was reported to be responsible for 9 anatomically distinct lesions out of 17 cases of cardiac disease, thus demonstrating its potent affinity for different cardiac tissues.[15]

One of the most important characteristics exhibited by thalidomide is its species-strain specificity. Nonhuman primates and rabbits exhibit markedly higher susceptibilities to thalidomide than mice or rats, which are far less sensitive to this drug.[16,17] The only animals showing the same degree of teratogenic sensitivity as that observed in humans exposed *in utero* are seven species of macaque monkeys, baboons, and marmosets.[18] Such specificity of action led to misinterpretation of the safety of this drug for pregnant women, resulting in disastrous consequences to their offspring.

Potential Teratogens Under Investigation
Radiation

Ionizing radiation was the first environmental agent shown to possess teratogenic properties. The effects of radiation exposure involve all organ systems. However, visceral organs such as the heart are only occasionally affected, whereas the central nervous system exhibits a much higher degree of susceptibility. Using the chick embryo as an animal model, Le Dourain and Le Dourain[19] demonstrated that the heart is less sensitive to radiation than many other embryonic tissues, confirmed by a lower percentage of pyknotic nuclei within irradiated cardiac tissues as compared to other irradiated tissues of the embryonic chick. However, internal and external cardiac morphologic changes were noted, as well as alterations in cardiac function. Irradiation during cardiogenesis at an intensity of 3000 rad (absorbed dose in ergs per gram of tissue) temporarily inhibits the mitotic index of the embryonic chick heart ([number of mitoses $\times$ 100]/[# of interphasic cells]) increases the heart rate, and decreases contractile ability. Vascular blood flow slows or ceases following irradiation, and ceases in all vessels at higher doses. Upon exposure to 3000 rads, blood flow changes result in altered aortic arch formation, primarily the third and fourth aortic arches. Irradiation inhibits endocardial cushion development, which may result in incomplete partitioning of the bulbus and arterial truncus, VSD, incomplete formation of the valves of the atrioventricular canal, absence or incomplete formation of the semilunar valves, and incomplete rotation and twisting of the bulbus.[19]

Alcohol

Among potential teratogens currently under investigation, alcohol is of particular concern as the most common drug abuse problem, affecting 1 to 2% of women of child-bearing age.[20] Congenital heart defects have been noted in 25 to 30% of infants afflicted with the fetal alcohol syndrome.[9] The most common lesions cited in a review by Nora and Nora[9] include VSD, patent ductus arteriosus (PDA), and ASD. However, results of independent epidemiologic investigations in France[21] and Seattle, WA[22,23] indicate ASD to be the most common cardiac lesion.

The adverse effect of alcohol on cardiac development in chick embryos is associated with increased dose and, within the time frame of cardiac morphogenesis, increased gestational age at time of exposure,[24-26] thus demonstrating the same concerns of stage specificity seen with rubella and thalidomide. A proposed mechanism for the action of alcohol on the developing heart will be discussed later in the chapter.

Hormones

Exogenous hormones are suspected of playing a teratogenic role, especially when in association with other environmental factors. Female hormonal agents reportedly have a nonspecific affinity for embryonic and fetal tissues.[27,28] Levy et al.[29] found that 7 of 76 mothers of babies with transposition of the great vessels had used estrogen and/or progestin contraceptives during pregnancy, whereas none of 76 women who had children with other defects had taken the hormones. The results suggest that exogenous sex hormones may play a role in a potential multifactorial causation of CHD. Robertson-Rintoul[30] found cardiovascular anomalies in two of five infants born to women who had taken oral contraceptives during pregnancy. However, Nishimura et al.[31] found no significant increase in congenital heart defects among aborted fetuses of women exposed to sex hormones as opposed to abortuses of women not exposed. Heinonen[32] found that the risk of CHD in patients of mothers who had been exposed to maternal sex hormones was 2.3 times greater than that of offspring whose mothers had not been exposed. Nora and Nora[33,34] found 20 patients with CHD among 224 examined in a retrospective study of the adverse effects of exposure to estrogen and progesterone contraceptive preparations early in gestation. In a prospective study, Nora and Nora[9] found a twofold to fourfold increase in anomalies in infants of mothers exposed to hormones during pregnancy. Common cardiovascular anomalies resulting from exposure to estrogenic hormones were VSD, tetralogy of Fallot, and TGA. In addition, Mitchell et al.[35] recognized an increase in TGA incidence in children of mothers who demonstrated estrogen deficiency.

Catecholamines

Much research has focused on catecholamines, naturally occurring hormones present in variable concentrations at different stages in gestation. Catecholamines have been associated with the production of aortic arch and cardiac anomalies in chick embryos when administered at specific points in gestation.[36] Catecholamines exhibit a positive dose response for chick embryos and have been shown to alter hemodynamics in animal studies.[37] In addition, there is a possible association between hemodynamic change and aortic arch maldevelopment[38] and a well-established relationship between aortic arch anomalies and cardiac defects.[36,39] A suggested mechanism of action will be discussed later.

Lithium

Among pharmacologic agents under investigation for their possible role in CHD, there has been a pattern of anomalies related to the use of lithium. Ebstein's anomaly is a cardiac defect which accounts for 1% of all CHD.[40] A statistical review performed by Mignot et al.[41] concerning lithium as a potential teratogen indicates that 7.8% of lithium-exposed embryos develop congenital cardiac defects. Nora and Nora[9] reported that lithium, used in the treatment of manic-depressive psychoses, is associated with cardiovascular disease in approximately 10% of all cases. In a retrospective study concerning the possible association between Ebstein's anomaly and maternal lithium exposure during pregnancy, Nora et al.[42] found 2 of 733 pregnancies to have involved maternal lithium exposure and both resulted in Ebstein's anomaly. In an investigation by Schou et al.[43] 6 of 118 infants of mothers who had taken lithium had congenital cardiovascular defects, a fivefold increase over the expected incidence. Furthermore, two of the six had Ebstein's anomaly, a 400-fold increase above the expected value. A subsequent report from the lithium registry revealed that 11 of 143 lithium-exposed pregnancies resulted in newborns with cardiovascular anomalies.[40]

Källén and Tandberg[44] conducted a cohort study on inpatient manic-depressive women, examining the possible relationship between the use of lithium and other psychotropic drugs in early pregnancy and perinatal death or survival with congenital malformations. They found that 7% of all women who had used lithium during early pregnancy gave birth to children with serious heart defects. Perinatal death rate was significantly increased (9 of 350 occurred; 4.3 of 350 expected — $p < 0.05$). In addition, the number of heart defects had increased significantly (6 of 350 occurred; 2.1 of 350 expected — $p < 0.05$). Length of gestation for 29 infants of women exposed to lithium was less than 36 weeks. The expected number of short pregnancies is 18.6 (no correction for maternal age parity). On examination of birth weights, it was found that 40 infants weighed less than 2500 g, whereas the expected number is 15.5 infants. However, smoking may influence weight values, as smoking has a known effect of birth weight, and women taking drugs for manic-depressive illness have been found to smoke more than healthy women and comparable women not using drugs. It is interesting to note that none of the 350 cases exhibited Ebstein's anomaly; however, the cohort study was relatively small.[44]

Hypoxia

Hypoxia is the oldest known environmental teratogen and has frequently been cited as a potential cause of human abnormalities.[45-50] The teratogenic effects of oxygen deprivation were observed as early as 1820 by St.-Hilaire.[50] Van Liere and Stickney[51] classified the major kinds of hypoxia: anemia, hypoxemia, stagnant hypoxia, and histotoxic hypoxia, and later Grabowski[52] summarized potential hypoxia-inducing conditions.

The stage of development and amount of hypoxia necessary to induce irreversible damage in humans is unknown, partly due to the many possibilities for synergistic action involving hypoxic agents and the difficulty in extrapolating from animals to humans. Epinephrine and other naturally occurring uterine-vasoconstrictive drugs have been investigated for their potential ability to induce hypoxia in the human fetus. Each has been shown to induce significant bradycardia in the human fetus,[53,54] as well as in monkeys,[55] rats and mice,[56] and rabbits and guinea pigs.[57] Chernoff and Grabowski[56] have investigated the amount of fetal bradycardia and resulting hypotension necessary to induce sufficient stagnant hypoxia to adversely affect the development of human tissue. In addition, Grabowski and Chernoff[58] investigated sodium nitrite ($NaNO_2$), a chemical found in certain foods and well water, and found that a single injection of $NaNO_2$ can induce bradycardia and hypotension in rat fetuses for periods up to 5 h.

Ruckman et al.[59,60] studied the effects of graded hypoxia on the embryonic chick during active cardiogenesis. Embryos with 65 or 84 h of prior incubation were placed in a humidified nitrogen-oxygen gas mixture. Following exposure to one of three hypoxic conditions or control condition (6% O_2, 11% O_2, 16% O_2, 21% O_2-room air), the four groups of 15 embryos were filmed hourly for 6 h and examined histologically. Results indicated that within 1 h, O_2 concentrations below room air produced significant dilatation of the primitive ventricle; and exposure to 6% O_2 was associated with tachycardia and loss of contractility ($p < 0.0001$). Within 3 h of hypoxic stress, the above mentioned changes in dimension and contractility were noted with mild hypoxia (16% O_2) ($p < 0.0001$). Heart rate was significantly altered with moderately severe hypoxia (6% O_2). In addition, although cellular damage

Table 3
FUNCTION CHANGE WITH TIME FOLLOWING EXPOSURE TO HYPOXIC AND/OR HYPOTHERMIC CONDITIONS

	Hypoxia	Hypothermia	Both
HR	↑*	↓*	N.S.
EDD	↑↑**	N.S.	N.S.
ESD	↑↑**	N.S.	N.S.
SF	↓↓**	N.S.	N.S.

Note: HR, heart rate; EDD, end diastolic dimension; ESD, end systolic dimension; SF, shortening fraction. $*p < 0.01$; $** p < 0.001$; N.S., not significant.

was not evident in mild hypoxia, it was observed with moderately severe hypoxia.[60] A possible mechanism of action will be discussed later.

Ruckman and Rademaker[61] studied the ability of the embryonic chick heart to recover cardiac function after hypoxic stress. After 18 h of recovery, embryos subjected to hypoxia (6% O_2, 11% O_2, 16% O_2) at 65 h of incubation demonstrated measurements of cardiac contractility identical to that of controls, confirming that the embryonic heart, even at a stage of active cardiogenesis, can recover normal function after hypoxic stress.[61]

In a study investigating the combined effect of hypoxia and hypothermia at the stage of active cardiogenesis, cinephotoanalysis was used to measure function change following induced hypoxia (6% O_2, 37.7°C), hypothermia (21% O_2, 32°C), and hypoxia with hypothermia (6% O_2, 32°C).[62] Function data obtained include heart rate (HR), end diastolic dimension (EDD), end systolic dimension (ESD), and shortening fraction (SF). Histologic analysis of embryos sacrificed following exposure to hypoxia showed myocardial edema, cytolysis, and pericardial hemorrhage, whereas no such cellular changes were observed following hypothermia or combined exposure.[63] Results confirm that hypoxia also induces marked impairment of embryonic function. However, simultaneous hypothermia appears to preserve form and function under conditions of acute hypoxia (Table 3).[62,63]

MATERNAL CONDITIONS

In addition to environmental agents, infectious or pharmacological, that may exert a teratogenic influence on the developing embryo, epidemiological studies have shown certain maternal conditions to be highly associated with cardiovascular defects. Maternal factors that Heinonen et al.[64] found associated with cardiovascular malformations in offspring were diabetes (duration greater than or equal to 5 years), thrombophlebitis during pregnancy, and chronic hypertension and/or eclampsia. Rowland et al.[65] reported a fivefold (4 to 5% risk) increase in cardiovascular anomalies in infants of diabetic mothers. These defects include structural abnormalities such as TGA and VSD. Since these mothers are being treated with hypoglycemic agents, however, it is difficult to separate the effect of the maternal illness from that of the mother's medications. Of the offspring of mothers wiht phenylketonuria, 25 to 50% had CHD.[66,67] Tetralogy of Fallot, VSD, ASD, and other cardiac defects were found in these infants. Offspring of mothers with systemic lupus erythematosus and other collagen vascular diseases often have complete heart block *in utero* and at birth.[68] Nora and Nora[3] reported the risk of complete heart block in infants of these mothers to be 20 to 40% in any given pregnancy. Further studies are needed to determine the bases for the associations between CHD in offspring and each of these maternal conditions.

Table 4
CHROMOSOMAL ABERRATIONS ASSOCIATED WITH CHD

	Common lesions	Incidence of CHD (%)
Trisomy 21	VSD or AV canal, ASD, PDA	50
Trisomy 18	VSD, PDA, pulmonary stenosis	99+
Trisomy 13	VSD, PDA	90
Trisomy 22	ASD complex, VSD, PDA	67
Turner's syndrome (XO)	Coarctation of the aorta, aortic stenosis, ASD	35

Note: CHD, congenital heart disease; VSD, ventricular septal defect; AV, atrioventricular; ASD, atrial septal defect; PDA, patent ductus arteriosus.

From Nora, J. J. and Nora, A. H., *Genetics and Counseling in Cardiovascular Diseases*, Charles C Thomas, Springfield, IL, 1978. With permission.

GENETIC ETIOLOGY OF CONGENITAL HEART DISEASE

Genetic Factors

Certain cardiac anomalies are known to be of genetic origin, associated with chromosomal aberrations, single mutant genes, or familial inheritance, but only a small percentage of all cardiac defects can be attributed to genetic factors. Nora and Nora[3] observed in their CHD clinic that only 8% of CHD patients had defects clearly attributable to genetic factors: approximately 5% to chromosomal aberrations and approximately 3% to single mutant genes. Those defects occurring through familial inheritance presumably involve an interaction of both genetic and environmental factors.

Down's syndrome, or mongolism, caused by an extra chromosome at the 21 position, is associated with single or multiple congenital heart defects. As high as 50% of patients with Down's syndrome also have CHD.[9] Of patients with trisomy 13, 90% have CHD, while more than 99% of patients with trisomy 18 have CHD.[9] These defects are primarily VSD and patent ductus arteriosus.[69] Table 4 lists chromosomal aberrations associated with CHD.[9]

After the discovery in 1959 by Lejeune and associates[70] that chromosomal aberrations resulted in mongolism, further chromosomal studies have attempted to define a chromosomal basis for certain cardiovascular anomalies. While some studies have shown associations between chromosomal aberrations and cardiovascular anomalies in a few patients,[71-73] chromosomal aberrations are not considered a major factor in most CHD. Emerit et al.[74] estimated the frequency of chromosomal aberrations as a cause of CHD to be 3 to 5%.

A small percentage of cardiac defects occurs as part of an established complex syndrome controlled by a single mutant gene. These syndromes are inherited either as autosomal dominant or autosomal recessive genes. Table 5 lists examples of these syndromes with associated congenital heart defects[9]

Investigations of a genetic basis for CHD encompass studies of occurrence of CHD in siblings. In a study of 1227 patients with CHD, Campbell[75] reported cardiovascular malformations in 1.7% of patients' sibs as compared to 0.6% in the general population. Lamy et al.,[76] in a study of over 2000 siblings of patients affected with CHD, noted cardiac defects in 1.47% of siblings. Campbell[75] found the same lesion in siblings and propositus 15 to 20 times more frequently than what would be expected by chance. However, Lamy et al.[76] found concordance in a relatively small number of cases.

Several authors have reported cases of a particular lesion passed through several generations. Warkany[77] cites several examples: a father with patent ductus arteriosus who had eight children, four of whom had the same defect;[78] an acyanotic mother with probable VSD

Table 5
SINGLE MUTANT GENE DISORDERS ASSOCIATED WITH CHD

Syndrome	Types of CHD	Risk of penetrance (%)
Autosomal dominant		
Marfan's syndrome	Mitral and aortic disease	60—80
Holt-Oram	ASD, VSD	~50
Ehlers-Danlos	A-V valve regurgitation, rupture of large blood vessels	~50
Noonan's syndrome	Pulmonary stenosis, ASD, left ventricular disease	50
Autosomal recessive		
Ellis-van Creveld	ASD, single atrium	50

Note: ASD, atrial septal defect; VSD, ventricular septal defect; A-V, atrioventricular.

From Nora, J. J. and Nora, A. H., *Genetics and Counseling in Cardiovascular Diseases,* Charles C Thomas, Springfield, IL, 1978. With permission.

who had, in each of two marriages, a child with the same defect;[79] atrial septal defect was established by autopsy in a woman who had transmitted the same defect to her daughter and granddaughter.[80] Patent ductus arteriosus and particularly ASD were among the specific lesions often described with repeated occurrence within families. While several studies conclude that heredity plays an important role,[81,82] the effects of environment and the interaction of multiple genes make determination of the specific role of genetic factors unclear.

Experimental Genetic Models

These observations of the possible role of genetics in human CHD have prompted animal studies. Detweiler and Patterson[83] performed preliminary studies of the distribution of cardiovascular disease in dogs and found the prevalence of CHD was 5 to 10 cases per 1000. In addition, Detweiler and Patterson[83] noted that certain defects apparently were distributed nonrandomly among the breeds, suggesting a genetic influence. Patterson[84] found an incidence of 6.8 dogs with cardiovascular malformations per 1000, confirming Detweiler's findings. He provided evidence that a variety of cardiac malformations in the dog can be caused by genotypes which have specific and localized effects on cardiac morphogenesis.[84] The high frequency of congenital heart defects found in the Keeshond breed of dogs provided Van Mierop et al.[85] the opportunity to study the pathogenesis of cardiac lesions, primarily involving the ventricular outflow tracts, through direct observations of embryologic material. Their studies confirmed the view that intracardiac anomalies in the Keeshond strain are due to maldevelopment of the embryonic conotruncal septum. Van Mierop and Patterson[86] considered the anomalies they found in the Keeshond dogs identical to anomalies in human cases of Eisenmenger complex; their findings support the view that Eisenmenger complex is due primarily to hypoplasia of conus cushions. Further embryologic animal studies similar to these will continue to elucidate mechanisms of cardiac maldevelopment in humans.

MULTIFACTORIAL INHERITANCE

Epidemiologic studies indicate that certain environmental agents, such as rubella or thalidomide, demonstrate high association rates with CHD, while certain genetic defects, such as Down or Ellis-van Creveld syndrome, also are highly associated with CHD. However, the large majority of congenital heart malformations appear to be due to a complex interaction of both environmental and genetic factors. Several investigators, including Falconer,[87] Edwards,[88] Morton et al.,[89] and Nora,[90] have proposed models of multifactorial inheritance which attempt to evaluate environmental and genetic influences to determine the risk of

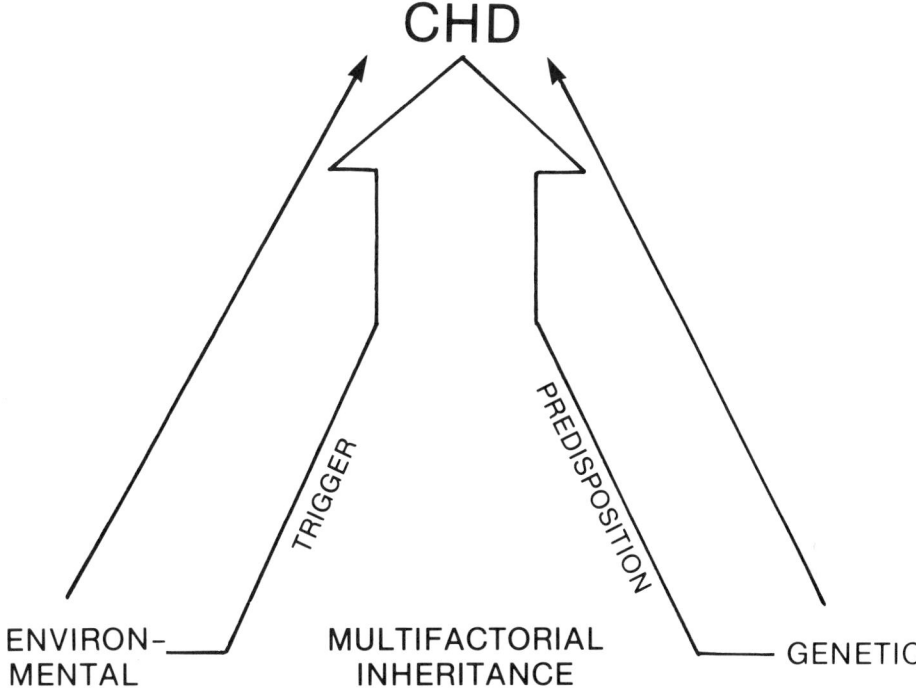

FIGURE 2. Few cases of congenital heart disease are due to environmental or genetic factors acting alone, while the large majority of cases are presumed to occur by "multifactorial inheritance". (From Nora, J. J. and Nora, A. H., *Circulation*, 57, 205, 1978. With permission.)

CHD recurrence. Nora[90] interpreted the genetic/environmental interaction in terms of a multifactorial inheritance hypothesis in which an environmental trigger produces congenital heart defects in those individuals most genetically predisposed (Figure 2). According to Nora's predictions, as high as 90% of congenital heart disease occurs by multifactorial inheritance. From Nora and Nora's counseling of low-risk ("type B") patients, the following general guidelines are proposed: "a higher risk for offspring than for sibs of a patient with congenital heart disease; a higher risk if the lesion is common (e.g., ventricular septal defects) than if the lesion is uncommon (e.g., tricuspid atresia); a two-fold to three-fold increase in risk if there are two affected first-degree relatives rather than one.[9]" If there are three affected first-degree relatives, especially if one is an affected parent ("type C" family),[91] then the risk is "comparable to or greater than Mendelian risks.[9]"

PROPOSED MECHANISMS OF TERATOGEN-INDUCED CONGENITAL HEART DISEASE

A number of studies have been conducted over the past several years to elucidate teratogenic mechanisms of action that cause CHD. However, determination of such mechanisms is complicated by several factors, some of which are listed in Table 6.[85]

Catecholamines

Sympathomimetic amines have been found to produce both aortic arch anomalies and ventricular septal defects.[36,39] Hodach et al.,[36] in an investigation concerning the potential

Table 6
LIMITING FACTORS IN DETERMINATION OF CARDIOVASCULAR PATHOGENIC MECHANISMS

1. Experimentally induced defects:
   Nonspecific agents
   "Artificial" anomalies produced by possibly alternate mechanisms in experimental animals
2. Time of exposure
3. Ability of organisms to compensate prior to observation
4. Development of secondary defects due to abnormal hemodynamic patterns
5. Separate defect-producing factors acting simultaneously
6. Inadequate numbers
7. Subjectivity of interpretation by investigators

teratogenic effect of epinephrine on the embryonic chick, directly exposed embryos to epinephrine during the period of cardiovascular embryogenesis: Hamburger-Hamilton (H-H)[92] stages 6 to 35 (24 to 190 h).[93] Results showed a spectrum of aortic arch anomalies produced during the period of aortic arch development (66 to 148 h). Exposure before or after this period did not lead to such anomalies (Figure 3). The largest percentage of aortic arch anomalies (94%) occurred on day 5 (105 to 112 h), during development of the third, fourth, and sixth aortic arches.[36]

The most common anomalies involved only the third pair of aortic arches (43.3%). Anomalies involving only the fourth pair of aortic arches comprised 36.6% of the total cases, and 4.6% of all anomalies involved solely the sixth pair of aortic arches. More than one pair of aortic arches were involved in 32.3% of total cases.

Hodach et al.[36] noted that anomalies involving the right fourth arch were consistently associated with VSD. Similarly, other investigators observed the simultaneous occurrence of aortic arch anomalies and VSD.[38,94,95] In addition, Hodach et al.[36] noted that if either of the sixth aortic arches or the arcus aorta (fourth right arch in chicks) failed to form, VSD consistently developed.

Investigations concerning the possible association between epinephrine exposure and chronotropism have revealed conflicting results.[96-100] However, in chick embryos exposed to epinephrine, Hoffman and Van Mierop[37] noted a marked increase in systolic and diastolic pressure following the second half of day 3 in gestation and continuing until hatching (Figure 3). The change in pressure was most pronounced in embryos exposed during the fourth to sixth day in gestation,[37] corresponding to the highest frequency of aortic arch anomaly incidence observed in the investigation of Hodach et al.[36] Determination of a potential relationship between hemodynamic change and morphologic alteration is a subject which deserves more study. However, it has been shown that exposure to epinephrine, a substance known to alter hemodynamics, causes aortic arch anomalies which then lead to cardiac defects[36]

Increased β-adrenergic activity has been proposed as a potential mechanism in production of aortic arch anomalies and cardiac defects (Figure 3, bottom). Exposure of chick embryos to pharmacologic agents, including isoproterenol, epinephrine, norepinephrine, and phenylephrine at H-H stages 20 to 27 produced cardiovascular anomalies proportional to the degree of β-adrenergic activity of each drug. Anomalies produced include ASD, VSD, double outlet right ventricle, aortic hypoplasia, and truncus arteriosus. Pretreatment with the β-antagonist propranolol blocked the action of the sympathomimetic amines, significantly reducing the frequency of anomalies.[39]

The methylxanthines theophylline and caffeine do not produce aortic arch anomalies in chicks. However, they potentiate effective doses of catecholamines. Theophylline increases the effect of norepinephrine more than 100-fold. Resulting anomalies include aneurysms of

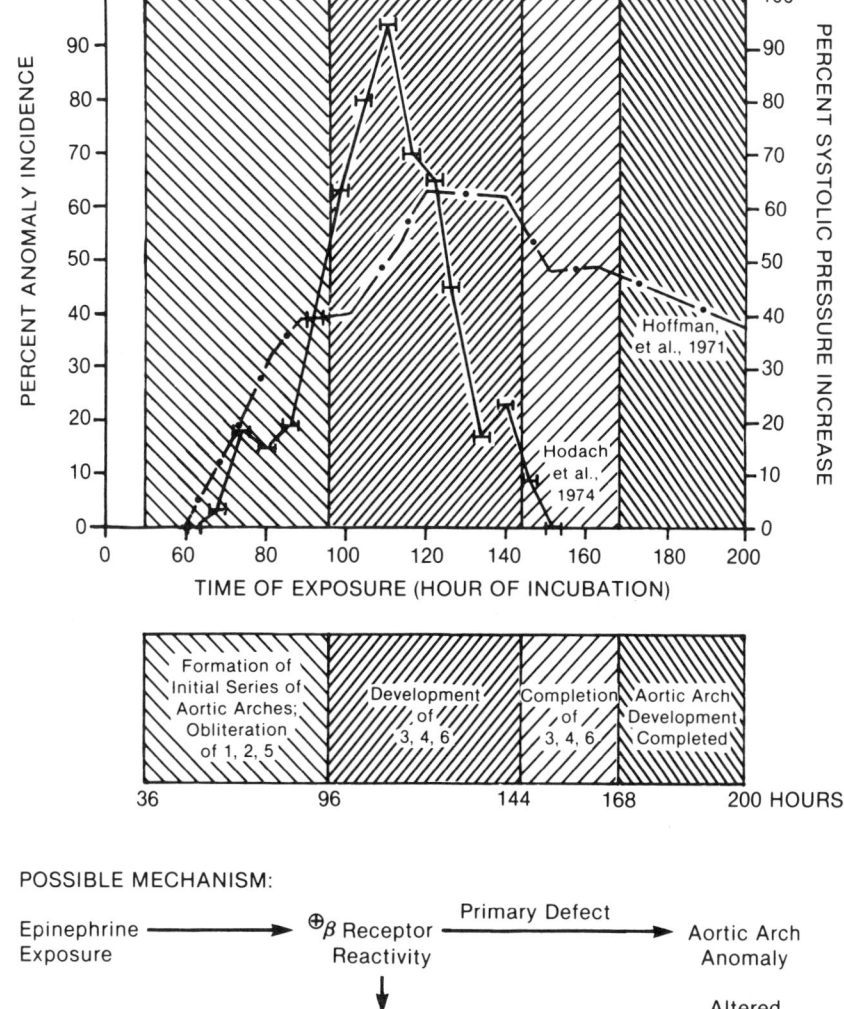

FIGURE 3. Percentages of aortic arch anomaly incidence (—) and arterial systolic pressure increase (—·—)) vs. time of exposure to epinephrine in chick embryos.

the ascending aorta and absence or virtually complete constriction of the right ductus arteriosus. Gilbert et al.[101] cited calcium mobilization and/or cyclic nucleotide phosphodiesterase (PDE) inhibition as potential causative factors (Figure 4). Methylxanthines have been shown to release ionized calcium from storage sites and increase the transmembranous influx of calcium in myocardial tissue.[102,103] Caffeine reportedly acts primarily by releasing intracellular calcium from subcellular compartments.[104-106]

Alcohol

The relationship between time of exposure and embryonic sensitivity to catecholamines is similarly exhibited in embryonic sensitivity to alcohol at specific stages in gestation. Chick embryos exposed to alcohol at H-H stage 18 (active cardiogenesis) showed impaired viability as compared to chick embryos exposed at H-H stage 1 (prestreak). In a study of 160 chick embryos, the LD_{50} for embryos exposed during active cardiogenesis was 0.06 ml absolute ethanol, whereas the LD_{50} for embryos exposed at H-H stage 1 was 0.60 ml.[24] In addition,

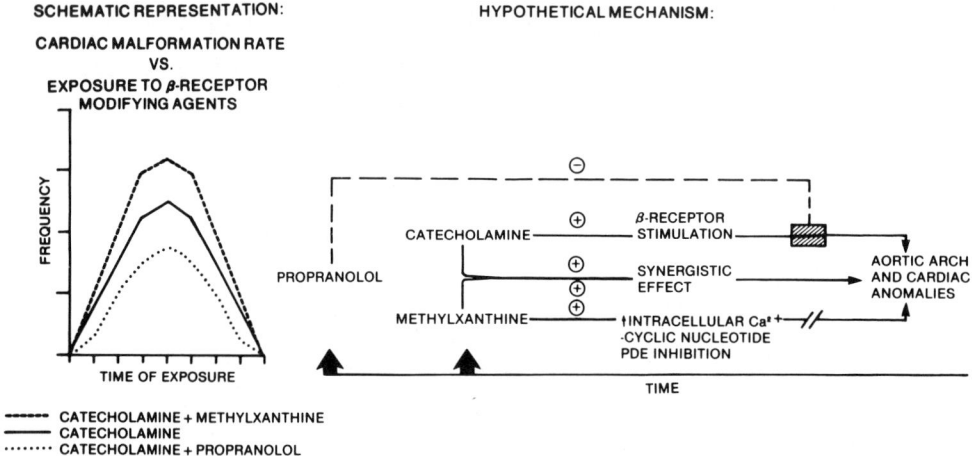

FIGURE 4. Proposed independent and combined effects of catecholamine and methylxanthine exposure. Pretreatment with propranolol appears to block the teratogenic effect.

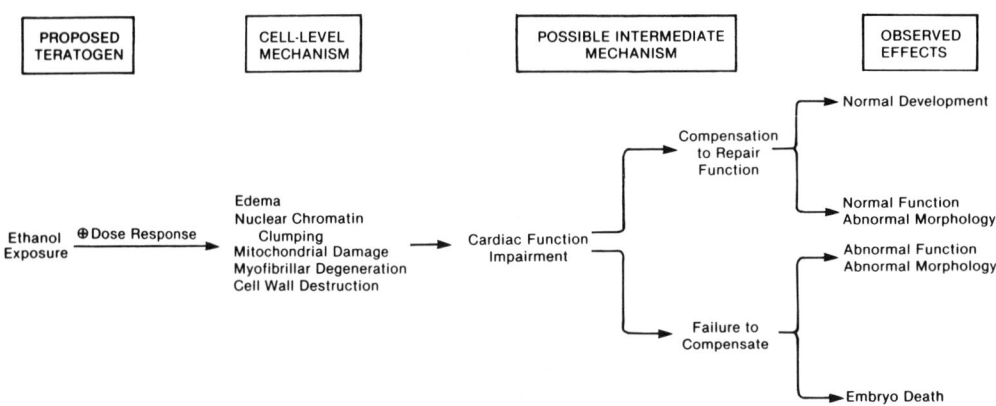

FIGURE 5. Proposed mechanisms of injury following ethanol exposure. A possible intermediate mechanism may be a key factor in the inability to observe function change following teratogenic exposure.

there was a positive association between dose injected and mortality among embryos injected at the same time in gestation.

Electron micrographs of cardiac tissue from sacrificed embryos show a positive association between the degree of cellular damage and the administered dose. Cardiac tissue of embryos exposed to low to moderate doses of ethanol (0.0 to 0.075 ml) show some degree of edema, nuclear chromatin clumping, and mitochondrial swelling, but little or no myofibril degeneration. Micrographs of embryos exposed to 0.1 to 0.2 ml ethanol show the same alterations seen at lower doses, but to a greater degree. In addition, tissue from embryos exposed to higher doses show a high degree of vacuole formation, myofibrillar breakdown, and cell wall degeneration. Furthermore, embryos exposed during active cardiogenesis showed much more severe cellular damage than that of embryos exposed at H-H stage 1.[24] Finally, among embryos injected at H-H stage 18 with equal doses of ethanol, embryos examined 24 postexposure showed more cellular damage than embryos observed 1 h postexposure.

The proposed mechanism of action (Figure 5) suggests myocardial cellular damage is the underlying cause of increased mortality and morphologic changes. There was no observable significant function change (heart rate and shortening fraction) among survivors

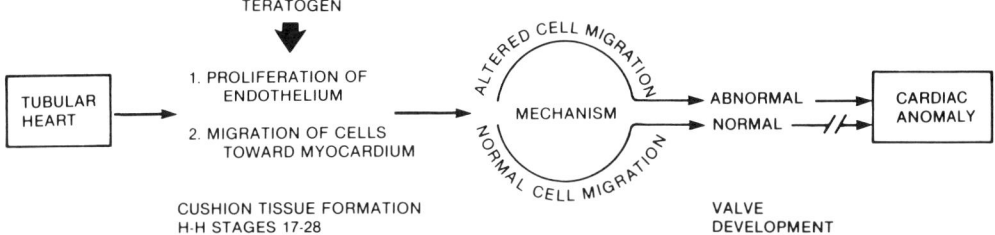

FIGURE 6. Teratogenic exposure during endocardial cushion development potentially causes valvular septal maldevelopment with subsequent cardiac defect.

filmed 1 and 24 h after exposure at H-H stage 18, possibly suggesting an intermediate mechanism leading either to recovery or to continued cellular degeneration and death.[25,26] Retention of myofibrillar integrity in chicks exposed to doses less than or equal to 0.075 ml ethanol may enable the heart to maintain function in spite of cellular damage, whereas higher doses (0.10 to 0.20 ml) may result in irreparable damage leading to death. The existence of such a mechanism would explain the inability to demonstrate function change.

The Role of Endocardial Cushions

Embryonic sensitivity to teratogenic exposure has been documented in studies concerned with endocardial cushion development in the embryonic chick. In a comparative study of developing rat and chick endocardial cushions (16 to 18 somites), Markwald et al.[107] proposed a mechanism for endocardial cushion origin and development. Evidence suggests that the source of endocardial cushion cells lies in the endocardium, or inner sleeve of the cardiac tube. After migrating centrifugally through the cardiac jelly matrix, the cushion cells make contact with the myocardium, or outer sleeve of the heart tube.[107,108] Modifying of cardiac jelly by endocardial cells is inferred from matrix uptake of ^{35}S, incorporation into chondroitin sulfate, and secretion of chondroitin sulfate-containing material contiguous with migrating cushion cells.[109] Overman and Beaudoin showed that teratogenic exposure potentially decreases incorporation of labeled sulfate into glycosaminoglycans.[110] Administration of salicylate, a glycosaminoglycan inhibitor, to chick embryos during cushion formation produced outflow obstruction, valvular atresia, and septal defects.[111] Markwald et al.[112] and Manasek[113,114] demonstrated that a defect in more than one location in the cardiac tube could potentially result in abnormal development of endocardial cushion tissue and, consequently, abnormal valves and septae (Figure 6).

Le Dourain and Le Dourain[19] note that a large number of internal abnormalities of the heart result from complete (or semicomplete) fusion of blood streams. Several investigators have shown that hemodynamic alteration potentially leads to subsequent change in cardiac structure.[38,95,115-122] Endocardial cushions, which are especially sensitive to irradiation,[19] may fail to develop or develop incompletely in response to teratogenic exposure, thus leading to incomplete septation and/or valvular development and, consequently changes in flow patterns and cardiac morphology.

Hypoxia

Grabowski[123] demonstrated that exposure to moderate hypoxia (10.5% for 5 h) in 5-d-old chick embryos produces malformations in approximately 25% of all embryos. The proposed mechanism of action is "the edema syndrome", characterized in part by generalized edema, hypervolemia, hemorrhages, and blisters. Results of an investigation by Jaffee[124] suggest the same mechanism of action. Exposure to hypoxia (5% O_2) and hypoxia + hypercapnea (5% O_2 + 2% CO_2) among H-H stage 17 (60 h) chick embryos produced a significant incidence of anomalies, including extracardiac as well as cardiovascular defects.

Observations of embryos exposed to hypoxia + hypercapnea showed edema, hemorrhages, hypervolemia, and bradycardia. Hemodynamic alterations produced by exposure to hypoxia + hypervolemia are potentially associated with gross cardiac malformations noted in the experimental chicks.

Upon analysis, blood serum shows marked changes in composition, suggesting that hypoxia interferes with normal physiologic maintenance of extracellular and intracellular fluids.[123,125] Accumulation of lactic acid as a result of anaerobiosis is a possible explanation for such serum changes. In a separate investigation Byerly[126] noted enlarged venous sinuses in suffocating chicks and proposed lactic acid accumulation to be the cause. In addition, Grabowski[125] demonstrated that lactic acid levels increase markedly with hypoxic stress, and injections of lactic acid mimic some of the effects seen in hypoxia.

Altered potassium levels may be another factor leading to observed anomalies in oxygen-deficient animals. Oxygen deficiency was induced in 14.5- to 19.5-d-old rats through one of five means; oxygen-deficient atmospheres, KCN, $NaNO_2$-induced methemoglobinemia, uterine vessel clamping, and maternal injection of adrenalin and vasopressin. As the degree of hypoxia increased, embryonic heart beat decreased and serum potassium levels increased to extremely high levels.[58] Increases in potassium levels may be explained by interference with the intracellular "sodium pump".[123]

A potential link between hypoxia-induced hypervolemia and rise in blood pressure with subsequent hemorrhage was addressed by Grabowski et al.[127] These investigators exposed 3- to 5-d-old chick embryos to 6 to 10% O_2 for 4 to 6 h. Results indicated that hypervolemia induced by hypoxia is responsible for blood pressure increases and appearance of hemorrhage.

Ruckman et al.[59-61] used cinephotoanalysis to measure the myocardial contractility of H-H stage 18 chick embryos following exposure to graded levels of hypoxia (16% O_2, 11% O_2, 6% O_2) and compared to controls (21% O_2). Results showed that myocardial contractility (measured as SF) was significantly depressed following hypoxic stress ($p < 0.05$). Furthermore, function impairment was related to the intensity of induced hypoxic stress. Finally, observation of films showed alterations in the pattern of contraction in 16, 11, and 6% O_2-exposed embryos. Electron microscopy revealed cellular edema and alteration of nuclear chromatin related to the severity and duration of hypoxic stress. Myocyte damage and death were noted after 5 h of exposure to moderate hypoxia.

The investigators found no association between rise in heart rate and severity of hypoxia. Exposure to the mildest hypoxia (16% O_2) resulted in immediate tachycardia, whereas no additional increase in heart rate was observed with more severe hypoxia. However, increased intraventricular pressure, an effect of the edema syndrome, may have been a causative factor leading to the observed rise in heart rate following mild hypoxic stress. Depression of contractility noted after more severe hypoxic exposure may serve to prohibit a rise in heart rate under these conditions.[59]

PROPOSED MECHANISMS OF ABNORMAL DEVELOPMENT IN TWO SPECIFIC LESIONS

Almost all congenital heart defects that occur in humans also occur in other vertebrates. To study the events which lead to the anomaly, investigators have used two types of approaches. In the first, environmental agents, either mechanical or pharmacological, which consistently produce cardiac defects, have been used to study embryologic pathogenesis. This approach is complicated by the nonspecificity of certain teratogens and the variable response of embryos depending on when the agent is introduced. Secondly, in a few animal models, hereditary cardiac defects occur in high proportions, permitting the embryologic study of naturally occurring anomalies. Both approaches, however, are complicated by the inherent ability of the embryo to compensate for abnormalities. Nonetheless, investigations of both environmental and genetic models have provided information which has helped improve our understanding of the development of CHD in humans. Approaches used to study the development of two specific lesions, VSD and TGA, will be discussed.

Ventricular Septal Defect

VSD occurs when one or all of three basic structures, the interventricular septum, endocardial cushions, and conus, fail to complete ingrowth of tissue which surrounds the interventricular opening and contributes to its closure. The fact that three separate areas of the heart contribute to closure, an event which is also dependent on proper timing, probably contributes to the frequent occurrence of VSD.[128]

Environmental Models of VSD

In a review of experimental studies, Wilson[129] found a high correlation between aortic arch anomalies and VSD. Demonstrating this association, Rychter,[38] with the use of silver microclips which stenosed both pulmonary arches (the sixth aortic arches), experimentally induced VSDs in chick hearts. Because the whole stroke volume could not pass through the narrowed ventricular outlet, blood was forced to escape into the neighboring ventricle through the ventricular septum, which was not yet closed. Consequently, fusion of the right and left proximal bulbar cushions and the tubercles from the fused atrioventricular cushions was disturbed by the altered bloodstream, and septation was not completed. Even if temporary, Rychter[38] states that disturbances in hemodynamic equilibrium may cause a VSD. In embryos in which the clip produced only transitory stenosis, VSDs still occurred.

After bilateral sixth arch elimination, which hemodynamically induced VSD, Pexieder[130] studied changes in cell death patterns inside the bulbar cushions using Nile blue supravital stain. In the transition zone between the proximal right and distal ventral bulbar cushions, where cell death almost never occurs, he found a new zone of intense cell death as early as 16 h after experimental intervention. In support of these observations, Rychter and Lemež,[131] in their investigations of VSD pathogenesis, showed that 32 h after intervention, the distal ventral bulbar cushion had a shorter proximal extension than in normals, and fusion of the distal ventral and proximal right bulbar cushions was prevented. Pexieder[130] proposed that fusion of these cushions was prevented by the newly formed cell death focus 16 h before the manifestations of delayed fusion at the organ level. With the same experimental conditions, Pexieder[132,133] investigated endocardial modifications using scanning electron microscopy. He observed crater-like formations on the endocardium proximal to the cushions and on the distal ventral cushion, which he interpreted as a sign of cell damage.[134] In addition, giant intercellular clefts observed on the proximal left bulbar cushions and distal ventral cushion were interpreted as a consequence of an increased washout of phagocytes. Pexieder[130] suggests that "both the localized intensification of cell death and the altered removal of phagocytes, dead cells, and cytosegresomes from the bulbar tissues contribute to the disturbance of normal fusion of the bulbar cushions, resulting in VSD." He proposes that this interpretation of cellular events provides an important connection between alterations in hemodynamics and morphogenesis that occur during the development of VSD.

Investigations of the effect hemodynamic changes on the ventricular septum developing between parallel bloodstreams have provided additional information regarding the formation of VSD. Following the administration of trypan blue to 44-h chick embryos, Jaffee[118] observed inhibition of formation of the left ventricular outflow tract. This inhibition resulted in increased vascular resistance to the left ventricle, leading to higher pressures in this chamber. Flow of blood between the ventricles was observed at 5.5 d, while it was not observed in normal hearts, indicating higher pressures in the left ventricle because blood flow was observed moving left to right.[118] Flow studies in the chick embryos demonstrated fusion of the two bloodstreams at the sites of the forming interventricular septum, interrupting normal septation which would occur between two well-defined bloodstreams.[118]

After lung bud excision in the chick embryo, Clark et al.[135] observed a high incidence of ventricular septal defects (13 of 18, or 72.2%), suggesting that a primary teratogenic influence on the developing lung may alter intracardiac morphology. Because vessel size is proportional to blood flow in embryonic systems, they suggested that smaller left pulmonary artery and vein diameters are related to decreased blood flow in these vessels. Increased vascular resistance due to removal of a portion of the lung bud could contribute to alterations in blood flow. The resultant hemodynamic alterations, as several investigators[36,38,39,118] have demonstrated, could account for the formation of ventricular septal defects.

Genetic Models of VSD

In embryos of Siller's strain of chicks, VSD is a heterogeneous inborn malformation either due to a primary disturbance of ventricular septal development of the consequence of more transient deviations in heart development.[38] Rychter and Lemež[136] used this animal model to study the embryologic development of VSD. Their speculation based on experimental findings that a temporary obstacle to blood flow existed inside the aortic outlet was confirmed by the presence of a supernumerary small proximal bulbar cushion in the left ventricular outflow tract. *In vivo* analysis of the colored bloodstream from the left ventricle demonstrated that the cushion deviated a portion of the bloodstream from the left into the right ventricle. Especially when there was a large supernumary cushion, they noted slight stenosis of the right fourth aortic arch, which they propose was due to altered hemodynamics. The interaction of changes in morphogenesis and hemodynamics could account for the high frequency of VSD in Siller's strain of chicks.

The genetically regulated development of cardiovascular defects in another animal model, the Keeshond strain of dogs, provided Van Mierop et al.[85] the opportunity to study embryologic pathogenesis. Their studies determined that intracardiac anomalies in the Keeshond dog result from abnormal development of the embryonic conotruncal septum. The most important disturbance in development was hypoplasia of the conus cushions, resulting in a hypoplastic or sometimes absent conal septum. Consequently, the interventricular communication remained widely patent.[85]

Transposition of the Great Arteries

Instead of the normal condition in which the aorta and the pulmonary trunk spiral after leaving the heart, in transposition of the great arteries, these two vessels arise from the right and left ventricles, respectively, and run parallel to each other.[137] The embryogenesis of this defect is more complex and obscure than other lesions. de la Cruz and da Rocha[138] attribute TGA to a "complete lack of rotation of the truncoconal septum". Van Praagh et al.[139] point out that the conotruncal wall itself must be involved in the failure of rotation. They continue that if there were no abnormality other than the lack of septal spiraling, the coronary arteries would be expected to arise in their usual position from the vessel overlying the left ventricle; however, the coronary ostia are routinely associated with the aorta.[139] They conclude that "the straight aortico-pulmonary septum is considered an effect of the conal malformation, no the cause of the transposition."[139]

After comparing abnormal and normal hearts in a number of mammalian species, Van Mierop[140] still considers TGA an enigma. Observations that TGA is often uncomplicated by other anomalies and that inversion of the ventricles (L-loop) is almost always associated with TGA have provided clues to the pathogenesis of TGA in a general sense. Van Mierop[140] suggests that because the ventricles anatomically are normal, the developmental abnormality occurs distal to the conal septum. He explains that the embryo "either uses the intercalated valve cushions to partition the truncus arteriosus (these cushions becoming the major ones, while the normal truncus cushions remain small and simply form a valve cusp each), or the aorticopulmonary septum develops on the left instead of the (normal) right side." Each of these mechanisms could result in TGA.

Environmental Models of TGA

Due to the complexity of this anomaly, it has been difficult to determine the pathogenesis of TGA from experimental animal models using environmental or pharmacological interventions. A few experiments, however, have provided information regarding the pathogenesis of TGA. After administration of trypan blue at the primitive streak stage in rat fetuses, Monie et al.[141] found shortening of the great arteries, abnormal looping of the heart tube, and a minimal degree of spiraling of the shortened truncus, which may have been a result of altered blood flow. TGA was commonly found. Dor and Corone[142] produced TGA in 3- to 3 $1/2$-embryonic-day chick embryos after a find lamella was taken from the shell membrane and applied to one segment of the conus. Adhesions subsequently formed around the conus, which were capable of altering conus development leading to the production of TGA.

In his experiments using glass models of the heart altered by distinct manipulations to resemble morphologically abnormal hearts with deficient myocardial growth, Goerttler[143] studied alterations in flow patterns which showed relationships to the conventional transposition of the great vessels. Goerttler's experiments provided support for a hemodynamic basis for TGA.

Genetic Models of TGA

Van Praagh et al.[144] studied both the pathologic anatomy of 221 cases of TGA in man and serial sections of 18 cases of TGA in 14-d fetal and newborn mice produced by the mating of animals homozygous for the situs inversus (iv) gene. Van Praagh et al.[144] believe this is the first time that typical TGA in man has been reproduced in a genetic model in mammals. Their data strongly supported the concept that abnormal development of the conus and of the ventricular loop are important in the production of abnormal atrioventricular relationships. They suggest that observations of "bizarre early cardiac loop formation may explain how it is possible for the ventricles to loop in one direction and for the conotruncus to twist in the opposite direction." The high frequency of TGA (18/80 or 22.5%) observed in this animal model of situs inversus supports the concept that TGA results from discordance between conal and ventricular development, specifically conal inversion.[144]

CONCLUSION

Whether due to environmental conditions, genetic factors, or an interaction of both, as is often the case, CHD has been investigated by both epidemiological studies in humans and by observations of pathogenesis in several experimental models. Both morphological alterations which impinge on developing hemodynamic patterns and altered blood flow patterns which affect morphological development contribute to the complexity of pathogenesis of these defects. Our ability to look at the interplay of form and function, however, has improved, and further investigations will elucidate mechanisms to help define the basis of malformations of the heart and blood vessels.

REFERENCES

1. **Zierler, S.**, Maternal drugs and congenital heart disease, *Obstet. Gynecol.*, 65, 155, 1985.
2. **Mitchell, S. C., Korones, S. B., and Berendes, H. W.**, Congenital heart disease in 56,109 births, *Circulation*, 43, 323, 1971.
3. **Nora, J. J. and Nora, A. H.**, Genetic epidemiology of congenital heart diseases, *Prog. Med. Genet.*, 5, 91, 1983.
4. **Gregg, N.**, Congenital cataracts following German measles in the mother, *Trans. Ophthalmol. Soc. Aust.*, 3, 35, 1941.

5. **Dekaban, A., O'Rourke, J., and Corman, T.,** Abnormalities in offspring related to maternal rubella during pregnancy, *Neurology,* 8, 387, 1958.
6. **Michaels, R. H. and Mellin, G. W.,** Prospective experience with maternal rubella and the associated congenital malformations, *Pediatrics,* 26, 200, 1960.
7. **Lundstrom, R.,** Rubella during pregnancy: a follow-up study of children born after an epidemic of rubella in Sweden, 1951, with additional information on prophylaxis and treatment of natural rubella, *Acta Paediatr.,* 51(Suppl. 133), 1, 1962.
8. **Campbell, M.,** Place of maternal rubella in the aetiology of congenital heart disease, *Br. Med. J.,* 1, 691, 1961.
9. **Nora, J. J. and Nora, A. H.,** The evolution of specific genetic and environmental counseling in congenital heart diseases, *Circulation,* 57, 205, 1978.
10. **Nora, J. J. and Nora, A. H.,** *Genetics and Counseling in Cardiovascular Diseases,* Charles C Thomas, Springfield, IL, 1978.
11. **Korones, S. B., Ainger, L. E., Monif, G. R. G., Roane, J., Sever, J. L., and Fuste, R.,** Congenital rubella syndrome: new clinical aspects with recovery of virus from affected infants, *J. Pediatr.,* 67, 166, 1965.
12. **Swan, C., Tosterin, A. L., Moore, B., Mayo, H., and Black, B. H. B.,** Congenital defects in infants following infectious diseases during pregnancy, with special reference to relationship between German measles and cataract, deaf-mutism, heart disease and microcephaly, and to period of pregnancy in which occurrence of rubella is followed by congenital abnormalities, *Med. J. Aust.,* 2, 201, 1943.
13. **Cooper, L. Z., Ziring, P. R., and Ockerse, A. B.,** Rubella: clinical manifestations and management, *Am. J. Dis. Child.,* 118, 18, 1969.
14. **Lenz, W.,** Chemicals and malformations in man, in *Congenital Malformations,* Fishbein, M., Ed., International Medical Congress, New York, 1964, 263.
15. **Weicker, J.,** Klinik und Epidemiologie der Thalidomid-Embryopathie, *Bull. Soc. R. Belge Gynecol. Obstet.,* 33, 21, 1963.
16. **Swinyard, C. A.,** *Limb Development and Deformity: Problems of Evaluation and Rehabilitation,* Scheltema and Holkema, Amsterdam, 1969.
17. **Wilson, J. G.,** Abnormalities of intrauterine development in non-human primates, in *The Use of Non-Human Primates in Research of Human Reproduction,* Diczfalusy, E. and Standley, C. C., Eds. WHO Research and Training Centre on Human Reproduction, Stockholm, 1972, 261.
18. **Wilson, J. G.,** An animal model of human disease: thalidomide embryopathy in primates, *Comp. Pathol. Bull.,* 5, 3, 1973.
19. **Le Dourain, N. and Le Dourain, G.,** The effects of radiation on cardiac development, in *Cardiac Development with Special Reference to Congenital Heart Disease,* Jaffe, O. C., Ed., University of Dayton Press, Dayton, OH, 1970, 121.
20. **Globus, M. S.,** Teratology for the obstetrician: current status, *Obstet. Gynecol.,* 55, 269, 1980.
21. **Lemoine, P., Harousseau, H., Borteyru, J. P., et al.,** Les enfants de parents alcooliques: anomalies observées à propos de 127 cas, *Ouest Med.,* 21, 476, 1968.
22. **Ulleland, C. N.,** The offspring of alcoholic mothers, *Ann. N.Y. Acad. Sci.,* 1971, 167, 1972.
23. **Jones, K. L., Smith, D. W., Ulleland, C. N., et al.,** Patterns of malformations in offspring of chronic alcoholic mothers, *Lancet,* i, 1267, 1973.
24. **Ruckman, R. N., Messersmith, D. J., O'Brien, S. A., and Morse, D. E.,** Alcohol sensitivity in the embryonic heart, *Pediatr. Cardiol.,* 5, 254, 1984.
25. **Ruckman, R. N., O'Brien, S. A., Messersmith, D. J., Boeckx, R. L., and Morse, D. E.,** Changes in myocardial structure and function in the embryonic chick exposed to ethanol, *Teratology,* 31, 31A, 1985.
26. **Ruckman, R. N., O'Brien, S. A., Messersmith, D. J., Boeckx, R. L., and Morse, D. E.,** Alterations of myocardial function and structure in a model of alcohol embryopathy, *Am. Heart J.,* 110, 705, 1985.
27. **Shapiro, S. and Slone, D.,** The effects of exogenous female hormones on the fetus, *Epidemiol. Rev.,* 1, 110, 1979.
28. **Schardein, J. L.,** Congenital abnormalities and hormones during pregnancy: a clinical review, *Teratology,* 22, 251, 1980.
29. **Levy, E. P., Cohen, A., and Fraser, F. C.,** Hormone treatment during pregnancy and congenital heart defects, *Lancet,* 1, 611, 1973.
30. **Robertson-Rintoul, J.,** Oral contraceptives: potential hazards of hormone therapy during pregnancy, *Lancet,* 2, 515, 1974.
31. **Nishimura, H., Uwabe, C., and Semba, R.,** Examination of teratogenicity of progestogens and/or estrogens by observation of the induced abortuses, *Teratology,* 10, 93, 1974.
32. **Heinonen, O. P., Slone, D., and Monson, R. R.,** Cardiovascular birth defects and antenatal exposure to female sex hormones, *N. Engl. J. Med.,* 296, 67, 1977.
33. **Nora, J. J. and Nora, A. H.,** Preliminary evidence for a possible association between oral contraceptives and birth defects, *Teratology,* 7, A-24, 1973.
34. **Nora, J. J. and Nora, A. H.,** Can the pill cause birth defects?, *N. Engl. J. Med.,* 291, 731, 1974.

35. **Mitchell, S. C., Sellman, E. H., and Westphal, M. C.**, Etiologic correlates in a study of congenital heart disease in 56,190 births, *Am. J. Cardiol.*, 28, 653, 1971.
36. **Hodach, R. J., Gilbert, E. F., and Fallon, J. F.**, Aortic arch anomalies associated with administration of epinephrine in chick embryos, *Teratology*, 9, 203, 1974.
37. **Hoffman, L. E. and Van Mierop, L. H. S.**, Effect of epinephrine on heart rate and arterial blood pressure of the developing chick embryo, *Pediatr. Res.*, 5, 472, 1971.
38. **Rychter, Z.**, Experimental morphology of the aortic arches and the heart loop in chick embryos, *Adv. Morphogenesis*, 2, 333, 1962.
39. **Hodach, R. J., Hodach, A. E., Fallon, J. F., Folts, J. D., Bruyere, H. J., and Gilbert, E. F.**, The role of β-adrenergic activity in the production of cardiac and aortic arch anomalies in chick embryos, *Teratology*, 12, 33, 1975.
40. **Weinstein, M. R. and Goldfield, M. D.**, Cardiovascular malformations with lithium use during pregnancy, *Am. J. Psychiatry*, 132, 529, 1975.
41. **Mignot, G., Devie, M., and Dummont, M.**, Lithium et grossesse, *J. Gynecol. Obstet. Biol. Reprod.*, 7, 1303, 1978.
42. **Nora, J. J., and Nora, A. H., and Toews, W. H.**, Lithium, Ebstein's anomaly and other congenital heart defects, *Lancet*, 2, 594, 1974.
43. **Schou, M., Goldfield, M. D., and Weinstein, M. R.**, Lithium and pregnancy. I. Report from the register of lithium babies, *Br. Med. J.*, 2, 135, 1973.
44. **Källén, B. and Tandberg, A.**, Lithium and pregnancy: a cohort study on manic-depressive women, *Acta Psychiatr. Scand.*, 68, 134, 1983.
45. **Dareste, M. C.**, *La Production Artificielle des Monstruosités*, C. Reinwald et Cie, Paris, 1877.
46. **Büchner, F.**, Experimentelle Entwicklungsstörungen durch allgemeinen Sauerstoffmangel, *Klin. Wochenschr.*, 26, 38, 1955.
47. **Rübasaamen, H.**, Uber die teratogenetische Wirkung des Missbildungen bei Mensch und Tier, *Beitr. Pathol. Anat. Allg. Pathol.*, 112, 336, 1950.
48. **Ingalls, T. H.**, Causes and prevention of developmental defects, *J. Am. Med. Assoc.*, 161, 1047, 1956.
49. **Windle, W. F.**, Role of respiratory distress in asphyxial brain damage of the newborn, *Cerebral Palsy J.*, 27, 3, 1966.
50. **St.-Hilaire, G.**, Des différents états de pesanteur des oeufs au commencement et à la fin de l'incubation, *J. Complemtaire Sci. Méd.* 7, 271, 1820.
51. **Van Liere, E. J. and Stickney, J. C.**, *Hypoxia, a Detailed Review of the Effects of Oxygen Want on the Body*, University of Chicago Press, Chicago, 1963.
52. **Grabowski, C. T.**, Atmospheric gases, in *Handbook of Teratology*, Vol. 1, Wilson, J. G. and Fraser, F. C., Eds., Plenum Press, New York, 1977, 405.
53. **Beard, R. W.**, Response of the fetal heart and maternal circulation to adrenalin and noradrenalin, *Br. Med. J.*, 1962, 443, 1962.
54. **Donarin, A.**, Noradrenalin and the foetal heart, *Lancet*, 1962, 756, 1964.
55. **Adamsons, K., Mueller-Heuback, E., and Myers, R. E.**, Production of fetal asphyxia in the rhesus monkey by administration of catecholamines to the mother, *Am. J. Obstet. Gynecol.*, 109, 248, 1971.
56. **Chernoff, N. and Grabowski, C. T.**, Responses of the rat foetus to maternal injections of adrenaline and vasopressin, *Br. J. Pharmacol.*, 43, 270, 1971.
57. **Dornhorst, A. C. and Young, I. M.**, The action of adrenaline and noradrenaline on the placental and foetal circulation in the rabbit and guinea pig, *J. Physiol.*, 118, 282, 1952.
58. **Grabowski, C. T. and Chernoff, N.**, Effects of hypoxia on the cardiovascular physiology of mammalian embryos, *Teratology*, 3, 201, 1970.
59. **Ruckman, R. N., Rosenquist, G. C., Rademaker, D. A., Morse, D. E., and Getson, P. R.**, The effects of graded hypoxia on the embryonic chick heart, *Teratology*, 32, 463, 1985.
60. **Ruckman, R. N., Rosenquist, G. C., and Rademaker, D. A.**, The effect of graded hypoxia on the embryonic heart, *Am. J. Cardiol.*, 49,(4), 939, 1982.
61. **Ruckman, R. N. and Rademaker, D. A.** Recovery of cardiac function after hypoxic stress, *Teratology*, 27, 73A, 1983.
62. **Ruckman, R. N., Rosenquist, G. C., and Rademaker, D. A.**, Functional change in the embryonic heart induced by hypoxia and/or hypothermia, *Teratology*, 23, 59A, 1981.
63. **Ruckman, R. N., Rosenquist, G. C., and Rademaker, D. A.**, Preservation of cardiac function in the stressed embryo, *Pediatr. Cardiol.*, 2, 165, 1982.
64. **Heinonen, O. P., Slone, D., and Shapiro, S.**, *Birth Defects and Drugs in Pregnancy*, Publishing Sciences Group, Littleton, MA, 1977, 124.
65. **Rowland, T. W., Hubbel, J. P., and Nadas, A. S.**, Congenital heart disease in infants of diabetic mothers, *J. Pediatr.*, 83, 815, 1973.
66. **Stevenson, R. E. and Huntlet, C. C.**, Congenital malformations in offspring of phenylketonuric mothers *Pediatrics*, 40, 33, 1967.

67. **Fisch, R. O., Doeden, D., and Lansky, L. L.,** Maternal phenylketonuria, *Am. J. Dis. Child.*, 118, 847, 1969.
68. **McCue, C. M., Mantakas, M. E., and Tingelstad, J. E.,** Congenital heart block in newborns of mothers with connective tissue disease, *Circulation*, 56, 82, 1977.
69. **Warkany, J., Passarge, E., and Smith, L. B.,** Congenital malformations in autosomal trisomy syndromes, *Am. J. Dis. Child.*, 112, 502, 1966.
70. **Lejeune, J., Gautier, M., and Turpin, R.,** Les chromosomes humains en culture de tissus, *C. R. Acad. Sci.*, 248, 602, 1959.
71. **Book, J. A., Santesson, B., and Zetterqvist, P.,** Association between congenital heart malformations and chromosomal variations, *Acta Paediatr.*, 50, 217, 1961.
72. **Anders, J. M., Moores, E. C., and Emanuel, R.,** Chromosome studies in 156 patients with congenital heart disease, *Br. Heart J.*, 27, 756, 1965.
73. **German, J., Ehlers, K. H., and Engle, M. A.,** Familial congenital heart disease. II. Chromosomal studies, *Circulation*, 34, 517, 1966.
74. **Emerit, I., de Grouchy, J., Vernant, P., and Corone, P.,** Chromosomal abnormalities and congenital heart disease, *Circulation*, 36, 886, 1967.
75. **Campbell, M.,** Causes of malformations of heart, *Br. Med. J.*, 2, 895, 1965.
76. **Lamy, M., de Grouchy, J., and Schweisguth, O.,** Genetic and non-genetic factors in etiology of congenital heart disease: study of 1188 cases, *Am. J. Hum. Genet.*, 9, 17, 1957.
77. **Warkany, J.,** Etiologic factors of congenital heart disease, in *Congenital Heart Disease*, Vol. 63, Bass, A. D. and Moe, G. K., Eds., American Association for the Advancement of Science, Washington, D.C., 1960.
78. **Walker, G. C. and Ellis, L. B.,** The familial occurrence of congenital cardiac anomalies, *Proc. N. Engl. Heart Assoc.*, 1940—41, 26.
79. **Apert, E. and Cambessédès,** Malformations cardiaques (maladie de Roger) chez une mère et deux de ses enfants, *Bull. Soc. Pediatr. Paris*, 28, 340, 1930.
80. **Campbell, M.,** Genetic and environmental factors in congenital heart disease, *Q. J. Med.*, 18, 379, 1949.
81. **Chelius, C. J., Rowe, G. G., and Crumpton, C. W.,** Familial aspects of congenital heart disease, *Am. J. Cardiol.*, 9, 508, 1962.
82. **Nora, J. J., McNamara, D. G., and Fraser, F. C.,** Hereditary factors in atrial septal defect, *Circulation*, 35, 448, 1967.
83. **Detweiler, D. K. and Patterson, D. F.,** Prevalence and types of cardiovascular disease in dogs, *Ann. N.Y. Acad. Sci.*, 127, 481, 1965.
84. **Patterson, D. F.,** Epidemiologic and genetic studies of congenital heart disease in the dog, *Circ. Res.*, 23, 171, 1968.
85. **Van Mierop, L. H. S., Patterson, D. F., and Schnarr, W. R.,** Hereditary conotruncal septum defects in Keeshond dogs: embryologic studies, *Am. J. Cardiol.*, 40, 936, 1977.
86. **Van Mierop, L. H. S. and Patterson, D. F.,** Pathogenesis of conotruncal defects and some other cardiovascular anomalies in the Keeshond dog, in *Etiology and Morphogenesis of Congenital Heart Disease*, Futura, Mount Kisco, NY, 1980, 177.
87. **Falconer, D. S.,** The inheritance of liability to certain diseases, estimated from the incidence among relatives, *Ann. Hum. Genet.*, 29, 51, 1965.
88. **Edwards, J. H.,** Familial predisposition in man, *Br. Med. Bull.*, 25, 58, 1969.
89. **Morton, N. E., Yee, S., Elston, R. C., and Lew, R.,** Discontinuity and quasicontinuity: alternative hypotheses of multifactorial inheritance, *Clin. Genet.*, 1, 81, 1970.
90. **Nora, J. J.,** Multifactorial inheritance hypothesis for the etiology of congenital heart diseases: the genetic-environmental interaction, *Circulation*, 38, 604, 1968.
91. **Nora, J. J.,** Etiologic factors in congenital heart diseases, *Pediatr. Clin. North Am.*, 18, 1059, 1971.
92. **Hamburger, V. and Hamilton, H. L.,** A series of normal stages in the development of the chick embryo, *J. Morphol.*, 88, 49, 1951.
93. **Sissman, N. J.,** Developmental landmarks in cardiac morphogenesis: comparative chronology, *Am. J. Cardiol.*, 25, 141, 1970.
94. **Wilson, J. G. and Warkany, J.,** Cardiac and aortic arch anomalies in the offspring of vitamin A deficient rats correlated with similar human anomalies, *Pediatrics*, 5, 708, 1950.
95. **Stephan, F.,** Contribution experimental a l'étude du developpment du système circulatoire chez l'embryon de poulet, *Bull. Biol. Fr. Belg.*, 86, 217, 1952.
96. **Hsu, F. T.,** The effect of adrenalin and acetylcholine on the heart rate of the chick embryo, *Clin. J. Physiol.*, 7, 243, 1933.
97. **Barry, A.,** The effect of epinephrine on the myocardium of the embryonic chick, *Circulation*, 1, 1362, 1950.
98. **Fingl, E., Woodbury, L. A., and Hecht, H. H.,** Effects of innervation and drugs upon direct membrane potentials of embryonic chick myocardium, *J. Pharmacol. Exp. Ther.*, 104, 103, 1952.

99. **Jones, D. S.,** Effects of acetylcholine and adrenaline on the experimentally uninnervated embryonic heart of the chick embryo, *Anat. Rec.,* 130, 253, 1958.
100. **McCarty, L. P., Lee, W. C., and Shideman, F. E.,** Measurement of the inotropic effects of drugs on the innervated and noninnervated embryonic chick heart, *J. Pharmacol. Exp. Ther.,* 129, 315, 1960.
101. **Gilbert, E. F., Bruyere, H. J., Ishikawa, S., Cheung, M. O., and Hodach, R. J.,** The effects of methylxanthines on catecholamine-stimulated and normal chick embryos, *Teratology,* 16(1), 47, 1977.
102. **Nayler, W. G.,** Effect of calcium on cardiac contractile activity and radiocalcium movement, *Am. J. Physiol.,* 204, 969, 1963.
103. **Nayler, W. G.,** Calcium exchange in cardiac muscle. A basic mechanism of drug action, *Am. Heart J.,* 73, 379, 1967.
104. **DeGubareff, T. and Sleator, W.,** Effects of caffeine on mammalian atrial muscle and its interaction with adenosine and calcium, *J. Pharmacol. Exp. Ther.,* 148, 202, 1965.
105. **Herz, R. and Weber, A.,** Caffeine inhibition of Ca uptake by muscle reticulum, *Fed. Proc., Fed. Am. Soc. Exp. Biol.,* 24, 208, 1965.
106. **Nayler, W. G. and Hasker, J. R.,** Effect of caffeine on calcium in subcellular fractions of cardiac muscle, *Am. J. Physiol.,* 211, 950, 1966.
107. **Markwald, R. R., Fitzharris, T. P., and Manasek, F. J.,** Structural development of endocardial cushions, *Am. J. Anat.,* 148, 85, 1977.
108. **Manasek, F. J.,** Determinants of heart shape in early embryos, *Fed. Proc., Fed. Am. Soc. Exp. Biol.,* 40(7), 201, 1981.
109. **Markwald, R. R. and Smith, W. N.,** Distribution of mucosubstances in the developing rat heart, *J. Histochem. Cytochem.,* 20, 896, 1972.
110. **Overman, D. O. and Beaudoin, A. R.,** Early biochemical changes in the embryonic rat heart after teratogenic treatment, *Teratology,* 3, 183, 1971.
111. **Gessner, I. H.,** Some biochemical and anatomic effects of sodium salicylate on the chick embryo heart, in *Pathophysiology of Congenital Heart Defects,* Adams, F. N., Swan, H. J. C., and Hall, V. E., Eds., University of California Press, Berkeley, 1970, 17.
112. **Markwald, R. R., Fitzharris, T. P., and Smith, W. N.,** Structural analysis of endocardial cytodifferentiation, *Dev. Biol.,* 42, 160, 1975.
113. **Manasek, F. J.,** The extracellular matrix: a dynamic component of the developing embryo, in *Current Topics in Developmental Biology,* Vol. 10, Moscona, A. A. and Monroy, A., Eds., Academic Press, New York, 1975, 35.
114. **Manasek, F. J.,** The extracellular matrix of the early embryonic hearts, in *Developmental and Physiological Correlates of Cardiac Muscle,* Lieberman, M. and Sano, T., Eds. Raven Press, New York, 1975, 1.
115. **Bremer, J. L.,** The presence and influence of two spiral streams in the heart of the chick embryo, *Am. J. Anat.,* 49, 409, 1932.
116. **Spitzer, A.,** *The Architecture of Normal and Malformed Heart. A Phylogenetic Theory of Their Development,* Charles C Thomas, Springfield, IL, 1951.
117. **Jaffee, O. C.,** Hemodynamics and cardiogenesis. I. The effects of altered vascular patterns on cardiac development, *J. Morphol.,* 110, 217, 1962.
118. **Jaffee, O. C.,** Hemodynamic factors in the development of the chick embryo heart, *Anat. Rec.,* 151, 69, 1965.
119. **Jaffee, O. C.,** The development of the arterial outflow tract in the chick embryo heart, *Anat. Rec.,* 158, 35, 1967.
120. **Gessner, I. H.,** Spectrum of congenital cardiac anomalies produced in the chick embryos by mechanical interference with cardiogenesis, *Circ. Res.,* 18, 625, 1966.
121. **Gessner, I. H. and Van Mierop, L. H. S.,** Experimental production of cardiac defects: the spectrum of dextroposition of the aorta, *Am. J. Cardiol.,* 25, 272, 1972.
122. **Jung, Y. H., Paul, M. H., Gailen, W. J., Friedberg, D. Z., and Kaplan, S.,** Experimental production of hypoplastic left heart syndrome in the chick embryo, *Am. J. Cardiol.,* 31, 51, 1973.
123. **Grabowski, C. T.,** Physiological changes in the bloodstream of chick embryos exposed to teratogenic doses of hypoxia, *Dev. Biol.,* 13, 199, 1966.
124. **Jaffee, O. C.,** The effects of moderate hypoxia and moderate hypoxia plus hypercapnea on cardiac development in chick embryos, *Teratology,* 10, 275, 1974.
125. **Grabowski, C. T.,** Lactic acid accumulation as a cause of hypoxia-induced malformations in the chick embryo, *Science,* 134, 1359, 1961.
126. **Byerly, T. C.,** Studies in growth, I. Suffocation effects in the chick embryo, *Anat. Rec.,* 32, 249, 1926.
127. **Grabowski, C. T., Tsai, E. T., and Toben, H. R.,** The effects of teratogenic doses of hypoxia on the blood pressure of chick embryos, *Teratology,* 2, 67, 1969.
128. **Jackson, B. T.,** Review article: the pathogenesis of congenital cardiovascular anomalies, *N. Engl. J. Med.,* 279, 80, 1968.

129. **Wilson, J. G.,** Experimental production of cardiac defects, in *Congenital Heart Disease,* Bass, A. D. and Moe, G. K., Eds., American Association for the Advancement of Science, Washington, D.C., 1960, 65.
130. **Pexieder, T.,** Cellular changes accompanying the pathogenesis of experimental hemodynamically induced ventricular septal defects in the chick embryo, in *Morphogenesis and Malformation of the Cardiovascular System,* Vol. 14, Rosenquist, G. and Bergsma, D., Eds., Alan R. Liss, New York, 1978, 452.
131. **Rychter, Z. and Lemež, L.,** Vascular system of the chick embryo. IV. Morphology of experimentally produced ventricular septal defects, *Česk. Morfol.,* 7, 21, 1959.
132. **Pexieder, T.,** Effet de l'hemodynamique sur la morphologie de l'endocarde embryonnaire, *Bull. Assoc. Anat. (Nancy),* 60, 399, 1976.
133. **Pexieder, T.,** SEM observations of the embryonic endocardium under normal and experimental hemodynamic conditions, *Bibl. Anat.,* 15, 531, 1977.
134. **Nelson, E., Sunaga, T., Shimamoto, T., Kawamura, J., Rennels, N. L., and Hebel, R.,** Ischemic carotid endothelium, *Arch. Pathol.,* 99, 125, 1975.
135. **Clark, E. B., Martini, R., and Rosenquist, G. C.,** Effect of lung bud excision on cardiopulmonary development in the chick, in *Morphogenesis and Malformation of the Cardiovascular System,* Vol. 14, Rosenquist, G. and Bergsma, D., Eds., Alan R. Liss, New York, 1978, 427.
136. **Rychter, Z. and Lemež, L.,** Development of hereditary ventricular septal defects in Siller's strain of chick embryos, in *Morphogenesis and Malformation of the Cardiovascular System,* Vol. 14, Alan R. Liss, New York, 1978, 377.
137. **Edwards, J. P.,** Congenital malformations of the heart and great vessels, in *Pathology of the Heart,* Gould, S. E., Ed., Charles C Thomas, Springfield, IL, 1960, 260.
138. **de la Cruz, M. V. and da Rocha, J. P.,** Ontogenetic theory for explanation of congenital malformations involving truncus and conus, *Am. Heart J.,* 51, 782, 1956.
139. **Van Praagh, R., Vlad, P., and Keith, J. D.,** Complete transposition of great arteries, in *Heart Disease in Infancy and Childhood,* Keith, J. D., Rowe, R. D., and Vlad, P., Eds., Macmillan, New York, 1967, 682.
140. **Van Mierop, L. H. S.,** Transposition of the great arteries: controversies concerning the nature and the pathogenesis of the anomaly, in *Embryology and Teratology of the Heart and the Great Arteries,* Van Mierop, L. H. S., Oppenheimer-Dekker, A., and Bruins, C. L. D. Ch., Eds., Leiden University Press, The Hague, 1978, 122.
141. **Monie, I. W., Takacs, E., and Warkany, J.,** Transposition of the great vessels and other cardiovascular abnormalities in rat fetuses induced by trypan blue, *Anat. Rec.,* 156, 175, 1966.
142. **Dor, X. and Corone, P.,** Cono-truncal torsions and transposition of the great vessels in the chick embryo, in *Perspectives in Cardiovascular Research, Vol. 5, Mechanisms of Cardiac Morphogenesis and Teratogenesis,* Raven Press, New York, 1981, 453.
143. **Goerttler, K.,** Embryology, teratology, and congenital heart disease: a correlation, in *Cardiac Development with Special Reference to Congenital Heart Disease,* Jaffee, O. C., Ed., University of Dayton Press, Dayton, OH, 1970, 81.
144. **Van Praagh, R., Layton, W. M., and Van Praagh, S.,** The morphogenesis of normal and abnormal relationships between the great arteries and the ventricles: pathologic and experimental data, in *Etiology and Morphogenesis of Congenital Heart Disease,* Futura, Mount Kisco, NY, 1980, 271.

RESPIRATORY DEVELOPMENT

FETAL LUNG GROWTH

R. Harding and A. D. Bocking

INTRODUCTION

Stimulated by a need to overcome problems associated with respiratory management during infancy, there has been a prodigious amount of basic and clinical research into pulmonary development in recent years. Much of this work has been directed toward an understanding of the Respiratory Distress Syndrome and has focused, therefore, on metabolic maturation of the lung, particularly aspects which concern the synthesis, storage, and release of surface active phospholipids. With the increasingly successful management of biochemical immaturity of the lungs, the problem of retarded morphological development is now becoming evident and may ultimately prove more difficult to deal with clinically. This chapter, therefore, is a review of current knowledge on the morphological development of the lung and of factors that influence it.

MORPHOLOGICAL DEVELOPMENT OF THE FETAL LUNG

Many comprehensive review articles have recently been published on mammalian lung growth.[9,10,17,19,25a,41,47,60,76,102,123] Drawing principally on these sources, only a brief outline of the major stages in pulmonary development will be given here. It is perhaps worth emphasizing that lung growth and development are not complete at birth, and many of the maturational processes occurring in the fetus proceed into infancy and childhood.

Based on an early description of growth of the human lung, the process of growth is now, by convention, divided into four phases: an embryonic period (24 d to 5 weeks), a pseudoglandular (or glandular) stage (5 to 17 weeks), a canalicular stage (17 to 26 weeks), and a saccular (or alveolar) stage (24 weeks to birth).[17,19,37] However, lung development proceeds in a smooth and continuous manner rather than in a series of steps.

Embryonic Period

The lung originates as an outgrowth of the primitive foregut, at about day 24 in the human embryo. This endodermal tissue gives rise to the entire epithelial lining of the lungs. The pulmonary outgrowth, by repeated divisions of the original paired bronchi, invades the splanchnic mesoderm and becomes invested with mesenchymal tissue. The major bronchopulmonary segments have formed by 34 d. During the embryonic phase the primitive airways are lined with tall columnar cells, rich in glycogen granules. The splanchnic mesodermal tissue surrounding the airways differentiates into cartilage, muscle, and connective tissue, and eventually gives rise to the pulmonary blood vessels and lymphatics.

Pseudoglandular Stage

During this period (5 to 17 weeks in the human) the developing lung resembles a tubuloacinar gland. Significant developments during this stage are the formation of all the major airways and the formation of demarcations between the acini.[17] Both of these outcomes are a consequence of continuous branching at the distal extremity of the epithelial tubes, which at this stage are blind ending.[19] Branching of the epithelial tubes is stimulated by their contact with mesenchymal tissue;[15] in this process, the presence of collagen is an important requirement.[121]

Throughout the pseudoglandular stage, the developing liquid-filled tubules are embedded in abundant mesenchyme which has only a modest blood supply. This, and the absence of terminal air sacs, rules out the possibility of extrauterine respiration at this stage.

The epithelium lining the "airways" begins to differentiate in a centrifugal manner, from hilum to periphery. The cell types are either simple tall columnar (proximally) or cuboidal (distally), with large intracellular quantities of glycogen. Differentiation of the epithelial cells leads to the appearance of goblet cells, mucous glands, and the appearance of cilia by 11 to 13 weeks. The development of identifiable acini is paralleled by the development of major blood vessels; there is a close relationship between the growing tubules and arteries, although there is little flow in the latter.[60,61]

Canalicular Stage

This stage of development, characterized by greatly increased vascularity of the mesenchyme, covers the period 17 to 24 weeks in the human fetus. There is some overlap with the pseudoglandular stage owing to the existence of a centrifugal gradient of maturation. The canalicular stage is also characterized by the establishment of the pulmonary acini (respiratory units), each composed of a terminal bronchiole, a number of prospective respiratory bronchioles, and a peripheral cluster of six to seven generations of closely branched buds, which later form terminal saccules.[17] With subsequent branching, the amount of mesenchyme is reduced as a result of growth (lengthening and widening, i.e., canalization) of the originally compact units.

There are also marked changes in the relationship between capillaries and the epithelium of the potential air spaces during this period. Principally, there develops an intimacy between capillaries and the epithelial layer, which itself becomes thinner. Cells destined to become Type I cells from progenitor "Type II" cells lose their lamellar bodies, and their cytoplasm becomes attenuated. This process represents the beginning of the formation of an air-blood interface. Fetal breathing movements are well established during this stage of development.

Terminal Sac (or Saccular) Stage

This period (24 weeks to birth in the human) sees the development of terminal saccules into alveolar ducts or alveolar sacs. Owing to the continuously changing morphology of these structures during this phase of fetal development, Burri[19] has proposed calling these ducts and saccules "transitory".

Major changes in the epithelial cells occur during this period. Following their earlier differentiation into two cell types, the production of surface active material by the progenitor Type II cells is increased. The capillary network (both circulatory and lymphatic) around the developing saccules proliferates during this period. With the increase in the amount of potential airspace and the subsequent reduction in the volume of mesenchyme, there are marked changes in the relationship between capillaries and developing airways.

At term in the human, the lung is still undergoing growth and maturation. The formation of definitive gas exchange units, the alveoli, is primarily a postnatal occurrence in the human,[19,60] and the absolute number of alveoli (or terminal sacs) greatly increases after birth. After birth there is further thinning of the septa between airspaces, resulting in a reduction in the separation between air and blood and a further increase in the proportion of airspaces in the lungs.

The major emphasis of this brief review of the stages of fetal lung growth has been on the development of the airways. Detailed accounts of the prenatal development of pulmonary circulation (e.g., References 60 and 61) have been reviewed by Polgar and Weng,[109] and the development of pulmonary innervation has been reviewed by Loosli and Hung.[87] There are also detailed accounts of epithelial cellular maturation within the lungs.[10,47,75,92,134]

PHYSICAL FACTORS IN LUNG DEVELOPMENT

Recent reviews on the regulation of fetal lung growth and structural maturation have emphasized the importance of physical, rather than humoral, factors.[70,81,82] It is now clear that the degree to which the lungs are expanded during fetal life has a major influence on their growth and maturational processes. The extent of fetal lung distension depends upon the balance of forces which are exerted on the developing pulmonary tissue both externally, via the pleura, and from within the liquid-filled airways. The existence of both an "adequate space" within the thorax and the presence of an adequate "distending pressure" within the airways have been identified as being critical to normal lung development.

Adequate Intrathoracic Space

The importance of a space within the thorax into which the lungs can grow is amply demonstrated in cases of congenital diaphragmatic hernia. More than half of the babies born alive with this condition die soon after birth.[111] In these instances, it is clear that lung growth is retarded as a consequence of the entry of abdominal contents into the thorax, highlighting the vulnerability of pulmonary tissue to small pressure gradients.

Diaphragmatic herniation has been performed experimentally in fetal sheep during the last third of gestation, resulting in hypoplasia of the lungs if uncorrected.[31,57,111] Surgical repair of the hernia leads to increased survival after birth.[57,111] The effects of diaphragmatic herniation have also been simulated in fetal sheep by progressive inflation of a balloon within the pleural cavity from 100 d of gestation.[57] Deflation of the balloon at 120 d, simulating "correction" of a diaphragmatic hernia, allowed postnatal survival at term (145 d), indicating that there may be compensatory growth and maturation following the period of retarded growth. There was a significant increase in lung weight, air capacity, compliance, and area of the pulmonary vascular bed when compared to lungs in lambs in which balloon inflation *in utero* was maintained.[57]

The principal effect of uncorrected diaphragmatic hernia on the fetal sheep lung has been described as hypoplasia, yet Pringle[111] considers that the lungs are more accurately described as dysplastic. He observed that large areas of undifferentiated mesenchyme are present between the terminal air sacs, with large distances between capillaries and terminal air sacs. Pulmonary growth had apparently been arrested during the period of diaphragmatic herniation.

Distension by Lung Liquid

Throughout late fetal life, the pulmonary airways contain a liquid (lung liquid) which is secreted across their epithelium at 1.5 to 4.9 ml·kg^{-1}·h.[33,71,91,99,104,107,115] At present there is little evidence on which type of epithelial cell is responsible for this fluid production, although it has been suggested that Type II cells may play a secretory role.[88] The volume of liquid held within the trachea and pulmonary airways of the fetal sheep during the latter half of gestation (term: 145 d) is 26 to 32 ml·kg^{-1} body weight.[33,99,100,115] There is, however, little information about pulmonary secretory activity and the dynamics of lung liquid turnover during the early stages of lung development in experimental animals. In the sheep fetus, a great increase in volume, proportional to body weight, occurs between 80 d, when the airways are short and narrow, and 110 d (i.e., during the canalicular period).[100]

There is now good experimental evidence that the degree of pulmonary distension with lung liquid has a major bearing on pulmonary growth and structural maturation. This comes primarily from studies in which lung liquid volume has either been increased or reduced for long periods.

Reduction in Lung Liquid Volume or Pressure

The significance to lung growth of the volume of liquid in the airways has been tested by two methods: continuous removal of the liquid and prevention of a distending pressure. In the most extreme technique,[7] the tracheas of three fetal sheep were drained under gravity for 21 to 25 d starting at 105 to 109 d of gestation (term is 145 d). At term, the lungs were greatly reduced in size and had remained at an early stage of development. In fetal sheep, a pulmonary luminal "distending pressure" is normally present (1 to 3 mmHg) in the absence of fetal breathing movements, i.e., about 50% of the time.[125] Tracheostomy prevents the establishment of this distending pressure, and, when performed at 117 to 122 d of gestation, resulted in reduced lung weight, volume, and distensibility.[43] The branching pattern of the terminal airways was not affected. Although these two experimental approaches[7,43] are similar in that they both prevent the development of distending pressures within the airways, they are not comparable in that fetal breathing movements are likely to have greatly differing effects on lung volume under the two sets of conditions. In fetuses with tracheal stomas, lung expansion is likely to be greatly increased during episodes of breathing movements as a result of high levels of influx of fluid,[52] whereas this is less likely to occur under conditions of tracheal fluid drainage.

Increase in Lung Liquid Volume or Pressure

In cases of congenital tracheal atresia, the lungs are found to be greatly distended. This abnormality has been modeled experimentally by tracheal ligation during the latter half of gestation. This procedure, which prevents the egress of pulmonary liquid, has been performed in a number of species. In fetal rabbits near term, tracheal ligation for up to 5 d led to an increase in the weight of the lungs and thinning of the walls of the terminal bronchioles and alveoli.[21] In similar experiments, however, there was no increase in DNA content.[130] Prolonged tracheal ligation in three fetal sheep[7] led to a large increase in lung liquid volume and lung weight. Alveolar walls were thinner, and the alveolar capillary membrane was thinner than in control fetuses. In contrast, tracheal ligation at 108 to 127 d in fetal rhesus monkeys[48] were not distended when delivered at 148 to 155 d, and lung weights were reduced. There is clearly a case for further controlled studies on the effects of increasing lung liquid volume (with measurement of liquid volume and pressure) in order to determine the relative importance of transpulmonary pressure and lung liquid volume.

The Influence of Fetal Breathing Movements

Following the suggestion that fetal breathing movements (FBM) may influence lung development,[129] there has been great interest in investigating this possibility. Experimental studies carried out to date have examined the effects of the abolition or impairment of FBM. There are, however, no published reports on the effects on the lung of enhancing FBM.

Phrenic Nerve Section

In fetal sheep, the abolition of FBM by bilateral phrenic nerve section in late gestation (103 to 113 d) resulted in reductions in lung liquid volume, tissue weight, and the number of "alveoli".[8] There was evidence that little pulmonary growth had occurred following the neurectomy. In similar studies in older fetal sheep (116 to 121 d), it was found that neither the dry weight of the lungs nor their surfactant content was affected.[44] A term human infant with phrenic nerve agenesis showed pulmonary hypoplasia with retarded bronchial branching.[50] Changes in lung growth following phrenic nerve damage are difficult to interpret owing to atrophy of the diaphragm, allowing it to move up into the thorax. Thus, any effect on lung growth may simply be attributable to displacement of thoracic volume, rather than to an absence of FBM. There is the added complication that the intercostal muscles, which normally play a minor role in the generation of FBM,[53] could compensate for the inactivity of the diaphragm. However, this does not appear to have occurred in the experiment of Fewell et al.[44]

Spinal Cord Section

The rationale for sectioning the spinal cord above the phrenic outflow, rather than the phrenic nerves themselves, is that fetal breathing movements would be abolished while atrophy of the diaphragmatic muscle would be avoided. High cervical cord section in late gestation fetal rabbits leads to reduced lung growth and maturation within 3 to 4 d.[130,132] The lungs of cord-sectioned fetuses, compared to controls, were poorly expanded and had thicker-walled terminal sacs. The lungs of fetuses with cord sections below the phrenic outflow were also affected, but to a lesser extent. This type of experiment, employing two levels of cord section, has been repeated in fetal sheep.[85] Section of the cervical spinal cord at 114 d led to both reduced lung weight (wet and dry) and reduced distensibility with air at term (142 to 149 d).

Effects of Diminished Fetal Breathing Movements

Another experimental approach to the question of the influence of FBM on lung growth and maturation has been to diminish the intrathoracic pressure fluctuations caused by FBM in fetal sheep by replacement of part of the chest wall by a more compliant material (silicone rubber sheet).[86] Recordings of tracheal pressure confirmed that FBM were reduced in amplitude. At term, the lungs were found to be lighter than controls and less distensible. It is possible, however, that in these experiments and in those involving spinal cord section there was a reduction in lung distension which may have contributed to the changes in lung development.

Recently, observations have been made on human fetuses in which, due to congenital abnormalities, FBM were either absent or abnormal (i.e., shallow and rapid or of low amplitude).[36] The lungs of these fetuses were of reduced wet weight, suggestive of hypoplasia. The mechanisms by which FBM affect lung development have not been identified. As discussed in the following sections, it is unlikely that sustained increases in pulmonary distension occur as a result of fetal breathing activity. One possibility that has not been explored is that FBM increase blood flow through the lungs, enhancing nutrient delivery.

Mechanical Effects on Nonpulmonary Tissue Growth

In a number of nonpulmonary tissues, there is evidence that mechanical stress, or distending force, has the effect of stimulating cell proliferation. Stretching the skin of mice over a period of 4 d produced hyperplasia due to increased mitotic activity and an increased progenitor cell population.[122] Similar findings have been made in *in vitro* studies. An increase in protein synthesis occurred in cranial sutures from newborn rabbits when they were subjected to a distending force for 1 to 2 d.[90] Intermittently applied distending forces have been shown to stimulate protein synthesis in cultured smooth muscle cells.[78] The stimulatory effects of stretching on cell division in chick embryo fibroblasts have been thought to be mediated by the cellular microfilament system.[29] If pulmonary tissue is also affected by stretching forces, then the beneficial effects of both tonic distension with lung liquid and phasic distension by FBM may be explained. Clearly, there is now a need for studies on the effects of stretch on pulmonary tissue *in vitro*.

REGULATION OF LUNG LIQUID VOLUME

The volume of liquid held within the airways of the fetal lungs is unlikely to be constant in either the short or long term. At any one time, this volume will depend on the net rate of formation of lung liquid and its rate of efflux into the pharynx and amniotic sac and on the influx of liquid via the pharynx. Studies in fetal sheep during the last third of gestation show that lung liquid volume is normally 26 to 32 ml·kg^{-1} (Table 1). Since there is very little direct information on variations in pulmonary liquid volume, factors influencing pulmonary secretion and the flow of liquid in the trachea will be discussed separately.

Secretion of Pulmonary Liquid

The fluid normally present in the developing pulmonary airways (and trachea) is primarily

Table 1
LUNG LIQUID SECRETION AND VOLUME IN FETAL SHEEP

Age range (d)	Secretion rate (ml·kg·h^{-1})	Volume (ml·kg^{-1})	Ref.
74 ± 0.7 (SEM)[a]	1.6	4.3	100
96 ± 0.9 (SEM)[a]	1.8	11.1	100
102—135	3.3[b]	—	107
128—134	3.8[b]	—	104
120—145	3.5[b]	26[b]	33
120—148	4.5[b]	—	91
125—145[a]	2.2	30	99
122—150	4.3[b]	—	71
134—142	4.9[b]	—	104
135—145[a]	1.5	—	115
103—133	9.6 ml·h^{-1}	—	5
120—124	11.7 ml·h^{-1}	—	18
117—127	6.6 ml·h^{-1}	57 ml	34
127—141	14.4 ml·h^{-1}	—	55
133—137	14.5 ml·h^{-1}	111 ml	34
135—139	17.0 ml·h^{-1}	—	18

Note: Age range: term is approximately 145 d.

[a] Anesthetized, exteriorized.
[b] Fetal weights estimated.

of pulmonary origin. Proof that the lung was capable of secreting this liquid, and that it was not simply inhaled amniotic fluid, was provided by experiments in which the fetal lungs became grossly distended following tracheal ligation.[21,66] Furthermore, the composition of lung liquid is quite distinct from that of amniotic fluid.[2,3,91] In particular, lung liquid has a higher concentration of chloride ion and has a lower pH than amniotic fluid, and the study of ion fluxes across the pulmonary epithelium has shown that active transport mechanisms are involved.[99]

The rate of secretion of pulmonary liquid has been measured only in the fetal ruminant. Several different methods have been used, each giving similar values in unanesthetized fetuses *in utero* (Table 1). In the sheep fetus, drainage of tracheal fluid into an external collection bag[4,5] yielded a mean figure of 9.6 ml·h^{-1} (103 to 133 d). Collection into a bag within the amniotic sac (120 to 148 d), thus eliminating the effects of a hydrostatic pressure, yielded values in the range 9 to 20 ml·h^{-1} (mean 4.5 ml·kg^{-1}·h,[91] but in a similar study on younger fetuses (94 to 115 d),[115] much lower figures were obtained. (0.9 to 2.2 ml·kg^{-1}·h).

Using a dye-dilution technique in anesthetized fetal goats (95 to 155 d), Cassin and Perks[22] measured a mean secretion rate of 6 ml·kg^{-1}·h (13.4 ml·h^{-1}). In unanesthetized fetal sheep *in utero* (120 to 140 d), an indicator dilution technique yielded secretion rates of 11.7 to 17.0 ml·h^{-1}.[18] A flowmeter sensitive to very low flow rates has been used in fetal sheep *in utero* (127 to 135 d) fitted with extracorporeal tracheal loops.[55] Mean production rates of pulmonary fluid were found to be 12.7 to 16.5 ml·h^{-1}. Between 69 and 142 d of gestation, the rate of production of lung liquid in fetal sheep increases with wet weight of lung tissue,[100] presumably due to increasing surface area. The production of pulmonary fluid has been shown to be an active process, and several factors, endocrine and physical, which influence this process have been identified.

Catecholamines

Administration of β-adrenergic agonists leads to a reduction in the production of pulmonary liquid in fetal lambs[128] and rabbits.[39] In fetal sheep, the effect of intravenous epinephrine in slowing secretion increased with gestational age, and after 130 d, it caused absorption of liquid.[18] It was proposed that this process may be involved in the clearance of liquid from the lungs around the time of birth (see the section on changes in pulmonary liquid dynamics). The mechanism by which catecholamines reduce, and eventually reverse, secretion of pulmonary liquid has not been clearly defined, but it is likely that the effect is mediated by an inhibition of chloride transport across the pulmonary epithelium[77] or by stimulation of sodium ion transport.[101a]

Vasopressin

Owing to the action of vasopressin on the movement of water across membranes, its effect on lung liquid secretion has recently received attention. Intravenous administration of arginine vasopressin in fetal goats produced reduced secretion or reabsorption of liquid.[103] This effect, like that of β-adrenergic agonists, appears to be age dependent in that only fetuses older than 131 d responded. The effect is also present in fetal sheep (115 to 132 d), in which arginine vasopressin and vasotocin were found to be of equal potency.[114]

Prolactin

Administration of prolactin into the lung fluid of exteriorized fetal goats led to a delayed increase in the rate of fluid secretion.[22] Sodium and chloride ion secretion increased in parallel. The increase in secretion could be due to stimulation of chloride, or sodium, transport.

Lung Expansion

Expansion of the lungs of fetal sheep[38] and goats[105] led to a reduction in the rate of pulmonary liquid secretion. The effect in goats was shown to be attributable to increased volume rather than to pressure. The changes in lung liquid volume created in these experiments are greater than changes which are likely to occur under physiological circumstances, such as during episodes of FBM.

Flow of Liquid in the Fetal Trachea

Attempts to measure the flux of liquid within the fetal trachea have until recently been thwarted by the absence of a technique suited to very low flow rates for use in long-term experimentation. The use of electromagnetic flow probes, designed for the measurement of blood flow, has led to the conclusion that very little flow occurs[51] or that flow occurs primarily in gushes.[30] In an early attempt to identify means by which the escape of lung liquid was controlled and to account for positive pressure within the trachea, the movement of tracheal fluid was studied by a radiographic technique.[1] This approach showed that the glottis, which was normally closed, dilated at times, allowing liquid to escape from the trachea. However, the exteriorized fetal sheep used in these studies were normally apneic, and it is likely that both the apnea and the reflex glottic closure were attributable to the experimental conditions.

In preliminary observations on lung liquid flow in fetal sheep *in utero*, which were fitted with an extracorporeal tracheal loop, flow was not uniform, and augmented efflux of liquid occurred in association with episodes of FBM.[4] However, in a subsequent study using the same technique, the authors failed to confirm their initial observations.[5]

Recently, two studies using different methods of flow detection in fetal sheep *in utero* have demonstrated that episodes of FBM exert a major influence on tracheal liquid fluxes. Tracing the movement of a bubble in an extracorporeal loop of fetal trachea showed that high rates of efflux were associated with the transition from slow-wave sleep (apnea) to REM sleep (fetal breathing activity) and during the first 5 min of episodes of FBM.[34] The

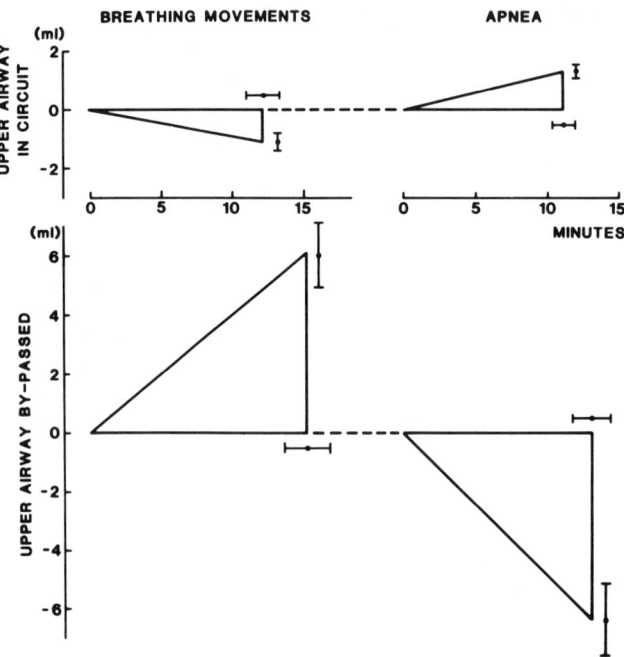

FIGURE 1. Mean changes in lung liquid volume in fetal sheep in relation to episodes of fetal breathing movements and apnea. Volume changes are based on data on tracheal fluxes in six fetuses (124 to 139 d of gestation) obtained when the upper airway was in circuit and when it was bypassed. When the upper airway is in circuit, there is a small net loss of liquid from the lungs which is restored during apneic periods. When the upper airway is bypassed, fetal breathing movements draw liquid into the lungs; a similar volume leaves again during apneic periods. The time course of changes in lung liquid volume are not necessarily as shown in this diagram. (Based on data from Reference 52a.)

introduction of a flowmeter, sensitive to low rates of flow, into a tracheal loop has enabled Harding et al.[55] to follow patterns of tracheal flow over long periods. They showed that mean rates of efflux were three to eight times greater during episodes of FBM than during intervening periods of apnea. This relationship was dependent upon the integrity of the recurrent laryngeal nerves and upon the functioning of skeletal muscles. These flow patterns suggest that, on average, the volume of pulmonary liquid diminishes slightly during episodes of FBM and becomes reestablished during apneic periods (Figure 1). Similar net changes in lung liquid volume have been reported by Dickson et al.[34] However, estimation of lung liquid volume by draining the lungs after periods of FBM and apnea has indicated that changes in the opposite direction occur.[95]

Very little information is available on lung liquid dynamics in the human fetus. A recent study, in which pulsed Doppler ultrasound was used near term, led to an estimated fluid movement of 6 ml per breath in the trachea and a maximum flow velocity of 40 cm·s^{-1}.[24] However, this technique suffers the same disadvantage as does the electromagnetic flowmeter in that only higher rates of flow are registered.

There are several possible explanations for the augmented efflux of tracheal fluid which accompanies episodes of FBM. One is that the secretion of fluid is increased by fetal breathing activity. This is unlikely, however, because neither phrenic nerve section[44] nor enhancement of FBM by inhibitors of prostaglandin synthesis[72] alters the rate of fluid production. Another possibility is that fluid accumulates in the lungs during periods of apnea because the resistance of the upper airway is raised; this is supported by the presence of elevated intratracheal pressure during apnea.[42,125] Measurements of resistance of the upper respiratory tract in fetal sheep have confirmed that its resistance is raised in the absence of FBM and have shown that the major site of resistance resides in the larynx.[52] Electromyographic recordings from laryngeal muscles in fetal sheep have shown that the major adductor muscles (thyroarytenoid) are usually tonically active during apnea, and inactive during episodes of FBM.[54] The resistance of the upper airway was found to be lowered during episodes of FBM,[52] probably as a result of rhythmic activation of laryngeal dilator muscles (posterior cricoarytenoid).[54]

The regulatory role of the fetal upper airway has been demonstrated in experiments in which the effect of its bypass has been assessed.[52a] When tracheal fluid was routed directly to the amniotic sac (i.e., bypassing the upper respiratory tract), there was a greatly increased efflux of fluid during apnea, particularly during periods of myometrial activity, presumably due to the absence of an appreciable resistance normally presented by the larynx (Figure 1). During episodes of FBM, fluid entered the lungs, rather than leaving them, probably due to the low resistance pathway between the trachea and the amniotic fluid. When the upper respiratory system is intact, there is only a small volume (1 to 2 ml) of fluid in the pharynx which can be drawn upon,[52] This may normally prevent the ingress of large amounts of liquid during episodes of FBM. The hypopharynx has recently been visualized by ultrasound in the human fetus, and has been observed to be partially distended with fluid.[26]

Changes in Pulmonary Liquid Dynamics Around the Time of Birth

Most of the data discussed above relate to events in the last third of gestation, when fetal ruminants are at a suitable stage of development for experimentation. Perhaps the period most critical to lung growth is the "canalicular" stage, between 100 and 120 d in the sheep fetus when the airways expand greatly. However, a critical period for postnatal viability is the perinatal period, when the lungs must make a smooth transition from being liquid-filled secretory organs to functioning as organs of gas exchange.

A reduction in the rate of secretion of lung liquid in fetal sheep has been detected during the 7 d preceding birth, at the same time as plasma cortisol concentrations are found to be rising.[71] This reduction in secretion accounts for the observation that extravascular water content of the lungs of fetal rabbits and lambs is greatly reduced before birth.[13,14] A recent study in fetal sheep has confirmed that the volume of liquid within the airways is significantly reduced during labor and that this may be partly due to a declining rate of secretion.[33]

The means by which lung liquid secretion is reduced around the time of birth has recently been the subject of investigation. There is good evidence that the elevation in circulating catecholamines which occurs before birth[64] may be implicated. Infusion of epinephrine into fetal sheep during late gestation caused a reduction in the rate of secretion of lung liquid, whereas close to term the effect was to cause absorption of liquid.[18,128] Catecholamines may also prepare the lung for extrauterine function by stimulating surfactant synthesis and release.[35,77,126]

PULMONARY HYPOPLASIA

Pulmonary hypoplasia is a pathological condition which has been reported to be the most common single abnormality identified in human infants dying in the early neonatal period.[80] Lung hypoplasia, defined as a decrease in lung/body weight ratio and retarded maturation, is present in as many as one in seven perinatal autopsies.[131] These cases represent one extreme of a spectrum of lung development, and milder degrees of pulmonary hypoplasia may lead to respiratory difficulties in the newborn period. With improved methods of treating conditions such as the Respiratory Distress Syndrome (RDS), meconium aspiration, and congenital pneumonia, pulmonary hypoplasia plays a much greater role in the pathogenesis of respiratory disorders following birth. Pulmonary hypoplasia may be classified as either primary or secondary, although primary pulmonary hypoplasia occurs only rarely and its cause is unknown.[80] Alternatively, impaired lung development may be secondary to such conditions as congenital diaphragmatic hernias or prolonged oligohydramnios.

Congenital Diaphragmatic Hernia

Congenital diaphragmatic hernia occurs in approximately 1 per 5000 live births and is associated with a mortality rate of 66 to 80%.[20,56] The cause of death in these infants is generally respiratory failure in the early neonatal period due to severe pulmonary hypoplasia.[16,69] Congenital diaphragmatic hernias can be diagnosed prenatally using ultrasound,[25] and therefore considerable experimental work has been conducted in animals, creating and repairing diaphragmatic defects (see the section titled Adequate Intrathoracic Space). Occasionally, other fetal pulmonary lesions such as cystic adenomatoid malformation may be erroneously diagnosed prenatally as a congenital diaphragmatic hernia. These lesions, however, are also space-occupying within the fetal chest and therefore may lead to pulmonary hypoplasia. Experiments in fetal lambs have shown that the correction *in utero* of surgically created diaphragmatic defects leads to an improvement in lung size and maturation and postnatal survival.[59,111] Whether or not a similar intervention *in utero* in humans will result in an improvement in neonatal outcome remains to be determined.

Oligohydramnios

Prolonged oligohydramnios (reduced amniotic fluid volume) is also associated with significant pulmonary hypoplasia in man and in other species. Hypoplastic lungs in association with oligohydramnios are characterized by the presence of narrow airways, retarded epithelial and interstitial growth, delay in the development of air-blood barriers, as well as a decrease in lung phospholipid concentrations.[133] Surface area within the lung is also likely to be diminished.

Amniotic fluid volume may be reduced either through fetal renal tract abnormalities or prolonged rupture of membranes.[110] In human infants with congenital bilateral renal agenesis, there is a reduction in total lung volume, number of airways, as well as acinar size.[63] These infants usually die of asphyxia secondary to pulmonary hypoplasia within the first few hours of life.[110] Oligohydramnios may also occur when there is bilateral obstruction to urine flow within the fetal urinary tract. This condition can be diagnosed *in utero* and, in selected cases, intervention *in utero* to relieve the obstruction and prevent the development of severe renal damage may be appropriate. The percutaneous placement of a catheter within the fetal bladder, allowing drainage of urine into the amniotic space when there is obstruction to urine flow by the presence of a posterior urethral valve, is an example of one possible intervention.[49] Relieving the obstruction to urine flow may correct the oligohydramnios associated with these conditions, thereby preventing the development of pulmonary hypoplasia.

Bilateral nephrectomy in fetal sheep only in the canalicular stage leads to retarded pulmonary development and cellular immaturity.[108] The outflow of urine from the bladder has been gradually reduced in fetal sheep *in utero*, leading to pulmonary hypoplasia and respiratory distress in the newborn.[6] This process can be reversed by performing a suprapubic cystotomy *in utero*, 3 weeks after the initial intervention with an improvement in postnatal survival. In pregnant guinea pigs, there is also a significant reduction in fetal lung weight when the fetal bladder outlet is obstructed, leading to oligohydramnios.[96] The reduction in lung/body weight ratio in these experiments was partially corrected by the infusion of saline into the amniotic space or by the creation of an abdominal wall hernia, allowing the peritoneal contents, including the distended bladder, to remain outside the fetus. Both of these procedures prevent the compression of the thoracic contents.

Pulmonary hypoplasia in the human may also be associated with oligohydramnios secondary to prolonged rupture of the fetal membranes.[12,74,98,106] In pregnant rats, oligohydramnios following needle puncture of the fetal membranes and uterus at day 15 leads to a significant decrease in the lung/body weight ratio as well as lung DNA content at term.[93]

The mechanisms by which oligohydramnios leads to pulmonary hypoplasia are not well understood, but there is sufficient experimental evidence to support a causal relationship between the two conditions. It is widely believed that a reduction in amniotic fluid volume leads to increased fetal "compression" *in utero*, as evidenced by limb facial deformities which are often associated with oligohydramnios. Thus, there may be a sustained reduction in fetal thoracic volume as a result of an increased degree of spinal flexion, particularly in association with nonlabor uterine contractions.[52b] Alternatively, the close apposition of the uterus to the chest and abdomen may affect FBM. It has recently been shown in both sheep[94] and humans[45] that oligohydramnios does not affect the incidence of FBM, but it is possible that the extent of thoracic and abdominal excursions and variations in lung expansion associated with episodes of FBM may be limited. In sheep it has been shown that reducing the effectiveness of FBM in causing intrathoracic pressure fluctuations leads to retarded lung development.[80,86]

Amniocentesis

In pregnant monkeys, amniocentesis between 47 and 95 d of gestation (equivalent to 14 to 25 weeks of human gestation) leads to a significant reduction in the number of respiratory airways and alveoli.[62] There is, however, a compensatory increase in the size of alveoli following birth at term (165 d). This study provides some insight into the possible explanations for the observation made by the Medical Research Council Working Party on Amniocentesis that there was a slight (1%) increase in the incidence of respiratory problems at birth in infants of mothers who underwent genetic amniocentesis, when compared to controls.[90] This finding was not confirmed, however, by two other groups of investigators.[97,117] In a study of 20 infants, there was a significant decrease in the crying vital capacity of neonates born to mothers who underwent amniocentesis when compared to those who did not.[127] There is therefore some evidence to suggest that a single needle puncture of the uterus and membranes during the second trimester or pregnancy in primates, without prolonged rupture of membranes and loss of amniotic fluid, may be associated with retarded pulmonary development. Further investigations into the possible mechanisms of this impairment are necessary.

ENDOCRINE CONTROL OF PULMONARY DEVELOPMENT

Glucocorticoids

The maturational effect of glucocorticoids on pulmonary maturation was first noted by Liggins[79] in fetal lambs born prematurely after fetal ACTH infusions. A considerable number of investigations have been conducted since then into the mechanisms whereby glucocorticoids affect lung growth and maturation. Fetal plasma cortisol concentration has been shown to be positively correlated with a number of indices of lung maturation in fetal sheep.[73] In addition, exogenous glucocorticoids are known to affect lung development by increasing branching and enlargement of glands as well as promoting the formation of alveoli.[68] The effect of cortisol on cell growth in the lung is dependent upon gestational age.[120] At later gestational ages, cortisol administration leads to an increase in the number of Type II alveolar epithelial cells which have the characteristic surfactant-containing lamellar bodies.[135] The effects of cortisol on lung maturation may be partially mediated through adrenergic mechanisms, since the number of pulmonary beta-adrenergic receptors is increased in fetal rabbit lung by cortisol administration.[23] A review of the effects of glucocorticoids on phospholipid metabolism as well as their use in attempts to reduce the severity of respiratory distress syndrome in humans is presented elsewhere in this volume.

Pituitary Hormones

Pulmonary maturation is delayed in the hypophysectomized ovine fetus.[83] This delay in maturation can be overcome by the infusion of $ACTH_{1-24}$ into the fetus, but not by cortisol administration, suggesting that the effect of ACTH is not mediated by cortisol alone. In contrast, the lungs of adrenalectomized sheep fetuses are brought to functional maturity by the infusion of cortisol, but not $ACTH_{1-24}$.[84] Despite these differential effects of cortisol and $ACTH_{1-24}$ on fetal pulmonary functional maturation, their administration to the hypophysectomized fetus leads to a morphological maturation.[27] It is possible that other hormones such as thyroxine or prolactin are required for cortisol to have its maturational effect on fetal pulmonary function.[84] In the intact ovine fetus, beta-adrenergic blockade has been shown to inhibit the maturational effect of ACTH on the lung, suggesting that adrenergic mechanisms are also involved.[126]

Thyroid Hormones

In fetal rabbits, thyroxine administration accelerates pulmonary maturation.[136] In addition, fetal lambs which have been thyroidectomized at 95 to 99 d (term is 145 d) have impaired lung growth as well as maturation.[40] The importance of thyroid hormones in human fetal development is not known, although neonates with RDS have lower cord serum thyroxine and triiodothyronine concentrations than healthy infants.[28] The effects of nephrectomy on lung development in fetal sheep may be partly mediated by a change in circulating levels of thyroid hormones. Plopper et al.[108] have shown that nephrectomy resulted in a reduction in plasma thyroxin (T_4), but not in cortisol.

Prolactin

The importance of fetal prolactin in lung maturation is controversial,[112] although fetal pulmonary tissue is known to contain prolactin receptors.[65] Its effect may be mediated by alterations in pulmonary fluid production.[22]

Insulin

There is an increased incidence of RDS in infants of diabetic mothers,[113] suggesting that insulin may be important in fetal pulmonary maturation. Insulin is known to impair the stimulatory effect of cortisol on surfactant synthesis in cultured fetal lung cells.[119] Careful control of maternal glucose concentrations in diabetic pregnancies may therefore minimize the antagonistic effect of insulin on fetal lung development.[46]

Growth Factors

Fibroblast pneumonocyte factor (FPF) is a peptide which is important in fetal lung maturation and is produced by human fetal lung fibroblasts following glucocorticoid administration.[116] Infusions of epidermal growth factor (EGF) into premature fetal lambs also accelerate pulmonary maturation.[124] EGF administration also has the effect of inhibiting lung liquid secretion, although this may be partly attributable to a simultaneous stimulation of catecholamine release.[67] Although fetal mouse lung tissue has been shown to generate somatomedin C, its importance in fetal lung growth has not been established.[32]

SUMMARY

The lung develops during fetal life as a liquid-filled organ connected to the amniotic sac by a pathway which favors efflux of liquid rather than influx. Although hormonal influences on pulmonary growth and cellular maturation have been demonstrated, like few other organs, physical factors seem to be of prime importance for morphological development of the lungs. As a result of the relatively compliant fetal rib cage and the frequent inactivity of the diaphragm muscle, there is considerable scope for pulmonary compression by external forces. The maintenance of the correct balance of forces across the developing pulmonary tissue appears to be important for the development of successive generations of airways, creation of an air-blood interface, and maturation of supporting tissue. As yet, however, the mechanisms by which pulmonary tissue growth is controlled at a cellular level are poorly understood.

ACKNOWLEDGMENTS

This work has been supported by the National Health and Medical Research Council of Australia (R.H.) and the Medical Research Council of Canada (A.D.B.).

REFERENCES

1. **Adams, F. A., Desilets, D. T., and Towers, B.,** Control of flow of fetal lung liquid at the laryngeal outlet, *Resp. Physiol.,* 2, 302, 1967.
2. **Adams, F. H., Moss, A. J., and Fagan, L.,** The tracheal fluid in the fetal lamb, *Biol. Neonate,* 5, 151, 1963.
3. **Adamson, T. M., Boyd, R. D. H., Platt, H. S., and Strang, L. B.,** Composition of alveolar liquid in the foetal lamb, *J. Physiol.,* 204, 159, 1969.
4. **Adamson, T. M., Brodecky, V., Lambert, T. F., Maloney, J. E., Ritchie, B. C., and Walker, A.,** The production and composition of lung liquid in the in-utero foetal lamb, in *Foetal and Neonatal Physiology,* Comline, R. S., Cross, K. W., Dawes, G. S., and Nathanielsz, P. W., Eds., Cambridge University Press, London, 1973, 208.
5. **Adamson, T. M., Brodecky, V., Lambert, T. F., Maloney, J. E., Ritchie, B. C., and Walker, A. M.,** Lung liquid production and composition in the "in utero" foetal lamb, *Aust. J. Exp. Biol. Med. Sci.,* 53, 65, 1975.
6. **Adzick, N. S., Harrison, M. R., Glick, P. L., and Flake, A. W.,** Fetal urinary tract obstruction: experimental pathophysiology, *Semin. Perinatol.,* 9, 79, 1985.
7. **Alcorn, D., Adamson, T. M., Lambert, T. F., Maloney, J. E., Ritchie, B. C., and Robinson, P. M.,** Morphological effects of chronic tracheal ligation and drainage in the fetal lamb lung, *J. Anat.,* 123, 649, 1977.
8. **Alcorn, D., Adamson, T. M., Maloney, J. E., and Robinson, P. M.,** Morphological effects of chronic bilateral phrenectomy or vagotomy in the fetal lamb lung, *J. Anat.,* 130, 683, 1980.
9. **Alcorn, D. G., Adamson, T. M., Maloney, J. E., and Robinson, P. M.,** A morphologic and morphometric analysis of fetal lung development in the sheep, *Anat. Rec.,* 201, 655, 1981.
10. **Alcorn, D. G., Alexander, I. G. S., Maloney, J. E., Ritchie, B. C., and Walker, A. M.,** Morphological development of the lung: a review, *Aust. Paediatr. J.,* 10, 189, 1974.
11. **Avery, M. E., Fletcher, B. D., and Williams, R. G.,** *The Lung and its Disorders in the Newborn Infant,* W. B. Saunders, Philadelphia, 1981.
12. **Bain, A. D., Smith, I. I., and Gould, I. D.,** Newborn born after prolonged leakage of liquor amnii, *Br. Med. J.,* 11, 598, 1964.
13. **Bland, R. D., Bressack, M. A., and McMillan, D. D.,** Labour decreases the lung water content of new born rabbits, *Am. J. Obstet. Gynecol.,* 135, 364, 1979.
14. **Bland, R. D., Hansen, T. N., Haberkern, G. M., Bressack, M. A., Hazinski, T. A., Raj, J. V., and Goldberg, R. B.,** Lung fluid balance in lambs before and after birth, *J. Appl. Physiol.,* 53, 992, 1982.
15. **Bluemink, J. G., van Maurik, P., and Lawson, K. A.,** Intimate cell contacts at the epithelial/mesenchymal interface in embryonic mouse lung, *J. Ultrastruct. Res.,* 55, 257, 1976.
16. **Boyden, E. A.,** The structure of compressed lungs in congenital diaphragmatic hernia, *Am. J. Anat.,* 134, 497, 1972.
17. **Boyden, E. A.,** Development and growth of the airways, in *Development of the Lung: Lung Biology in Health and Disease,* Vol. 6, Hodson, W. A., Ed., Marcel-Dekker, New York, 1977, 3.
18. **Brown, M. J., Olver, R. E., Ramsden, C. A., Strang, L. B., and Walters, D. V.,** Effects of adrenaline and of spontaneous labour on the secretion and absorption of lung liquid in the fetal lamb, *J. Physiol.,* 334, 137, 1983.
19. **Burri, P. H.,** Fetal and postnatal development of the lung, *Annu. Rev. Physiol.,* 46, 617, 1984.

20. Butler, N. and Claireaux, A. E., Congenital diaphragmatic hernia as a cause of perinatal mortality, *Lancet*, 1, 659, 1962.
21. Carmel, J. A., Friedman, F., and Adams, F. H., Fetal tracheal ligation and lung development, *Am. J. Dis. Child.*, 109, 452, 1965.
22. Cassin, S. and Perks, A. M., Studies of factors which stimulate lung fluid secretion in fetal goats, *J. Dev. Physiol.*, 4, 311, 1982.
23. Cheng, J. B., Goldfien, A., Ballard, P. L., and Roberts, J. M., Glucocorticoids increase pulmonary beta-adrenergic receptors in fetal rabbit, *Endocrinology*, 107, 1646, 1980.
24. Chiba, Y., Utsu, M., Kanzaki, T., and Hasegawa, T., Changes in venous flow and intra-tracheal flow in fetal breathing movements, *Ultrasound Med. Biol.*, 11, 43, 1985.
25. Chinn, D. H., Filly, R. A., Callen, P. W., Nakayama, D. K., and Harrison, M. R., Congenital diaphragmatic hernia diagnosed prenatally by ultrasound, *Radiology*, 148, 119, 1983.
25a. Collins, M. H., Kleinerman, J., Moessinger, A. C., Collins, A. H., James, S., and Blanc, W. A., Morphometric analysis of the growth of the normal fetal guinea pig lung, *Anat. Rec.*, 216, 381, 1986.
26. Cooper, C., Mahoney, B. S., Bowie, J. D., Albright, T. O., and Callen, P. W., Ultrasound evaluation of the normal fetal airway and esophagus, *J. Ultrasound Med.*, 4, 343, 1985.
27. Crane, R. K., Davies, P., Liggins, G. C., and Reid, L., The effect of hypophysectomy, thyroidectomy and replacement therapy by cortisol and ACTH on ovine fetal lung structure, *J. Dev. Physiol.*, 5, 281, 1983.
28. Cuestas, R. A., Lindell, A., and Engel, R. R., Low thyroid hormones and respiratory distress syndrome of the newborn: studies on cord blood, *N. Engl. J. Med.*, 295, 297, 1976.
29. Curtis, A. S. G. and Seehar, G. M., The control of cell division by tension or diffusion, *Nature (London)*, 274, 52, 1978.
30. Dawes, G. S., Fox, H. E., Leduc, B. M., Liggins, G. C., and Richard, R. T., Respiratory movements and rapid eye movement sleep in the foetal lamb, *J. Physiol.*, 220, 119, 1972.
31. de Lorimier, A. A., Tierney, O. F., and Parker, H. R., Hypoplastic lungs in fetal lambs with surgically produced congenital diaphragmatic hernia, *Surgery*, 62, 12, 1967.
32. D'Ercole, A. J., Applewhite, G. T., and Underwood, L. E., Evidence that somatomedin is synthesized by multiple tissues in the fetus, *Dev. Biol.*, 75, 315, 1980.
33. Dickson, K. A., Maloney, J. E., and Berger, P. J., Decline in lung liquid volume prior to labour in fetal lambs, *J. Appl. Physiol.*, 61, 2266, 1986.
34. Dickson, K. A., Maloney, J. E., and Berger, P. J., State-related changes in lung liquid secretion and tracheal flow rate in fetal lambs, *J. Appl. Physiol.*, 62, 34, 1987.
35. Dobbs, L. G. and Mason, R. J., Pulmonary alveolar type II cells isolated from rats. Release of phosphatidylcholine in response to beta-adrenergic stimulation, *J. Clin. Invest.*, 63, 378, 1979.
36. Dornan, J. C., Ritchie, J. W., and Meban, C., Fetal breathing movements and lung maturation in the congenitally abnormal human fetus, *J. Dev. Physiol.*, 6, 367, 1984.
37. Dubreuil, G., Lacoste, A., and Raymond, R., Observations sur le developpement du poumon humain, *Bull. Histol. Appl. Physiol. Pathol. Microsc.*, 13, 235, 1936.
38. Egan, E. A., Effect of lung inflation on alveolar permeability to solutes, in *Ciba Foundation Symposium 38*, Elsevier/North-Holland, Amsterdam, 1976, 101.
39. Enhorning, G., Chamberlain, D., Contraras, C., Burgoyne, R., and Robertson, B., Isoxuprine-induced release of pulmonary surfactant in the rabbit fetus, *Am. J. Obstet. Gynecol.*, 129, 197, 1977.
40. Erenberg, A., Rhodes, M. L., Weinstein, M. M., and Kennedy, R. L., The effect of fetal thyroidectomy on ovine fetal lung maturation, *Pediatr. Res.*, 13, 230, 1979.
41. Farrell, P. M., Morphological aspects of lung maturation, in *Lung Development: Biological & Clinical Perspectives, Vol. 1, Biochemistry & Physiology*, Farrell, P. M., Ed., Academic Press, New York, 1982, 13.
42. Fewell, J. M. and Johnson, P., Upper airway dynamics during and breathing apnoea in fetal lambs, *J. Physiol.*, 339, 495, 1983.
43. Fewell, J. E., Hislop, A. A., Kitterman, J. A., and Johnson, P., Effect of tracheostomy on lung development in fetal lambs, *J. Appl. Physiol.*, 55, 1103, 1983.
44. Fewell, J. G., Lee, C. C., and Kitterman, J. A., Effects of phrenic nerve section on the respiratory system of fetal lambs, *J. Appl. Physiol.*, 51, 293, 1981.
45. Fox, H. E. and Moessinger, A. C., Fetal breathing movements and lung hypoplasia: preliminary human observations, *Am. J. Obstet. Gynecol.*, 15, 531, 1985.
46. Frantz, I. D. and Epstein, M. F., Fetal lung development in pregnancies complicated by diabetes, *Semin. Perinatol.*, 2, 347, 1978.
47. Gail, D. B. and Lenfant, C. J. M., Cells of the lung: biology and clinical implications, *Am. Rev. Respir. Dis.*, 127, 366, 1983.
48. Gluck, L., Chez, R. A., Kulovich, M. V., Hutchinson, D. L., and Niemann, W. H., Comparison of phospholipid indicators of fetal lung maturity in the amniotic fluid of the monkey (*Macaca mulatta*) and baboon (*Papio papio*), *Am. J. Obstet. Gynecol.*, 120, 524, 1974.

49. **Golbus, M. S., Filly, R. A., Callen, P. W., Glick, P. L., Harrison, M. R., and Anderson, R. L.**, Fetal urinary tract obstruction: management and selection for treatment, *Semin. Perinatol.*, 9, 91, 1985.
50. **Goldstein, J. D. and Reid, L. M.**, Pulmonary hypoplasia resulting from phrenic nerve agencies and diaphragmatic amyoplasia, *J. Pediatr.*, 97, 282, 1980.
51. **Goodlin, R. C. and Rudolph, A. M.**, Tracheal flow and function in fetuses in utero, *Am. J. Obstet. Gynecol.*, 106, 597, 1970.
52. **Harding, R., Bocking, A. D., and Sigger, J. N.**, Upper airway resistances in fetal sheep: the influence of breathing activity, *J. Appl. Physiol.*, 60, 160, 1986.
52a. **Harding, R., Bocking, A. D., and Sigger, J. N.**, Influence of upper respiratory tract on liquid flow to and from fetal lungs, *J. Appl. Physiol.*, 60, 160, 1986.
52b. **Harding, R., Dickson, K. A., and Hooper, S. B.**, Fetal breathing, tracheal fluid movement and lung growth, in *Advances in Fetal Physiology,* Gluckman, P. D., Johnston, B. M., and Nathanielsz, P. W., Eds., Perinatology Press, Ithaca, NY, 1989, 153.
53. **Harding, R., Johnson, P., McClelland, M. E., McLeod, C. N., and Whyte, P. L.**, Laryngeal function during breathing and swallowing in fetal and newborn lambs, *J. Physiol.*, 272, 14, 1977.
54. **Harding, R., Johnson, P., and McClelland, M. E.**, Respiratory function of the larynx in developing sheep and the influence of sleep state, *Respir. Physiol.*, 40, 165, 1980.
55. **Harding, R., Sigger, J. N., Wickham, P. J. D., and Bocking, A. D.**, The regulation of flow of pulmonary fluid in fetal sheep, *Respir. Physiol.*, 57, 47, 1984.
56. **Harrison, M. R., Bjordal, R. I., Langmark, F., and Knutrud, O.**, Congenital diaphragmatic hernia: the hidden mortality, *J. Pediatr. Surg.*, 13, 227, 1978.
57. **Harrison, M. R., Bressack, M. A., Chury, A. M., and de Lorimier, A. A.**, Correction of diaphragmatic hernia in utero. II. Simulated correction permits fetal lung growth with survival at birth, *Surgery,* 88, 260, 1980.
58. **Harrison, M. R., Jester, J. A., and Ross, N. A.**, Correction of congenital diaphragmatic hernia in utero. I. The model: intrathoracic balloon produces fatal pulmonary hypoplasia, *Surgery,* 88, 174, 1980.
59. **Harrison, M. R., Ross, N. A., and de Lorimier, A. A.**, Correction of congenital diaphragmatic hernia in utero. III. Development of a successful surgical technique using abdominoplasty to avoid compromise of umbilical blood flow, *J. Pediatr. Surg.*, 16, 934, 1981.
60. **Hislop, A. and Reid, L. M.**, Growth and development of the respiratory system — anatomical development, in *Scientific Foundations of Paediatrics,* Davis, J. A. and Dobbing, J., Eds., Heinemann, London, 1974, 214,
61. **Hislop, A. and Reid, L. M.**, Formation of the pulmonary vasculature, in *Development of the Lung,* Hodson, W. A., Ed., Marcel Dekker, New York, 1977, 37.
62. **Hislop, A., Fairweather, D. V. I., Blackwell, R. J., and Howard, S.**, The effect of amniocentesis and drainage of amniotic fluid on lung development in Macaca fascicularis, *Br. J. Obstet. Gynecol.*, 91, 835, 1984.
63. **Hislop, A., Hey, E., and Reid, L.**, The lungs in congenital bilateral renal agenesis and dysplasia, *Arch. Dis. Child.*, 54, 32, 1979.
64. **Jones, C. T. and Robinson, R. O.**, Plasma catecholamines in foetal and adult sheep, *J. Physiol.*, 248, 15, 1975.
65. **Josimovich, J. B., Merisko, K., Boccella, L., and Tobin, H.**, Binding of prolactin by fetal rhesus cell membrane fractions, *Endocrinology,* 100, P557, 1977.
66. **Jost, A. and Policard, A.**, Contribution experimentale a l'etude du development prenatal du poumon chez le lapin, *Arch. Anat. Microsc.*, 37, 323, 1948.
67. **Kennedy, K. A., Wilton, P., Mellander, M., Rojas, J., and Sundell, H.**, Effect of epidermal growth factor on lung liquid production and catecholamine blood levels in fetal lambs, *Pediatr. Res.*, 18, 329A, 1984.
68. **Kikkawa, Y., Kaibora, M., Motoyama, E. K., Orzolesi, M. M., and Cook, C. D.**, Morphologic development of the fetal rabbit lung and its acceleration with cortisol, *Am. J. Pathol.*, 64, 423, 1971.
69. **Kitagawa, M., Hislop, A., Boyden, E. A., and Reid, L.**, Lung hypoplasia in congenital diaphragmatic hernia, a quantitative study of airway, artery and alveolar development, *Br. J. Surg.*, 58, 342, 1971.
70. **Kitterman, J. A.**, Physical factors and fetal lung growth, in *Respiratory Control and Lung Development in the Fetus and Newborn,* Johnston, B. M. and Gluckman, P. D., Eds., Perinatology Press, Ithaca, NY, 1986, 63.
71. **Kitterman, J. A., Ballard, P. L., Clements, J. A., Mescher, E. J., and Tooley, W. H.**, Tracheal fluid in fetal lambs: spontaneous decrease prior to birth, *J. Appl. Physiol.*, 47, 985, 1979.
72. **Kitterman, J. A., Liggins, G. T. G., Clements, J. A., Campos, G. A., Lee, C. H., and Ballard, P. L.**, Inhibitors of prostaglandin synthesis, tracheal fluid and surfactant in fetal lambs, *J. Appl. Physiol.*, 51, 1562, 1981.
73. **Kitterman, J. A., Liggins, G. C., Campos, G. A., Clements, J. A., Forster, C. S., Lee, C. A., and Creasy, R. K.**, Prepartum maturation of the lung in fetal sheep: relation to cortisol, *J. Appl. Physiol.*, 51, 384, 1981.

74. **Kuhn, C.**, The cytology of the lung: ultrastructure of the respiratory epithelium and extracellular lining layers, in *Lung Development: Biological & Clinical Perspectives*, Vol. 1, Farrell, P. M., Ed., Academic Press, New York, 1982, 27.
76. **Langston, C. and Thurlbeck, W. M.**, Lung growth and development in late gestation and early postnatal life, *Perspect. Pediatr. Pathol.*, 7, 203, 1982.
77. **Lawson, E. E., Brown, E. R., Torday, J. S., Manasky, D. L., and Taeusch, H. W.**, The effect of epinephrine on tracheal fluid flow and surfactant efflux in fetal sheep, *Am. Rev. Respir. Dis.*, 118, 1023, 1978.
78. **Leung, D. V. M., Glagov, S., and Mathew, M. B.**, Cyclic stretching stimulates synthesis of matrix components by arterial smooth muscle cells in vitro, *Science*, 191, 475, 1976.
79. **Liggins, G. C.**, Premature delivery of fetal lambs infused with glucocorticoids, *J. Endocrinol.*, 45, 515, 1969.
80. **Liggins, G. C.**, Growth of the fetal lung, *J. Dev. Physiol.*, 6, 237, 1984.
81. **Liggins, G. C. and Kitterman, J. A.**, Development of the fetal lung, in *The Fetus and Independent Life*, Ciba Foundation Symp. 86, Pitman Publishing, London, 1981, 308.
82. **Liggins, G. C. and Schellenburg, J. C.**, Aspects of fetal lung development, in *The Physiological Development of the Fetus & Newborn*, Jones, C. T. and Nathanielsz, P. W., Eds., Academic Press, London, 1985, 179.
83. **Liggins, G. C., Kitterman, J. A., Campos, G. A., Clements, J. A., Forster, C. S., Lee, C. H., and Creasy, R. K.**, Pulmonary maturation in the hypophysectomized ovine fetus. Differential responses to adrenocorticotrophin and cortisol, *J. Dev. Physiol.*, 3, 1, 1981.
84. **Liggins, G. C., Schellenberg, J. C., Finberg, K., Kitterman, J. A., and Lee, C. H.**, The effect of $ACTH_{1-24}$ or cortisol on pulmonary maturation in the adrenalectomized ovine fetus, *J. Dev. Physiol.* 7, P105, 1985.
85. **Liggins, G. C., Vilos, G. A., Campos, G. A., Kitterman, J. A., and Lee, C. H.**, The effect of spinal cord transection on lung development in fetal sheep, *J. Dev. Physiol.*, 3, 267, 1981.
86. **Liggins, G. C., Vilos, G. A., Kitterman, J. A., and Lee, C. H.**, The effects of bilateral thoracoplasty on lung development in fetal sheep, *J. Dev. Physiol.*, 3, 275, 1981.
87. **Loosli, C. G. and Hung, K. S.**, Development of pulmonary innervation, in *Development of the Lung*, Hodson, W. A., Ed., Marcel Dekker, New York, 1977, 269.
88. **Mason, R. J., Williams, M. C., and Widdicombe, J. H.**, Secretion and fluid transport by alveolar type II epithelial cells, *Chest*, Suppl. 81, 61, 1982.
89. Medical Research Council, An assessment of the hazards of amniocentesis, *Br. J. Obstet. Gynecol.*, 85(Suppl. 2), 1, 1978.
90. **Meikle, M. C., Reynolds, J. J., Sellers, A., and Dingle, J. T.**, Rabbit carnial sutures *in vivo*: a new experimental model for studying the response of fibrous joints to mechanical stress, *Calcif. Tissue Int.*, 28, 137, 1979.
91. **Mescher, E. J., Platzker, A. C. G., Ballard, P. L., Kitterman, J. A., Clements, J. A., and Tooley, W. H.**, Ontogeny of tracheal fluid, Pulmonary surfactant and plasma corticoids in the fetal lamb, *J. Appl. Physiol.*, 39, 1017, 1975.
92. **Meyrick, B. and Reid, L. M.**, Ultrastructure of alveolar lining and its development, in *Development of the Lung*, Hodson, W. A., Ed., Marcel Dekker, New York, 1977, 135.
93. **Moessinger, A. C., Bossi, G. A., Ballantyne, G., Collins, M. H., James, L. S., and Blanc, W. A.**, Experimental production of pulmonary hypoplasia following amniocentesis and oligohydramnios, *Early Hum. Dev.*, 8, 343, 1983.
94. **Moessinger, A. C., Fewell, J. E., Stark, R. I., Collins, M. H., Daniel, S. S., Singh, M., Blanc, W. A., Kleineman, J., and James, L. S.**, Lung hypoplasia and breathing movements following oligohydramnios in fetal lambs, in *The Physiological Development of the Fetus and Newborn*, Jones, C. T. and Nathanielsz, P. W., Eds., Academic Press, London, 1985, 193.
95. **Murai, D. T., Lee, C. H. Y., and Kitterman, J. A.**, Breathing movements intermittently increase lung volumes in the fetal sheep, *Clin. Res.*, 31, 173A, 1983.
96. **Nakayama, D. K., Glick, P. L., Harrison, M. R., Villa, R. L., and Noall, R.**, Experimental pulmonary hypoplasia due to oligohydramnios and its reversal by relieving thoracic compression, *J. Paediatr. Surg.*, 18, 347, 1983.
97. National Institute of Child Health and Development (NICHD), National Registry for Amniocentesis Study Group, Midtrimester amniocentesis for prenatal diagnosis, *J. Am. Med. Assoc.*, 236, 1471, 1976.
98. **Nimrod, C., Varela-Gittings, F., Machin, G., Campbell, D., and Wesenberg, R.**, The effect of very prolonged membrane rupture on fetal development, *Am. J. Obstet. Gynecol.*, 148, 540, 1984.
99. **Normand, I. C. S., Olver, R. E., Reynolds, E. O. R., Strang, L. B., and Welch, K.**, Permeability of lung capillaries and alveoli to non-electrolytes in the foetal lambs, *J. Physiol.*, 219, 303, 1971.
100. **Olver, R. E., Schneeburger, E. E., and Walters, D. V.**, Epithelial solute permeability, ion transport and tight junction morphology in the developing lung of the fetal lamb, *J. Physiol.*, 315, 385, 1981.

101. **Olver, R. E., and Strang, L. B.,** Ion fluxes across the pulmonary epithelium and the lung liquid in the foetal lamb, *J. Physiol.*, 241, 327, 1974.
101a. **Olver, R. E., Ramsden, C. A., Strang, L. B., and Walter, D. V.,** The role of amiloride-blockable sodium transport in adrenaline-induced lung liquid reabsorption in the fetal lamb, *J. Physiol.*, 376, 321, 1986.
102. **Perelman, R. H., Engle, M. J., and Farrell, P. M.,** Perspectives on fetal lung development, *Lung*, 159, 53, 1981.
103. **Perks, A. M. and Cassin, S.,** The effects of arginine vasopressin and other factors on the production of lung fluid in fetal goats, *Chest*, Suppl. 81, 63S, 1982.
104. **Perks, A. M. and Cassin, S.,** The effects of arginine vasopressin on lung liquid secretion in chronic fetal sheep, in *The Physiological Developments of the Fetus and Newborn*, Jones, C. T. and Nathanielsz, P. W., Eds., Academic Press, London, 1985, 253.
105. **Perks, A. M. and Cassin, S.,** The rate of production of lung liquid in fetal goats, and the effect of expansion of the lungs, *J. Dev. Physiol.*, 7, 149, 1985.
106. **Perlman, M., Williams, J., and Hirsch, M.,** Neonatal pulmonary hypoplasia after prolonged leakage of amniotic fluid, *Arch. Dis. Child.*, 51, 349, 1976.
107. **Platzker, A. C. G., Kitterman, J. A., Mescher, E. J., Clements, J. A., and Tooley, W. H.,** Surfactant in the lung and tracheal fluid of the fetal lamb and acceleration of its appearance by dexamethasone, *Pediatrics*, 56, 554, 1975.
108. **Plopper, C. G., Kendall, J. Z., Saldana-Gautier, L. R., Arias-Bravo, J. W., Rivera-Alcini, M. E., Szmyd, S., and Webb, P. D.,** Lung development in the nephrectomized ovine fetus, *J. Dev. Physiol.*, 6, 313, 1984.
109. **Polgar, G. and Weng, T. R.,** The functional development of the respiratory system, *Am. Rev. Respir. Dis.*, 120, 625, 1979.
110. **Potter, E. L.,** Bilateral absence of ureters and kidneys, *Obstet. Gynecol.*, 25, 3, 1965.
111. **Pringle, K. C.,** Fetal lamb and fetal lamb lung growth following creation and repair of a diaphragmatic hernia, in *Animal Models in Fetal Medicine*, Nathanielsz, P. W., Ed., Perinatology Press, Ithaca, NY, 1984, 109.
112. **Quirk, J. G., MacDonald, P. C., and Johnston, J. M.,** Role of fetal pituitary prolactin in fetal lung maturation, *Semin. Perinatol.*, 6, 238, 1982.
113. **Robert, M. F., Neff, R. K., Hubbel, J. P., Taeusch, H. W., and Avery, M. E.,** The association between maternal diabetes and the respiratory distress syndrome in the newborn, *N. Engl. J. Med.*, 294, 357, 1976.
114. **Ross, M. G., Ervin, G., Leake, R. D., Fu, P., and Fisher, D. A.,** Fetal lung liquid regulation by neuropeptides, *Am. J. Obstet. Gynecol.*, 150, 421, 1984.
115. **Scarpelli, E. M., Condorelli, S., and Cosmi, E. V.,** Lamb fetal pulmonary fluid. I. Validation and significance of method for determination of volume and volume change, *Pediatr. Res.*, 9, 190, 1975.
116. **Shermeta, D. W. and Oesch, I.,** Characteristics of fetal lung fluid production, *J. Pediatr. Surg.*, 16, 943, 1981.
117. **Simpson, N. E., Dallaire, L., Miller, J. R., Siminovitch, L., Homerton, J. L., Miller, J., and McKeen, C.,** Prenatal diagnosis of genetic disease in Canada: report of a collaborative study, *Can. Med. Assoc. J.*, 115, 739, 1976.
118. **Smith, B.,** Lung maturation in the fetal rat: acceleration by injection of fibroblast-pneumonocyte factor, *Science*, 204, 1094, 1979.
119. **Smith, B. T., Giroud, C. J. P., Robert, M., and Avery, M. E.,** Insulin antagonism of cortisol action on lecithin synthesis by cultured fetal lung cells, *J. Pediatr.*, 87, 953, 1975.
120. **Smith, B. T., Torday, J. S., and Giroud, C. J. P.,** Evidence for different gestation-dependent effects of cortisol on cultured fetal lung cells, *J. Clin. Invest.*, 53, 1518, 1974.
121. **Spooner, B. S. and Faubion, J. M.,** Collagen involvement in branching morphogenesis of embryonic lung and salivary glands, *Dev. Biol.*, 77, 84, 1980.
122. **Squier, C. A.,** The stretching of mouse skin *in vivo*: effect on epidermal proliferation and thickness, *J. Invest. Dermatol.*, 74, 68, 1980.
123. **Strang, L. B.,** Growth and development of the lung: fetal and postnatal, *Annu. Rev. Physiol.*, 39, 253, 1977.
124. **Sundell, H., Serenius, F. S., and Borthe, P.,** The effect of EGF on fetal lamb maturation, *Pediatr. Res.*, 9, 371, 1975.
125. **Vilos, G. A. and Liggins, G. C.,** Intrathoracic pressures in fetal sheep, *J. Dev. Physiol.*, 4, 247, 1982.
126. **Vilos, G. A., Challis, J. R. G., Pliagas, G. A., Lye, S. J., Possmayer, F., and Harding, P. G.,** Propranolol inhibits the maturational effect of adrenocorticotropin in the fetal sheep lung, *Am. J. Obstet. Gynecol.*, 153, 472, 1985.
127. **Vyas, H., Milner, A. D., and Hopkin, I. E.,** Amniocentesis and fetal lung development, *Arch. Dis. Child.*, 57, 627, 1982.

128. **Walters, D. V. and Olver, R. E.,** The role of catecholamines in lung liquid absorption at birth, *Pediatr. Res.,* 12, 239, 1978.
129. **Wigglesworth, J. S.,** The effects of placental insufficiency on the fetal lung, *J. Clin. Pathol.,* 29(Suppl. 10), 27, 1976.
130. **Wigglesworth, J. S. and Desai, R.,** Effect of lung growth of cervical cord section in the rabbit fetus, *Early Hum. Dev.,* 3, 51, 1979.
131. **Wigglesworth, J. S. and Desai, R.,** Is fetal respiratory function a major determinant of perinatal survival?, *Lancet,* 1, 264, 1982.
132. **Wigglesworth, J. S., Winston, R. M. C., and Bartlett, K.,** Influence of the central nervous system on fetal lung development, *Arch. Dis. Child.,* 52, 965, 1977.
133. **Wigglesworth, J. S., Desai, R., and Guerrini, P.,** Fetal lung hypoplasia: biochemical and structural variations and their possible significance, *Arch. Dis. Child.,* 56, 606, 1981.
134. **Williams, M. C.,** Development of the alveolar structure of the fetal rat in late gestation, *Fed. Proc., Fed. Am. Soc. Exp. Biol.,* 36, 2653, 1977.
135. **Wong, N. S., Kotas, R. V., Avery, M. E., and Thurlbeck, W. M.,** Accelerated appearance of osmiophilic bodies in fetal lungs following steroid injection, *Appl. Physiol.,* 30, 362, 1971.
136. **Wu, B., Kikkawa, Y., Orzalesi, O. M., Motoyama, E. K., Kaibora, M., Zigas, C. J., and Cook, C. D.,** The effect of thyroxine on the maturation of fetal rabbit lungs, *Biol. Neonate,* 22, 161, 1973.

DEVELOPMENT OF BREATHING MOVEMENTS AND THEIR REGULATION BEFORE BIRTH

David W. Walker

SPECIES

Breathing movements during intrauterine life in mammals have been described in sheep,[1] goats,[2] cats,[3] guinea pigs,[4] monkeys,[5,6] and humans.[7] They have been observed in chickens during development in the egg.[8] A review of the subject, with historical detail, was given by Wilds.[10]

RESPIRATORY MOVEMENTS

Muscles

Fetal breathing movements (FBMs) involve the diaphragm and laryngeal adductor muscles.[11,12] The intercostal muscles are largely silent during normal, quiet breathing in fetal lambs from 120 d gestation, but they become active when breathing is stimulated by CO_2,[11] certain drugs,[13] and probably also by changes of temperature. Inspiratory intercostal activity has been described during normal quiet breathing in fetal lambs up to 110 d gestation.[14]

Detection and Measurement

In experimental animals, FBMs are measured from diaphragm and/or laryngeal adductor electromyograms (EMGs), or from the negative change of intrathoracic (intratracheal or intrapleural) pressure. Changes in the lateral dimensions of the fetal chest during FMBs have been measured with ultrasound.[15] FBMs cause phasic changes in superior vena caval pressure and flow.[2,16] In birds, FBMs cause changes of pressure within the airspace of the egg.[9] Techniques for the measurement and analysis of FBMs have been discussed by Harding and Poore.[17] Computer-based system of analysis have been described.[11,18]

In humans, FBMs are detected by real-time ultrasound methods. Real-time ultrasound methods using linear array and sector scanners have largely supplanted A-scan and M-mode devices for visualizing FBMs in the humans.[19] The resolution is sufficient even for small FBMs, since in fetal monkeys, the smallest change of intratracheal pressure was always associated with observable movement of the chest and abdominal walls.[5] Graphic display of human FBMs have been achieved using a combination of B- and M-mode systems[20,21] or electronic methods to continuously track movement of selected echoes.[22] The automatic tracking devices give an analog output of both amplitude and frequency of chest and abdominal wall movements. Doppler methods have also been used, since the modulation of blood flow in the descending aorta, umbilical vein, and inferior vena cava by FBMs can be detected as a Doppler shift of ultrasound.[23-26] Difficulties concerning the use of Doppler techniques have been discussed by Cousin.[22] Perhaps the greatest limitation is that the signal is due to modulation of blood flow, and there may not be a strict relationship of this to the respiratory movement itself.

Types of FBMs

At least two types of FBMs occur in fetal sheep. The most common are low-amplitude FBMs, which are irregular in amplitude and rate.[1,27] They occur in episodes lasting up to about 30 min. The mean amplitude (in terms of intrathoracic pressure) during such episodes is 3.12 ± 0.19 mmHg, and the mean inspiratory time (T_I) is 0.45 ± 0.02 ms.[11] The frequency distribution of breath amplitude, T_I, and breath interval is asymmetric, being skewed to the right.[11]

The second type of breathing consists of isolated, single breaths of much greater amplitude (>20 mmHg) and shorter T_I.[1,28] These "deep inspiratory efforts",[28] sometime improperly termed "gasps", occur occasionally within an episode of low-amplitude, irregular FBMs, but they are more notable for their appearance at times when this type of breathing is absent, when they occur as a series of single breaths at 1 to 5 per minute.

Short episodes of more rapid, regular, and augmented breathing have been seen to occur within an episode of low-amplitude, irregular FBMs.[29] FBMs of greatly augmented amplitude and rate also occur during bouts of fetal swallowing and body movement.

During FBMs in both man and experimental animals there is "paradoxical" movement of the chest and abdominal walls. In sheep it has been shown that the point along the trunk which separates the opposite movements of the thorax and abdomen is not constant, but may move upward or downward with changes of fetal posture.[15] In the human fetus, inspiration causes a 2- to 5-mm inward movement of the chest wall with a 3- to 8-mm outward movement of the abdomen.[19] The amplitude of thoracic and abdominal displacements correlates well with gestational age (but not with estimated fetal weight), and a nomogram has been published.[30] Abdominal displacement is usually greater than the thoracic displacement.

In the human fetus, breathing rates between 47 and 58 breaths per minute have been reported,[7] with a coefficient of variability of 28.6 ± 13.2%.[31] Apnea has been defined as a breath-breath interval >6 s, since 97% of all breath intervals are less than this value from at least 30 weeks of gestation.[32] Breathing rates in excess of 100 breaths per minute have been reported and are regarded as a sign of fetal compromise.[33-35]

"Hiccoughs" are large inward-outward movements of the chest wall of short duration which occur in episodes lasting 1.7 to 13.4 min.[7] They may occur during episodes of FBM or during apnea.

AFFERENT NERVE ACTIVITY

Tactile stimulation elicits respiratory movements in fetal lambs 40 to 60 d gestation, although usually there is also general activation of body movements which obscure specific respiratory movements.[36] Stimulation of the face region evokes movements in human fetuses after at least 12 weeks of gestation, in which a gasp can often be distinguished.[37] Electrical stimulation of the sciatic nerve in fetal lambs sometimes provokes a train of breathing movements,[38,39] an effect which probably relies upon activation of small-diameter pain fibers, since sciatic nerve stimulation at levels just sufficient to evoke reflex contraction of the flexor muscles did not alter breathing movements.[40]

Increase of lung volume in fetal lambs decreased breathing rate, an effect presumably mediated by pulmonary stretch receptors.[41] Pulmonary stretch receptors were observed to fire continuously in studies conducted on anesthetized fetal lambs.[42] Activity increased and decreased when lung volume was decreased and increased by withdrawal or injection of fluid. It is not known if the pulmonary stretch receptors are active in the unanesthetized fetus *in utero* and if they are sensitive to the small changes of lung volume which occur during breathing movements and changes of posture. Section of the vagus nerve did not change the pattern of FBMs in unanesthetized fetal lambs.[1]

Carotid chemoreceptor activity has been recorded from the carotid sinus nerve of anesthetized fetal lambs, and it increased in response to hypoxemia, hypercapnia, and umbilical cord ligation.[43] Fetal lambs *in utero* in which the carotid sinus and vagal nerves had been sectioned made FBMs of lower amplitude and incidence compared to intact, control fetuses.[44] These observations indicate that the carotid chemoreceptors are active and have effects on FBMs *in utero*.

GESTATIONAL AGE

In humans, the mean incidence and amplitude of FBMs increase with gestational age. Patients studied serially from 20 weeks gestation showed that FBMs occurred as isolated breaths or in brief groups of four to ten breaths.[45] By 30 weeks gestation, longer episodes of FBMs were observed. At 24 weeks, the percent of time spent breathing ranged from 5 to 10% without the periodicities that characterize FBMs after 30 weeks.[46] After 30 weeks the percent of time spent breathing is approximately 31% (range 17 to 65%) with no significant change until term.[19] Breathing becomes more regular, with the peak of the breath-to-breath interval distribution shifting from the range 0.5 to 1 s at 30 weeks to 1 to 5 s at 38 weeks.[19,32] There are changes in the effect of maternal meals on FBM with gestational age (see below).

In sheep, FBMs are almost continuous before 100 d gestation, but become organized into episodes lasting 5 to 30 min by 125 d.[14,37] Inspiratory intercostal muscle activity is present at the earlier age, but is rarely observed after 125 d.[14] These changes occur in association with the appearance of sleep cycles at about 120 d gestation.[1,14,37]

BEHAVIORAL STATES

In fetal lambs after 125 d gestation the episodes of low-amplitude, irregular FBMs occur during periods when there is increased extraocular activity, low-amplitude electrocortical activity, and reduced activity in the muscles of the neck, back, and limbs. Such a period is regarded as a form of rapid eye movement (REM) or "paradoxical" sleep, and accounts for 40 to 50% of the recording time.[1,12,47] The episodes of low-amplitude, irregular FBMs occupy from 30 to 80% of the REM period, but this varies greatly from time to time and between animals. Periods of REM sleep lasting up to 30 min alternate with periods in which there is relatively little electroocular activity, electrocortical activity of higher amplitude, and the muscles of the neck, back, and limbs show increased, and sometimes sustained, activity.[1,12,47] Such periods are regarded as a form of nonrapid eye movement (NREM) or "quiet" sleep. Low-amplitude, irregular FBMs do not usually occupy more than 6% of the NREM period, whereas the deep inspiratory efforts often occur at this time.[1,12] Low-amplitude, irregular FBMs may continue during the 1 to 5 min taken for the transition from REM to NREM sleep, when they become deeper and more regular.[48] Much of the occurrence of FBMs in NREM is accounted for by their persistence during the transition phase.

In human pregnancy, fetal behavioral states have been identified from approximately 36 weeks of gestation.[49] Four different states were identified using the same parameters (eye movements, breathing and body movements, heart rate variability) and definitions employed to describe sleeping and waking states in the newborn infant. After 36 weeks gestation, continuous regular FBMs occurred during the "1F" state when there were few body movements, no eye movements, and a stable heart rate with low-amplitude oscillations. During other states, when eye movements were present (e.g., states 2F and 3F), FBMs were irregular in rate and length of each episode. Although it is clear that FBMs come to be associated with other physiological markers of behavioral states in both the human[49] and other experimental animals,[1,14] it is important to note that patterns within FBMs are present before fully coherent behavioral states are established. This suggests that brainstem mechanisms modify respiratory output before maturation of full control over other motor output has been attained.[14]

An aroused or wakeful state has been identified in fetal lambs.[50,52] This occupies 5 to 10% of total recording time, and FBMs are usually of deeper amplitude. Some workers consider that hypercapnia or somatic nerve stimulation induces wakefulness, which then causes increase of FBMs.[38,47,50,53]

DIURNAL RHYTHMS

A diurnal rhythm in the incidence of FBMs has been described in the sheep and man.[54,55] In the human, the incidence of FBMs is lowest near midnight, but increases throughout the remainder of the night to reach a peak at about 0700 h. This rhythm can be discerned from at least 30 weeks gestation.[7] In contrast to the peaks and troughs which occur during the day in relation to meals, the nocturnal increase of FBMs occurs when maternal glucose concentrations are stable or falling.[55,56] There is a nocturnal increase of glucocorticoids in maternal plasma, and it has been suggested that these may increase fetal glucose concentrations.[56] Administration of synthetic glucocorticoids to pregnant women suppressed the endogenous diurnal change of plasma cortisol and removed the overnight increase of FBMs.[57,58]

A diurnal rhythm of FBM was not observed in insulin-dependent diabetic women at 33 to 37 weeks gestation.[59] In sheep, the diurnal rhythm in the incidence of FBMs has been associated with increased duration of low-voltage electrocortical activity,[54] but the metabolic or neurochemical changes which determine these rhythms are not understood.

UTERINE CONTRACTIONS AND LABOR

FBMs in sheep diminish 2 to 3 d before the onset of spontaneous labor,[60] and in rhesus monkeys decreased incidence of breathing movements was reported during early labor.[6] In the human, the incidence of fetal breathing movements is reduced during induced[61,62] and spontaneous labor.[62-64] The decrease of breathing movements precedes spontaneous labor by 2 to 3 d,[64] and appears to be greater if labor begins before rupture of the membranes. During active labor, fetal breathing movements are almost entirely absent.[61] The most likely explanation for the decrease of breathing movements is the increased release of prostaglandin E_2 (PGE_2) into the fetal circulation, since the substance is known to reduce the incidence of breathing movements.[65]

Spontaneous uterine activity present before the onset of active labor also influences FBMs. In sheep, spontaneous uterine contractions occurring at one to four per hour, and causing an increase of intrauterine pressure of 3 to 8 mmHg, are commonly associated with cessation of FBMs.[66] The cause of this cessation may be a brief reduction of uterine blood flow during the contraction, leading to a transient hypoxemia.[67] In human pregnancies with normal fetal growth, Braxton-Hicks contractions are associated with a decrease in the rate of FBMs during the onset phase of the contraction, followed by an increased rate of breathing as the contraction subsides.[68] These changes of breathing rate appear to precede any change of fetal oxygenation which could occur because of the contraction, and it was suggested that physical distortion of the fetal trunk could have altered the sensory input from the chest wall and lung afferents.[68] In sheep, small increases of central venous pressure have been shown to increase intracranial pressure and cause FBMs to stop momentarily.[69] Uterine contractions may increase central venous pressure and lead to increased intracranial pressure, and so contribute to changes observed in the fetuses of both man and sheep. In human pregnancies with fetal growth retardation, nonlabor uterine contractions were not associated with changes of FBMs.[70]

TEMPERATURE

Decrease of body temperature of fetal lambs *in utero* resulted in the initiation of continuous and deeper FBMs.[71] In partly exteriorized fetal lambs, cooling caused an increase in the sensitivity and decrease in the threshold to CO_2.[72] These effects have been related to stimulation of cutaneous thermoreceptors.[71]

Increase of body temperature of fetal lambs *in utero* by 0.7 to 1.5°C resulted in an increase in rate, depth, and incidence of FBMs, but this effect was sustained for only 2 to 3 h.[73] Maternal (and fetal) body temperatures may be raised during fever, exercise,[74,75] or hot weather,[76] and so cause changes in FBM.

BLOOD GASES AND pH

In fetal lambs, the overall incidence of FBMs is less when the resting arterial PO_2 is low, but for any given PaO_2, the incidence of FBM is higher if the resting $PaCO_2$ is high.[77] An unusual form of FBM has been noted in instances where the pH has been abnormally low for some days.[78-80] The FBMs were fixed at a relatively slower rate and low amplitude. This form of FBM was observed to precede intrauterine death, and has not been reproduced by experimental manipulation of blood gases or pH.

Oxygen

In fetal lambs, the overall incidence of FBMs is decreased when the PaO_2 is lowered abruptly by 8 to 10 mmHg, and they may cease altogether if the PaO_2 is sufficiently low (generally less than 12 mmHg).[27] The principal effect of hypoxia seems to be to decrease the number of FBMs, since any that continue to be made are of normal amplitude and T_I.[14] The incidence of FBMs are decreased to a lesser extent by hypoxia in fetal lambs less than 110 d gestation compared to lambs after 120 d gestation.[14]

The cause of the decrease of FBMs during hypoxia is not known. Mechanisms may include one or more of the following:

1. Increased cerebral blood flow, leading to decrease of H^+ concentration in the region of the central chemoreceptors
2. Decrease in metabolic rate
3. Specific inhibition of FBMs by a central neuronal pool sensitive to changes of cerebral PO_2 or O_2 delivery

Evidence for (3) above comes from the observation that transection of the brainstem of fetal lambs above the pons results in the augmentation and persistence of FBMs during hypoxia.[81] More restricted lesions in an area of the mesencephalon are also associated with persistence of FBMs during hypoxia.[82] It has been proposed that a suprapontine pool of neurons is excited by low PO_2, causing the inhibition of FBMs and probably also decreasing the excitability of lumbar motoneurons.[81,83,84] The neurotransmitter involved in mediating this effect has not been identified. Transections and lesions probably interfere with brainstem blood flow and its change during hypoxia, and it is possible that (1) above is the principal mechanism involved in the decrease of FBMs. In human pregnancies, the effect of induced hypoxia has not been studied for ethical reasons. However, it has been noted that FBMs were reduced during and after nonlabor uterine contractions where deceleration of fetal heart rate occurred, a phenomenon normally associated with fetal hypoxemia.[70]

In contrast to the observations that FBMs decrease in incidence when the PaO_2 is lowered experimentally, short episodes of FBMs of increased amplitude have been observed in association with spontaneous falls of PaO_2.[85] It is considered that O_2 consumption is increased during the short episodes of augmented breathing.[85]

Carbon Dioxide

Increase of $PaCO_2$ leads to an increase in breath amplitude and a decrease in the variability of the breath-to-breath intervals, although mean rate of FBMs does not always change.[11,27,29,39,87] Several of the parameters of the diaphragm EMG (rate of rise during inspiration, integrated EMG amplitude, and area) increase, indicating increased central respiratory drive and neural output.[11,39,53] There is an increase in inspiratory intercostal muscle activity during hypercapnia.[11]

The "sensitivity" of FBMs to CO_2 (increase of breath amplitude vs. increase of $PaCO_2$) has been calculated from steady-state changes of $PaCO_2$ and from progressive increases of $PaCO_2$ using rebreathing techniques.[11,39] The breathing parameter used has not been the same when the two techniques have been employed, and therefore a comparison of the results is not possible. Using steady-state changes of $PaCO_2$, tracheal pressure amplitude and inspiratory slope (of tracheal pressure change) increased linearly over the range of $PaCO_2$ from 47 to 80 mmHg, yielding values of 0.23 ± 0.08 mmHg·mmHg CO_2^{-1} and 0.75 ± 0.19 mmHg·s^{-1}·mmHg CO_2^{-1}, respectively. Examination of Figure 3C of Moss and Scarpelli's 1979 study[39] using the CO_2-rebreathing technique shows that the tracheal pressure amplitude/$PaCO_2$ relationship would have a value of approximately 2.5 mmHg·mmHg CO_2^{-1}. The relationship of inspiratory slope (respiratory drive) to $PaCO_2$ yielded values of 2.4 to 3.7 mmHg·s^{-1}·mmHg CO_2.[39] No systematic study has been made to examine the different estimates of CO_2-sensitivity given by the two techniques.

Using the rebreathing technique in fetal lambs 140 to 144 d prepared for study under local anesthesia, it has been shown that several experimental treatments alter the CO_2 response curve. Theophylline increased the slope, but did not change the position of the response line.[88] Sciatic nerve stimulation and intravenous (i.v.) injection of naloxone shifted the response line upwards, but did not apparently change the slope of the line (although the authors have concluded otherwise).[39,53] Under the conditions of these experiments (local anesthesia, uterus exposed from the abdomen, but fetus within), fetuses spontaneously making FBMs had a *lower* sensitivity to CO_2 than those apneic before the administration of CO_2.[39] No comparable data are available for the effect of CO_2 during apnea and breathing in fetal animals that are *in utero* and have recovered from surgery.

Hydrogen Ion

I.v. infusion of HCl or NH_4Cl into fetal lambs resulted in an increase in the incidence of FBM after a delay of 3 to 7 h.[89,90] The delay was attributed to slow penetration of H^+ into the brain extracellular space. Episodes of FBMs lasting 4 to 8 h were noted, and the stimulation of breathing persisted for up to 24 h.[90]

Bicarbonate

Using ventriculocisternal perfusion, it has been shown that the overall incidence and amplitude of FBMs is altered by changes of cerebrospinal fluid (CSF) HCO_3^- concentration.[48] FBMs decrease in incidence and amplitude cerebrospinal fluid when CSF HCO_3^- is raised, and increase when HCO_3^- is lowered. Comparison with similar experiments in adult goats under isocapnic conditions shows that the depth of breathing has approximately the same relationship to changes of CSF HCO_3^- in the adult and in the fetus.[91] These studies show that H^+ ion concentration in the fetal brainstem is an important determinant of the level of respiratory activity and that the functioning and response of the central chemoreceptor-respiratory generator complex is probably very similar in the fetus and adult, if sheep and goats can be considered comparable.[91]

METABOLITES AND HORMONES

Glucose

In man and sheep, changes in plasma glucose induce changes in the incidence and depth of FBMs, with no consistent change of rate or breath-to-breath variability.[56,92-95] The current explanation for this effect is that increased availability of glucose allows increased cerebral metabolism, and the resulting increase of metabolic CO_2 in the vicinity of the central chemoreceptors causes increased respiratory output. In humans, the changes of plasma glucose which occur during the day and after meals are followed by changes of FBMs.[7]

Meals containing only fat and protein and which did not change maternal plasma glucose concentrations were not followed by an increase of FBMs.[56] FBMs are decreased after an overnight fast, and the increase of FBMs following oral glucose is proportionately the same in fasting and nonfasting patients.[96] In sheep, a fasting or induced hypoglycemia is required to decrease FBMs, but FBMs are not consistently increased by hyperglycemia.[93,94]

Although hypoglycemia in fetal lambs, whether induced by fasting of the ewe[97] or by infusion of insulin to the ewe or fetus,[94] resulted in decreased FBMs, the changes of plasma glucose concentration did not always reduce cerebral metabolic rate or glucose uptake.[97] It has been suggested that changes of glucose may affect FBM only when the O_2 delivery to the brain is suboptimal or because changes of cerebral glucose uptake influence turnover of serotonin in the brain.[94,97,98]

In man, the glucose-mediated increase of FBMs occurs after 30 to 32 weeks, but not before this.[46,99] This may reflect a change in the response of the fetal central nervous system (CNS) to glucose, or indicate a difference in the placental acquisition and delivery of glucose to the fetus at the two gestational ages.

Catecholamines

I.v. infusion of isoprenaline into unanesthetized fetal monkeys increased the incidence and amplitude of FBMs, but noradrenaline and adrenaline had little or no effect.[100] Infusion of either of the three catecholamines was reported to increase FBMs slightly in fetal lambs,[60] although others have reported that adrenaline had no such effect.[100a] Noradrenaline increased O_2 consumption in fetal lambs,[101] and this might be the reason for the increased amplitude and incidence of FBM attributed to catecholamines.

Opioids

Opioid-containing neurons within the brainstem may act to "brake" FBMs. Naloxone increased the amplitude of FBMs in fetal lambs at rest[102] and during hypercapnia.[103] Pretreatment with naloxone increased the respiratory sensitivity of fetal lambs to CO_2.[39,53] Naloxone injected into the subarachnoid space of the ventral medulla increased FBMs in incidence and amplitude within 5 to 15 min.[104] In these experiments the central administration of naloxone also caused changes of electrocortical activity, and FBMs were present during non-REM sleep and wakefulness, an effect not seen when naloxone was given peripherally.[102,103,105] Chronic exposure to methadone is associated with a reduced incidence of FBMs and a decreased response to 5% CO_2 in humans.[107] Each daily dose of methadone also decreased the incidence of FBMs in these patients. Pethidine abolished FBMs in sheep.[106] Immunoreactive β-endorphin, Leu-enkephalin, dynorphin-A, and pro-opiomelanocortin fibers have been visualized in fetal rat brain,[108] but not yet in species where FBMs have been measured. I.v. injection of Met-enkephalin (20 μg) arrested FBMs for 30 to 180 s in fetal lambs.[105]

In contrast to the observations that endogenous opiates may decrease FBMs, acute exposure of fetal lambs to morphine increased the incidence and amplitude of FBMs.[109-112] This response developed after a delay of approximately 30 min and was blocked by naloxone.[110] It is possible that the hyperglycemic effects of acute morphine exposure[113] contributed to the stimulation of FBMs. Others have found i.v. infusion of morphine first increased and then decreased FBMs in sheep, suggesting that stimulation occurs only at low concentrations of the drug.[114] Intracerebroventricular administration of β-endorphin led to an increase of FBMs in sheep 2 to 4 h after conclusion of a 2-h infusion.[115] I.v. administration of β-endorphin had no effect on FBMs.[115]

Prostaglandins

Prostaglandin (PG) synthesis within the brain may also diminish breathing movements tonically. I.v infusion of PGE$_2$ decreased the incidence and amplitude of FBMs in sheep.[65] Inhibition of PG synthetase by indomethacin or meclofenamate resulted in a prolonged increase in incidence and amplitude of FBMs.[116,117] PG synthesis within the brainstem appears to be responsible for tonic suppression of FBMs in sheep because ventriculocisternal administration of meclofenamate resulted in augmented FBMs, and i.v. meclofenamate was also effective after midcollicular transection of the brainstem in fetal lambs.[118] The decrease of FBMs during labor may be caused by the increase of circulating PGE$_2$ that occurs at this time. Whereas PGs influence both the incidence and amplitude of FBMs, opioids appear to affect only amplitude. PGs may suppress FBMs by diminishing the sensitivity or output from the central chemoreceptors.[117]

DRUGS

Alcohol

Ethanol administered orally or intravenously to pregnant women caused suppression of FBMs lasting at least 3 h.[119,120] The effect of glucose in enhancing FBMs was blocked in the presence of alcohol.[121] In sheep, i.v. infusion of ethanol to the ewe for 1 h suppressed FBMs for up to 9 h.[122] This effect outlasted the changes of electrocortical and electroocular activities which ethanol also caused, suggesting that alcohol suppressed FBMs by an action on pontine and medullary brain centers, rather than because of effects on higher brain centers.[122]

Cigarette Smoking and Nicotine

Cigarette smoking by pregnant women decreased the incidence of FBMs for about 1 h.[123,124] The effect appeared to be greater in pregnancies complicated by fetal growth retardation. Other studies have found that smoking increased breathing rate without change of incidence.[125] The effects of smoking have been attributed to the vasoconstrictive effects of nicotine on the uteroplacental circulation, causing transient fetal hypoxemia. Injection of nicotine into pregnant sheep caused a prompt decrease of fetal PaO$_2$ and in the incidence of FBMs,[126] although the doses given were large relative to plasma nicotine concentrations produced by smoking a cigarette. Nicotine is also implicated by the findings that smoking of herbal cigrettes (containing no nicotine) do not alter FBMs, and that chewing of gum containing nicotine does decrease FBMs.[124] Smoking also causes fetal circulatory changes (increased heart rate, aortic and umbilical vein blood flow) consistent with activation of the fetal sympathoadrenal system, either by nicotine entering the fetal circulation or by hypoxic changes of blood gases.[25,127] The effects of carbon monoxide in cigarette smoke has not been fully evaluated.

Anesthetics and Analgesics

Barbiturates given to sheep in doses which caused only slight sedation of the ewe resulted in fetal apnea.[41,108,128] This effect appears to be exerted at brain centers above the pons because pentobarbitone had no effect on FBMs when the fetal brainstem had been transected.[81]

In human pregnancies, meperidine decreased the incidence of FBM when given in doses of approximately 0.2 to 1 mg/kg.[129,130] Meperidine given to pregnant sheep did not alter FBMs, but abolished the response to hypercapnia.[108] Diazepam injections into pregnant sheep decreased the incidence of FBMs,[131] whereas chronic infusion caused an increased incidence.[132]

Other Drugs

Caffeine injected intravenously or into the carotid artery of fetal sheep increased the incidence and amplitude of FBMs.[131] Theophyline given to pregnant ewes decreased the threshold and increased the slope of the fetal CO_2-FBM response curve.[88] The analeptic drug doxapram increased the incidence and amplitude of FBMs in fetal sheep.[131,133] The muscarinic agonist pilocarpine also increased FBM incidence and amplitude in fetal lambs.[133a] Drugs which either increase[13] or decrease[134] serotonin turnover in the brain have the effect of increasing the incidence and amplitude of FBMs. It is not clear if the effect of any of these drugs is primarily upon the respiratory centers, the peripheral and central chemoreceptors, or the reticular formation.[47,52,88]

INTRAUTERINE GROWTH RETARDATION

The incidence of FBMs is slightly decreased in growth-retarded human fetuses.[135,135a] In growth-retarded infants, there was also a reduction of FBMs and body movements when nonlabor uterine contractions caused delayed (late) deceleration of heart rate.[70] There was no change of FBMs when contractions did not cause heart rate decelerations, and this observation contrasts with the situation in normally grown fetuses where nonlabor uterine contractions are associated with changes of breathing rate.[68]

Growth retardation also appears to alter the effects of changes of blood gases on FBMs. The incidence of FBMs increased from 29.8 ± 3.9 to $68.2 \pm 3.6\%$ in growth-retarded fetuses when the mother breathed 50% O_2.[136] No such response was observed in normally grown fetuses. The increased incidence of FBMs is at the upper limit of the range normally observed, and the response may be due to the vasoconstrictive effect of increased O_2 on cerebral blood vessels when the preexisting hypoxemia is relieved. In sheep, maternal hyperoxia increases the incidence of FBMs only when the fetus is already hypoxemic.[27]

In growth-retarded fetal lambs (produced by reduction of placental size), FBMs were reduced in incidence and rate from at least 100 d gestation until term (145 d).[137] The integrated activity of the diaphragm electromyogram was reduced in the growth-retarded lambs until 120 d gestation, but not different from controls after this time. This form of experimental growth retardation also reduced the number of fatigue-resistant fibers in the diaphragm and reduced relative lung size.[137] In another study, the incidence of FBMs in growth-retarded fetal lambs was one half that of normally grown lambs.[138]

CONCLUDING REMARKS

FBMs begin early in fetal life when they may be one component of a general activation of motor outflow from the central nervous system. FBMs become organized in conjunction with development of sleep states, but they are also regulated to some degree by metabolic substrates, the central and peripheral chemoreceptors, and by production of opioids and PGs within the brainstem. FBMs are important in allowing full structural and functional maturation of the lung, since in those instances where FBMs are absent,[139,140] or the fall of intrathoracic pressure cannot take place,[141,142] the lungs may be hypoplastic and the surfactant content of lung liquid less than normal.[139] (This topic is discussed fully elsewhere in this volume.) The value of observing FBMs in the human for assessing the clinical condition of the fetus has been discussed.[143-145] They appear to have greatest value in clinical evaluation when used in conjunction with other tests.[144,145]

REFERENCES

1. **Dawes, G. S., Fox, A. E., Leduc, B. M., Liggins, G. C., and Richards, R. T.,** Respiratory movements and rapid eye movement sleep in the fetal lamb, *J. Physiol.*, 220, 119, 1972.
2. **Towell, M. and Salvador, H. S.,** Intrauterine asphyxia and respiratory movements in the fetal goat, *Am. J. Obstet. Gynecol.*, 8, 1124, 1974.
3. **Snyder, F. F. and Rosenfeld, M.,** Direct observation of intrauterine respiratory movements of the fetus and the role of carbon dioxide and oxygen in their regulation, *Am. J. Physiol.*, 119, 153, 1937.
4. **Kendall, J. Z.,** Respiratory movements in the fetal guinea pig in utero, *J. Appl. Physiol.*, 42, 661, 1977.
5. **Fox, H. E. and Hohler, C. W.,** Fetal evaluation by real-time imaging, *Clin. Obstet. Gynecol.*, 20, 339, 1977.
6. **Martin, C. B., Murata, Y., Petrie, R. H., and Parer, J. T.,** Respiratory movements in fetal rhesus monkeys, *Am. J. Obstet. Gynecol.*, 119, 939, 1974.
7. **Patrick, J., Campbell, K., Carmichael, L., Natale, R., and Richardson, B.,** Patterns of human fetal breathing during the last 10 weeks of pregnancy, *Obstet. Gynecol.*, 56, 24, 1980.
8. **Windle, W. F., Monnier, M., and Steele, A. G.,** Fetal respiratory movements in the cat, *Physiol. Zool.*, 11, 425, 1938.
9. **Baggott, G. K., Dawes, C. M., Nair, G., and Osman, H.,** The measurement of lung ventilation in the embryo of the domestic fowl, *J. Physiol.*, 349, 4P, 1984.
10. **Wilds, P. L.,** Observation of intrauterine fetal breathing movements — a review, *Am. J. Obstet. Gynecol.*, 131, 315, 1978.
11. **Dawes, G. S., Gardner, W. N., Johnston, B. M., and Walker, D. W.,** Effects of hypercapnia on tracheal pressure, diaphragm and intercostal electromyograms in unanaesthetized fetal lambs, *J. Physiol.*, 326, 461, 1982.
12. **Harding, R.,** State-related and developmental changes in laryngeal function, *Sleep*, 3, 307, 1980.
13. **Quilligan, E. J., Clewlow, F., Johnston, B. M., and Walker, D. W.,** Effect of 5-hydroxytryptophan on electrocortical activity and breathing movements of fetal sheep, *Am. J. Obstet. Gynecol.*, 141, 271, 1981.
14. **Clewlow, F., Dawes, G. S., Johnston, B. M., and Walker, D. W.,** Changes in breathing, electrocortical and muscle activity in unanaesthetized fetal lambs with age, *J. Physiol.*, 341, 463, 1983.
15. **Poore, E. R. and Walker, D. W.,** Chest wall movements during fetal breathing in sheep, *J. Physiol.*, 301, 307, 1980.
16. **Rudolph, A. M.,** Effects of fetal breathing on superior and inferior vena caval blood flow in the fetal lamb, in *Proc. 6th Conf. on Fetal Breathing*, Paris, 1979, Abstr. 1.
17. **Harding, R. and Poore, E. R.,** Techniques for the measurement and analysis of fetal breathing, in *Animal Models in Fetal Medicine (II)*, Nathanielsz, P. W., Ed., Elsevier Biomedical Press, Amsterdam, 1982, 220.
18. **Wickham, P. J. D. and Walker, D. W.,** Analysis of fetal breathing in real time using a microprocessor, *J. Appl. Physiol.*, 62, 1733, 1987.
19. **Patrick, J. and Challis, J. H. G.,** Measurement of human fetal breathing movements in healthy pregnancies using a real time scanner, *Semin. Perinatol.*, 4, 257, 1980.
20. **Bots, R. S. G. M., Broeders, G. H. B., and Farman, D. J.,** The dynamics of human fetal breathing movements: a multiscan echofetographic approach, *Eur. J. Obstet. Gynecol. Reprod. Biol.*, 6, 339, 1976.
21. **Marsal, K., Gennser, G., and Lindstrom, K.,** Real-time ultrasonography for quantified analysis of fetal breathing movements, *Lancet*, 2, 718, 1976.
22. **Cousin, A. J.,** Advanced methods for measurement of fetal breathing movements in the human fetus, *Semin. Perinatol.*, 4, 261, 1980.
23. **Chiba, Y., Utsu, M., Kanzaki, T., and Hasegawa, T.,** Changes in venous flow and intra-tracheal flow in fetal breathing movements, *Ultrasound Med. Biol.*, 11, 43, 1985.
24. **Goodman, J. D. S. and Mantell, C. D.,** A second means of identifying fetal breathing movements using Doppler ultrasound, *Am. J. Obstet. Gynecol.*, 126, 73, 1980.
25. **Marsal, K., Lindblad, A., and Lingman, G.,** Blood flow in the fetal descending aorta: intrinsic factors affecting fetal blood flow, i.e., fetal breathing movements and cardiac arrhythmia, *Ultrasound Med. Biol.*, 10, 339, 1984.
26. **Eik-Nes, S. H., Marsal, K., Brubakk, A. O., and Ulstein, M.,** Ultrasonic measurements of human fetal blood flow in aorta and umbilical vena cava: influence of fetal breathing movements, in *Recent Advances in Ultrasonic Diagnosis 2*, Kurjak, A., Ed., Excerpta Medica, Amsterdam, 1980, 233.
27. **Boddy, K., Dawes, G. S., Fisher, R., Pinter, S., and Robinson, J. S.,** Fetal respiratory movements, electrocortical and cardiovascular responses to hypoxaemia and hypercapnia in sheep, *J. Physiol.*, 243, 599, 1974.
28. **Harding, R., Johnson, P., and McClelland, M. E.,** Respiratory function of the larynx in developing sheep and the influence of sleep state, *Resp. Physiol.*, 40, 165, 1980.

29. **Chapman, R. L. K., Dawes, G. S., Rurak, D. W., and Wilds, P. L.,** Breathing movements in fetal lambs and the effect of hypercapnia, *J. Physiol.*, 302, 19, 1980.
30. **Neldham, S.,** Fetal respiratory movements: a nomogram for fetal thoracic and abdominal respiratory movements, *Am. J. Obstet. Gynecol.*, 142, 867, 1982.
31. **Andrews, J., Shime, J., Gave, D., Salgado, J., and Whillans, G.,** The variability of fetal breathing movements in normal fetuses at term, *Am. J. Obstet. Gynecol.*, 151, 280, 1985.
32. **Patrick, J., Campbell, K., Carmicheal, L., Natale, R., and Richardson, B.,** A definition of human apnea and the distribution of fetal apneic intervals during the last 10 weeks of pregnancy, *Am. J. Obstet. Gynecol.*, 136, 471, 1980.
33. **Manning, F. A., Heaman, M., Boyce, D., et al.,** Intrauterine fetal tachypnea, *Obstet. Gynecol.*, 58, 398, 1981.
34. **Romero, R., Chervanak, F. A., Berkowitz, R. L., and Hobbins, J. C.,** Intrauterine fetal tachypnea, *Am. J. Obstet. Gynecol.*, 144, 356, 1982.
35. **Duff, P., Sanders, R. S., and Hayashi, R. H.,** Intrauterine tachypnea — a sign of fetal distress?, *Am. J. Obstet. Gynecol.*, 142, 1054, 1982.
36. **Barcroft, J. and Barron, D. H.,** The genesis of respiratory movements in the fetus of the sheep, *J. Physiol.*, 88, 56, 1937.
37. **Bowes, G., Adamson, T. M., Ritchie, B. C., Dowling, M., Wilkinson, M. H., and Maloney, J. E.,** Development of patterns of respiratory activity in unanaesthetized fetal sheep in utero, *J. Appl. Physiol.*, 50(14), 693, 1981a.
38. **Ioffe, S. E., Jansen, A. H., Russell, B. J., and Chernick, V.,** Respiratory response to somatic stimulation in fetal lambs during sleep and wakefulness, *Pfluegers Arch.*, 388, 143, 1980a.
39. **Moss, I. R. and Scarpelli, E. M.,** Generation and regulation of breathing in utero: fetal CO_2 response test, *J. Appl. Physiol.*, 47, 527, 1979.
40. **Rigatto, H., Blanco, C. E., and Walker, D. W.,** The response to stimulation of the hindlimb nerves in fetal lambs, in utero, during different phases of electrocortical activity, *J. Dev. Physiol.*, 5, 15, 1982.
41. **Maloney, J. E., Adamson, T. M., Brodecky, V., Dowling, M. H., and Ritchie, B. C.,** Modification of respiratory center output in the unanaesthetized fetal sheep "in utero", *J. Appl. Physiol.*, 39, 552, 1975.
42. **Ponte, J. and Purves, M. J.,** Types of afferent nervous activity which may be measured in the vagus nerve of the sheep fetus, *J. Physiol.*, 229, 51, 1973.
43. **Blanco, C. E., Dawes, G. C., Hanson, M. A., and McCooke, H. B.,** The response to hypoxia of arterial chemoreceptors in fetal sheep and new-born lambs, *J. Physiol.*, 351, 25, 1984.
44. **Murai, D. T., Chu-Ching, H. L., Wallen, L. D., and Kitterman, J. A.,** Denervation of peripheral chemoreceptors decreases breathing movements in fetal sheep, *J. Appl. Physiol.*, 59, 575, 1985.
45. **Trudinger, B. J. and Knight, P. C.,** Fetal age and patterns of human fetal breathing movements, *Am. J. Obstet. Gynecol.*, 137, 724, 1980.
46. **Natale, R., Patrick, J., and Richardson, B.,** Effects of maternal plasma venous glucose on concentrations on fetal breathing movements, *Am. J. Obstet. Gynecol.*, 132, 36, 1978.
47. **Jansen, A. H., Ioffe, S., Russell, B. J., and Chernick, V.,** Influence of sleep state on the response to hypercapnia in fetal lambs, *Respir. Physiol.*, 48, 125, 1982.
48. **Hohimer, A. R., Bissonnette, J. M., Richardson, B. S., and Machida, C. M.,** Central chemoreceptor regulation of breathing movements in fetal lambs, *Respir. Physiol.*, 52, 99, 1983.
49. **Nijhuis, J. G., Prechtl, H. F. R., Martin, C. B., and Bots, R. S. G. M.,** Are there behavioural states in the human fetus?, *Early Hum. Dev.*, 6, 177, 1982.
50. **Ioffe, S., Jansen, A. H., Russell, B. J., and Chernick, V.,** Sleep, wakefulness and the monosynaptic reflex in fetal and newborn lambs, *Pfluegers Arch.*, 388, 149, 1980.
51. **Ruchebusch, Y.,** Development of sleep and wakefulness in the foetal lamb, *Electroencephalogr. Clin. Neurophysiol.*, 32, 119, 1972.
52. **Jansen, A. H., Ioffe, S., and Chernick, V.,** Drug induced changes in fetal breathing activity and sleep state, *Can. J. Physiol. Pharmacol.*, 61, 315, 1982.
53. **Moss, I. R. and Scarpelli, E. M.,** CO_2 and naloxone modify sleep/wake state and activate breathing in the acute fetal lamb preparation, *Respir. Physiol.*, 55, 325, 1984.
54. **Boddy, K., Dawes, G. S., and Robinson, J. S.,** A 24 hour rhythm in the foetus, in *Foetal and Neonatal Physiology*, Proc. Sir Joseph Bacroft Symp., Comline, R. S., Cross, K. S., Dawes, G. S., and Mathanielsz, P. W., Eds., Cambridge University Press, Cambridge, England, 1973, 63.
55. **Patrick, J., Natale, R., and Richardson, B.,** Patterns of human fetal breathing activity at 34 to 35 weeks gestational age, *Am. J. Obstet. Gynecol.*, 132, 507, 1978.
56. **Patrick, J.,** Fetal breathing movements, *Clin. Obstet. Gynecol.*, 25, 787, 1982.
57. **Patrick, J., Challis, J., Campbell, K., Carmicheal, L., Richardson, B., and Tevaarwerk, G.,** Effects of synthetic glucocorticoid administration on human fetal breathing movements at 34 to 35 weeks gestational age, *Am. J. Obstet. Gynecol.*, 139, 324, 1981.

58. **Challis, J., Patrick, J., Richardson, B., and Tevaarwerk, G.**, Loss of diurnal rhythm in plasma estrone, estradiol and estriol in women treated with synthetic glucocorticoids at 34 to 35 weeks gestation, *Am. J. Obstet. Gynecol.*, 139, 338, 1981.
59. **Wladimiroff, J. W. and Roodenburg, P. S.**, Human fetal breathing and gross body activity relative to maternal meals during insulin-dependent pregnancy, *Acta Obstet. Gynecol. Scand.*, 61, 65, 1982.
60. **Dawes, G.**, The central control of fetal breathing and skeletal muscle movements, *J. Physiol.*, 346, 1, 1984.
61. **Richardson, B., Natale, R., and Patrick, P.**, Human fetal breathing activity during electively induced labour at term, *Am. J. Obstet. Gynecol.*, 133, 247, 1979.
62. **Boylan, P. and Lewis, P. J.**, Fetal breathing in labor, *Obstet. Gynecol.*, 56, 35, 1980.
63. **Wittman, B. K., Davison, B. M., Lyons, E., Frolich, J., and Towell, M. E.**, Real-time ultrasound observation of fetal activity in labor, *Br. J. Obstet. Gynaecol.*, 86, 278, 1979.
64. **Carmichael, L., Campbell, K., and Patrick, J.**, Fetal breathing, gross fetal body movements, and maternal and fetal heart rates before spontaneous labor at term, *Am. J. Obstet. Gynecol.*, 148, 675, 1984.
65. **Kitterman, J. A., Fewell, J. E., and Tooley, W. H.**, Inhibition of breathing movements in fetal sheep by prostaglandins, *J. Appl. Physiol.*, 54, 687, 1983.
66. **Nathanielsz, P. W., Bailey, A., Poore, E. R., Thorburn, G. D., and Harding, R.**, The relationship between myometrial activity and sleep state and breathing in fetal sheep throughout the last third of gestation, *Am. J. Obstet. Gynecol.*, 138, 653, 1980.
67. **Harding, R., Poore, E. R., and Cohen, G. L.**, The effect of brief episodes of diminished uterine blood flow on breathing movements, sleep states and heart rate in fetal sheep, *J. Dev. Physiol.*, 3, 231, 1981.
68. **Wilkinson, C. and Robinson, J.**, Braxton-Hicks contractions and fetal breathing movements, *Aust. N.Z. J. Obstet. Gynaecol.*, 22, 212, 1982.
69. **Walker, D. W. and Harding, R.**, The effects of raising intracranial pressure on breathing movements, eye movements and electrocortical activity in fetal sheep, *J. Dev. Physiol.*, 8, 105, 1985.
70. **Beckedam, D. J. and Visser, G. H. A.**, Effects of hypoxemic events on breathing, body movements and heart rate variations: a study of growth-retarded human fetuses, *Am. J. Obstet. Gynecol.*, 153, 52, 1985.
71. **Gluckman, P. D., Gunn, T. R., and Johnston, B. M.**, The effect of cooling on breathing and shivering in unanaesthetized fetal lambs *in utero*, *J. Physiol.*, 343, 495, 1983.
72. **Moss, I. R., Mantone, A. J., and Scarpelli, E. M.**, Effect of temperature on regulation of breathing and sleep/wake state in fetal lambs, *J. Appl. Physiol.*, 54, 536, 1983.
73. **Walker, D. W. and Davies, A. N.**, Prolonged effect of maternal hyperthermia on breathing movements of fetal lambs, *J. Dev. Physiol.*, 8, 485, 1986.
74. **Clapp, J. F.**, Acute exercise stress in the pregnant ewe, *Am. J. Obstet. Gynecol.*, 136, 489, 1980.
75. **Metcalfe, J., Hohimer, A. R., and Morton, M. J.**, Maternal exercise and fetal development, in *The Physiological Development of the Fetus and Newborn*, Jones, C. T. and Nathanielsz, P. W., Eds., Academic Press, London, 1985, 711.
76. **Davies, A. N., Walker, D. W., McMillen, I. C., and Thorburn, G. D.**, The acute effects of maternal hyperthermia on the foetus, in *Reproduction in Sheep*, Findlay, J. K. and Lindsay, D. F., Eds., Australian Academy of Science, Canberra, 1985, 153.
77. **Bissonnette, J. M., Hohimer, A. R., Cronan, J. Z., and Paul, M. S.**, Effect of oxygen and of carbon dioxide tension on the incidence of apnea in fetal lambs, *Am. J. Obstet. Gynecol.*, 135, 575, 1980.
78. **Chapman, R. L. K., Dawes, G. S., Rurak, D. W., and Wilds, P. L.**, Intermittent breathing before death in fetal lambs, *Am. J. Obstet. Gynecol.*, 131, 894, 1978.
79. **Manning, F. A., Martin, C. B., Murata, Y., et al.**, Breathing movements before death in the primate fetus, *(Macaca mulatta)*, *Am. J. Obstet. Gynecol.*, 135, 71, 1979.
80. **Patrick, J., Dalton, K. J., and Dawes, G. S.**, Breathing patterns before death in fetal lambs, *Am. J. Obstet. Gynecol.*, 125, 73, 1976.
81. **Dawes, G. S., Gardner, W. N., Johnston, B. M., and Walker, D. W.**, Breathing in fetal lambs: the effect of brain stem section, *J. Physiol.*, 335, 535, 1983.
82. **Johnston, B. M., Gluckman, P. D., and Parsons, Y.**, The effects of midbrain lesions on breathing in fetal lambs, in *The Physiological Development of the Fetus and Newborn*, Jones, C. T. and Nathanielsz, P. W., Eds., Academic Press, London, 1985, 621.
83. **Blanco, C. E., Dawes, G. S., and Walker, D. W.**, Effects of hypoxia on polysynaptic hindlimb reflexes of unanaesthetized fetal and newborn lambs, *J. Physiol.*, 339, 453, 1983.
84. **Walker, D.**, Brain mechanisms, hypoxia and fetal breathing, *J. Dev. Physiol.*, 6, 225, 1984.
85. **Rurak, D. W. and Cooper, C.**, The effect of relative hypoxemia on the pattern of breathing movements in fetal lambs, *Respir. Physiol.*, 55, 23, 1984.
86. **Rurak, D. W. and Gruber, N. C.**, Increased oxygen consumption associated with breathing activity in fetal lambs, *J. Appl. Physiol.*, 54, 701, 1983.
87. **Bowes, G., Wilkinson, M. H., Dowling, M., Ritchie, B. C., Brodecky, V., and Maloney, J. E.**, Hypercapnic stimulation of respiratory activity in unanaesthetized fetal sheep in utero, *J. Appl. Physiol.*, 50, 701, 1981.

88. **Moss, I. R. and Scarpelli, E. M.,** Stimulatory effect of theophylline on regulation of fetal breathing movements, *Pediatr. Res.,* 15, 870, 1981.
89. **Hohimer, A. R. and Bissonnette, J. M.,** Effect of metabolic acidosis on fetal breathing movements, in utero, *Respir. Physiol.,* 43, 99, 1981.
90. **Molteni, R. A., Melmed, M. H., Sheldon, R. E., Jones, M. D., and Meschia, G.,** Induction of fetal breathing by metabolic acidemia and its effect on blood flow to the respiratory muscles, *Am. J. Obstet. Gynecol.,* 136, 609, 1981.
91. **Bissonnette, J. M. and Hohimer, A. R.,** Central neurogenesis of respiration in the fetus, in *Respiratory Control and Lung Development in the Fetus and Newborn. Reproductive and Perinatal Medicine III,* Johnston, B. M. and Gluckman, P. D., Eds., Perinatology Press, Ithaca, NY, 1986, 237.
92. **Natale, R., Nasello-Paterson, C., and Turliuk, R.,** Longitudinal measurement of fetal breathing, body movements, heart rate, and heart rate accelerations and decelerations at 24 to 32 weeks of gestation, *Am. J. Obstet. Gynecol.,* 151, 256, 1985.
93. **Richardson, B. S., Hohimer, A. R., Mueggler, P., and Bissonnette, J. M.,** Effect of glucose concentration on fetal breathing movements and electrocortical activity on fetal lambs, *Am. J. Obstet. Gynecol.,* 142, 678, 1982.
94. **Richardson, B. S., Hohimer, A. R., Bissonnette, J. M., and Machida, C. M.,** Insulin hypoglycemia, cerebral metabolism and neural function in fetal lambs, *Am. J. Physiol.,* 248, R72, 1985.
95. **Adamson, S. L., Bocking, A., Cousin, A., Rapaport, I., and Patrick, J.,** Ultrasonic measurement of rate and depth of human fetal breathing: effect of glucose, *Am. J. Obstet. Gynecol.,* 147, 288, 1983.
96. **Fox, H. E., Hohler, C. W., and Steinbrecher, M.,** Human fetal breathing movements after carbohydrate ingestion in fasting and nonfasting subjects, *Am. J. Obstet. Gynecol.,* 144, 213, 1982.
97. **Richardson, B. S., Hohimer, A. R., Bissonnette, J. M., and Machida, C. M.,** Cerebral metabolism in hypoglycemic and hyperglycemic fetal lambs, *Am. J. Physiol.,* 245, R730, 1983.
98. **Bissonnette, J. M., Hohimer, A. R., Richardson, B. S., and Machida, C. M.,** Effect of acute hypoglycemia on cerebral metabolic rate in fetal sheep, *J. Dev. Physiol.,* 7, 421, 1985.
99. **Meis, P. M., Rose, J. C., Swain, M., and Nelson, L. H.,** Gestational age alters the breathing response to intravenous insulin and intravenous glucose administration, *Am. J. Obstet. Gynecol.,* 151, 438, 1985.
100. **Murata, Y., Martin, C. B., Miyake, K., Socal, M., and Druzin, M.,** Effect of catecholamine on fetal breathing activity in rhesus monkeys, *Am. J. Obstet. Gynecol.,* 139, 942, 1981.
100a. **Jansen, A. H. and Chernick, V.,** Development of respiratory control, *Physiol. Rev.,* 63, 437, 1983.
101. **Lorign, R. H. W. and Longo, L. D.,** Norepinephrine elevation in the fetal lamb: oxygen consumption and cardiac output, *Am. J. Physiol.,* 239, R115, 1980.
102. **Adamson, S. L., Patrick, J. E., and Challis, J. R. G.,** Effects of naloxone on the breathing, electrocortical, heart rate, glucose and cortisol response to hypoxia in the sheep fetus, *J. Dev. Physiol.,* 6, 495, 1984.
103. **Joseph, S. A., McMillen, I. C., and Walker, D. W.,** Effect of naloxone on fetal breathing movements during hypercapnia in fetal lambs, *J. Appl. Physiol.,* 62, 673, 1987.
104. **Ioffe, S., Jansen, A. H., and Chernick, V.,** Blockade of endogenous opiates in the ventral medulla induces wakefulness and continuous fetal breathing in exteriorized fetal sheep, in *The Physiological Development of the Fetus and Newborn,* Jones, C. T. and Nathanielsz, P. W., Eds., Academic Press, London, 1985, 663.
105. **Hofmeyr, G. J., Bamford, O. S., Howlett, T., and Parkes, M. J.,** Methonine-enkephalin and the arrest of fetal breathing, in *The Physiological Development of the Fetus and Newborn,* Jones, C. T. and Nathanielsz, P. W., Eds., Academic Press, London, 1985, 653.
106. **Boddy, K., Dawes, G. S., Fischer, R., Pinter, S., and Robinson, J. S.,** The effects of pentobarbitone and pethidine of fetal breathing movements in sheep, *Br. J. Pharmacol.,* 57, 311, 1976.
107. **Richardson, B. S., O'Grady, J. P., and Olsen, G.,** Fetal breathing movements and the response to carbon dioxide in patients on methadone maintenance, *Am. J. Obstet. Gynecol.,* 150, 400, 1984.
108. **Khachaturian, H., Alessi, N. E., Munfakh, N., and Watson, S. J.,** Ontogeny of opioid and related peptides in the rat CNS and pituitary: an immunocytochemical study, *Life Sci.,* 33, 61, 1983.
109. **Olsen, G. D., Hohimer, A. R., and Mathis, G. D.,** Cerebral blood flow and metabolism during morphine-induced stimulation of breathing movements in fetal lambs, *Life Sci.,* 33(Suppl. 1), 751, 1983.
110. **Sheldon, R. E. and Toubas, P. L.,** Morphine stimulates rapid, regular, deep and sustained breathing efforts in fetal lambs, *J. Appl. Physiol. Respir. Exercise Physiol.,* 57, 40, 1984.
111. **Umans, J. G. and Szeto, H. H.,** Effects of opiates on fetal behavioural activity in utero, *Life Sci.,* 33(Suppl.I), 639, 1983.
112. **Rigatto, H., Lee, D., Caces, R., Alberscheim, S., and Moore, M.,** The effect of morphine on breathing and behaviour in fetal sheep, in *The Physiological Development of the Fetus and Newborn,* Jones, C. T. and Nathanielsz, P. W., Eds., Academic Press, London, 1985, 643.
113. **Feldberg, W. and Shaligram, S. W.,** The hyperglycaemic effect of morphine, *Br. J. Pharmacol.,* 46, 602, 1972.

114. **Olsen, G. D. and Dawes, G. S.**, Morphine-induced depression and stimulation of breathing movements in the fetal lamb, in *The Physiological Development of the Fetus and Newborn*, Jones, C. T. and Nathanielsz, P. W., Eds., Academic Press, London, 1985, 633.
115. **McMillen, I. C. and Walker, D. W.**, Effect of intravascular or intra-cerebroventricular infusion of β-endorphin on fetal breathing movements, *J. Appl. Physiol.*, 61, 1005, 1986.
116. **Kitterman, J. A., Liggins, G. C., Clements, J. A., and Tooley, W. H.**, Stimulation of breathing movements in fetal sheep by inhibitors of prostaglandin synthesis, *J. Dev. Physiol.*, 1, 453, 1979.
117. **Hohimer, A. R., Richardson, B. S., Bissonnette, J. M., and Machida, C. M.**, The effect of indomethacin on breathing movements and cerebral blood flow and metabolism in the fetal sheep, *J. Dev. Physiol.*, 7, 217, 1985.
118. **Koos, B. J.**, Central stimulation of breathing movements in fetal lambs by prostglandin synthetase inhibitors, *J. Physiol.*, 362, 455, 1985.
119. **Fox, H. E., Steinbrecher, M., Pessel, D., Inglis, J., Medvid, L., and Angel, E.**, Maternal ethanol ingestion and the occurrence of human fetal breathing movements, *Am. J. Obstet. Gynecol.*, 132, 354, 1978.
120. **McLeod, W., Brien, J., Loomis, C., Carmichael, L., Probert, C., and Patrick, J.**, Effect of maternal ethanol ingestion of fetal breathing movements, gross fetal body movements, and heart rate at 37-40 weeks gestation, *Am. J. Obstet. Gynecol.*, 145, 251, 1983.
121. **McLeod, W., Brien, J., Carmichael, L., Probert, C., Steenart, N., and Patrick, J.**, Maternal glucose injections do not alter the suppression of fetal breathing following ethanol administration, *Am. J. Obstet. Gynecol.*, 148, 634, 1983.
122. **Patrick, J., Richardson, B., Hasen, G., Clarke, D., Wlodek, M., Bonsquet, J., and Brien, J.**, Effects of maternal ethanol infusion on fetal cardiovascular and brain activity in lambs, *Am. J. Obstet. Gynecol.*, 151, 859, 1985.
123. **Gennser, G. and Marsal, K.**, Influence of smoking on intra-uterine fetal breathing in man, *Acta Obstet. Gynecol. Scand. Suppl.*, 47, 27, 1974.
124. **Manning, F., Wyn-Pugh, E., and Boddy, K.**, Effect of cigarette smoking on fetal breathing movements in normal pregnancies, *Br. Med. J.*, 1, 552, 1975.
125. **Thaler, I., Goodman, J. D. S., and Dawes, G. S.**, Effects of maternal cigarette smoking on fetal breathing and fetal movements, *Am. J. Obstet. Gynecol.*, 138, 282, 1980.
126. **Manning, F., Walker, D., and Feyerabend, C.**, The effect of nicotine on fetal breathing movements in conscious pregnant ewes, *Obstet. Gynecol.*, 52, 563, 1978.
127. **Sindberg Eriksen, P. and Gennser, G.**, Acute responses to maternal smoking of the pulsatile movements in fetal aorta, *Acta Obstet. Gynecol. Scand.*, 63, 647, 1984.
128. **Condorelli, S. and Scarpelli, E. M.**, Fetal breathing: induction in utero and effects of vagotomy and barbiturates, *J. Pediatr.*, 88, 94, 1976.
129. **Gennser, G., Marsal, K., and Lindstrom, K.**, Influence of external factors on breathing movements in the human fetus, in *Proc. 5th Eur. Congr. Perinatal Medicine*, Rooth, G. and Bratteby, L. E., Eds., Almqvist & Wiskell, Uppsala, Sweden, 1976, 181.
130. **Boddy, K.**, The influence of maternal drug administration on human fetal breathing movements in utero, in *Therapeutic Problems in Pregnancy*, Lewis, P. J. Ed., MTP Press, Lancaster, England, 1977, 153.
131. **Piercy, W. N., Day, M. A., Neims, A. H., and Williams, R. L.**, Alteration of ovine respiratory-like activity by diazepam, caffeine and doxapram, *Am. J. Obstet. Gynecol.*, 127, 43, 1977.
132. **Worthington, D., Piercy, W. N., and Smith, B. T.**, Modification of ovine fetal respiratory-like activity by chronic diazepam administration, *Am. J. Obstet. Gynecol.*, 131, 749, 1978.
133. **Hogg, M. I., Golding, R. H., and Rosen, M.**, The effect of doxapram on fetal breathing in sheep, *Br. J. Obstet. Gynaecol.*, 84, 48, 1977.
133a. **Brown, E. R., Lawson, E. E., Jansen, A., Chernick, V., and Taeusch, H. W.**, Regular fetal breathing induced by pilocarpine infusion in the near term fetal lamb, *J. Appl. Physiol.*, 50, 1348, 1981.
134. **Johnston, B. M., Walker, D. W., and Green, A. R.**, Lack of long term effect of p-chlorophenylalanine on brain 5-hydroxytryptamine and electrocortical activity in conscious fetal sheep, *Experientia*, 40, 291, 1984.
135. **Boddy, K.**, Fetal circulation and breathing movements, in *Fetal Physiology and Medicine*, Beard, R. W. and Nathanielsz, P. W., Eds., W.B. Saunders, London, 1976, 302.
135a. **Trudinger, B. J., Lewis, P. J., and Petit, B.**, Fetal breathing patterns in intrauterine growth retardation, *Br. J. Obstet. Gynaecol.*, 86, 432, 1979.
136. **Dornan, J. C. and Ritchie, J. W. K.**, Fetal breathing movements and maternal hyperoxia in the growth retarded fetus, *Br. J. Obstet. Gynaecol.*, 90, 210, 1983.
137. **Maloney, J. E., Bowes, G., Brodecky, V., Dennett, X., Wilkinson, M., and Walker, A.**, Function of the future respiratory system in the growth retarded fetal sheep, *J. Dev. Physiol.*, 4, 279, 1982.
138. **Worthington, D., Piercy, W. N., and Smith, B. T.**, Effects of reduction of placental size in sheep, *Obstet. Gynecol.*, 58, 215, 1981.

139. **Dornan, J. C., Ritchie, J. K. W., and Meban, C.**, Fetal breathing movements and lung maturation in the congenitally abnormal human fetus, *J. Dev. Physiol.*, 6, 367, 1985.
140. **Alcorn, D., Adamson, T. M., Maloney, J. E., and Robinson, P. M.**, Morphological effects of either chronic bilateral phrenectomy or vagotomy in the fetal lamb lung, *J. Anat.*, 130, 683, 1980.
141. **de Lorimer, A. A., Tierney, O. F., and Parker, H. R.**, Hypoplastic lungs in fetal lambs with surgically produced congenital diaphragmatic hernia, *Surgery*, 62, 12, 1967.
142. **Liggins, G. C., Campos, G. A., Kitterman, J. A., and Lee, C. H.**, The effects of bilateral thoracoplasty on lung development in fetal sheep, *J. Dev. Physiol.*, 3, 275, 1981.
143. **Kaplan, M.**, Fetal breathing movements. An update for the pediatrician, *Am. J. Dis. Child.*, 137, 177, 1983.
144. **Patrick, J. and Richardson, B.**, Clinical significance of fetal breathing and other movements, in *The Physiological Development of the Fetus and Newborn*, Jones, C. T. and Nathanielsz, P. W., Eds., Academic Press, London, 1985, 669.
145. **Manning, F. A., Platt, L. D., and Sipos, L.**, Antepartum fetal evaluation: development of a biophysical profile, *Am. J. Obstet. Gynecol.*, 136, 787, 1980.

THE DEVELOPMENT OF THE SURFACTANT SYSTEM IN THE LUNG

Terence E. Nicholas

INTRODUCTION

The gas-filled alveolus is at least partially[1] lined with a substance, known as pulmonary surfactant (S), which reduces surface tension.[2] Although most accept that an adequate amount of functional S is essential for a viable lung, its precise role is debated.[2] The S-system is the last fetal system to develop that is vital for extrauterine survival, and the ability of the neonatal lung to supply adequate S to the gas-liquid interface during the first breaths is critical for the fetal-neonatal transition. This transition involves both the reabsorption of fetal lung fluid and establishment of a new fluid balance, and the stabilization of the neonatal lung. Therefore, the fetal lung must be able to supply adequate S for the transition, and then have sufficient reserves to service a rapid turnover of alveolar S. Neonatal Respiratory Distress Syndrome (RDS) is associated with a reduced amount of tissue and alveolar S.

The aim of this short review is to outline our present knowledge of the development of the S-system. Due to space constraints, the references quoted are not always the first published, but rather the most informative available, and are often other reviews. Readers are referred to excellent recent reviews for fuller accounts of this huge, rapidly advancing area.[2-6]

SURFACTANT IN THE MATURE LUNG

In order to appreciate the fetal development of a system, it is necessary to understand the mature system.

Overall Turnover

Based primarily on a single autoradiographic study,[9] the picture has emerged that S is synthesized within the endoplasmic reticulum of the alveolar type II cell, modified in the Golgi apparatus and then stored dehydrated in vesicles known as lamellar inclusion bodies (lb). Unique proteins appear to be involved in the intracellular transport of S.[2] In response to a poorly understood signal(s), the lb are extruded into the aqueous hypophase lining the alveolar compartment, where they are converted to a square-lattice form of S, known as tubular myelin.[1] This liquid-crystal form may act as a reservoir for the monolayer of S at the interface, where it modifies surface tension and stabilizes the structure and fluid balance of the alveolus. The concentration of calcium ions appears crucial in determining the form of S in the hypophase. Alveolar S is "inactivated" and as much as 90% may be taken back into the alveolar type II cell, where some may be reutilized as whole molecules. The meaning of "inactivation" is far from clear. Likewise, we do not know whether or not reuptake only involves "inactivated" S. Finally, we have no indication of what controls the reuptake. If the process obeys first-order kinetics, then the more S in the alveolar compartment, the faster the reuptake. Certainly the rate of removal following a rapid increase in alveolar S does appear to follow first-order kinetics (unpublished observations).

Composition

Mature S is a complex mixture with the approximate composition:[7] phospholipids (77%), neutral lipids (glycerides, cholesterol, free fatty acids; (13%), protein (8%), and carbohydrates (2%). The phospholipids comprise: phosphatidylcholine (PC; 75%, 50% saturated), phosphatidylethanolamine (PE; 5%), phosphatidylglycerol (PG; 11%), phosphatidylinositol

(PI; 3%), phosphatidylserine (PS; 2%), and sphyngomyelin (Sph; 2%). Although the same components appear in all mammalian S, the relative amounts reported vary markedly,[6,7] possibly due to the methods used. The chemical nature of S is not fully elucidated: this is due both to problems in isolating pure S, and because not all compounds isolated are identified. This problem may be resolved with further evolution of HPLC methods. Finally, the complexity is compounded by each lipid having its own distinct complement of fatty acids.

Physicochemical Nature

Little is known of the interactions between, or the importance of, the different components of S. Rather than a true lipoprotein complex, S may be a loose collection of lipids and proteins that modify the behavior of dipalmitoylphosphatidylcholine (DPPC), the principal phospholipid. DPPC imbues S with the ability to lower surface tension of a monolayer towards zero during compression. However, at 37°C DPPC is in the solid state and can spread only very slowly at an interface. Other components such as cholesterol, free fatty acids, and unsaturated phospholipids render it fluid. Although there are unusually high concentrations of PG present in mature S, its ability to modify the behavior of DPPC is probably related to its complement of unsaturated fatty acids.

Two principal nonserum proteins are found in alveolar S under reducing conditions: one of 35 kDa and one of 10 kDa.[8] The former is found in all species examined and can be further defined into three to five proteins. There is some unconfirmed evidence that the 35- and 10-kDa apoproteins bear a precursor-product relationship. In turn, it is possible that they are derived from a 250-kDa protein found both in amniotic fluid and in alveolar wash of patients with proteinosis. These apoproteins may be important in maintaining the structure of lb and tubular myelin and in the spreading and reuptake of alveolar S. Although many apparently different apoproteins have been associated with S, it is not clear whether or not different laboratories are dealing with the same proteins. Until we know the precise composition of the alveolar hypophase and whether or not it is constant, we will not be able to describe accurately the interactions between the extruded lb, tubular myelin, and monomolecular alveolar S.

BIOSYNTHESIS OF SURFACTANT-TYPE PHOSPHOLIPIDS

PC is mainly synthesized via the cytidine diphosphocholine (CDP)-choline pathway, and PG and PI are synthesized via the CDP-diacylglycerol path.[2-6] There are basically three steps in the CDP-choline process: the supply and esterification of fatty acids, the supply and modification of the base units, and the deacylation and reacylation of β-fatty acids. Dihydroxyacetone phosphate, an intermediate in glycolysis, is converted principally to acylhydroxyacetone phosphate and 1-acylglycerol 3-phosphate,[10] which is then acylated to phosphatidic acid. This is dephosphorylated to diacylglycerol. Choline is incorporated as follows: the activated form of choline, CDP-choline, is formed by choline kinase converting choline to choline phosphate. Choline phosphate cytidylytransferase then catalyzes the combination of choline phosphate and cytidine triphosphate (CTP) to form CDP-choline. This in turn contributes phosphorylcholine, which together with the diacylglycerol and the enzyme choline phosotransferase, forms PC. Possibly as little as 25% DPPC is synthesized *de novo*. More often, a β-unsaturated fatty acid is removed by phospholipase A_2, and the position is reacylated with a saturated fatty acid, usually palmitate, via the action of lysophosphatidylcholine acyltransferase.

ALVEOLAR EPITHELIUM IN THE FETUS

The mature epithelium comprises two cell types. Alveolar type I cells occupy 97% of the surface and mediate gas exchange. Each cell has multiple, attenuated cytoplasmic plates which may line adjacent alveoli. The cells have few cytoplasmic inclusions, are incapable of mitotic division, and are very susceptible to damage. In contrast, the alveolar type II cell is a squat cell that occupies about 3% of the surface, and is sited primarily in the "corners" of the alveolus. It is characterized by apical microvilli, a large nucleus, and abundant endoplasmic reticulum, Golgi apparatus, and mitochondria. It also contains multivesicular bodies and many lb. The lb first appear during the canalicular stage of fetal development, that is, at 80 to 85% of gestation in most mammals, but at 60% in humans. The type II cell is the progenitor cell of the type I cells, and, in the fetal lung, lb actually appear prior to the attenuation of the cytoplasm. The fetal type II cells contain large amounts of glycogen which disappear as the lb become more abundant.

APPEARANCE OF SURFACTANT IN THE FETUS

The apoproteins first appear in human amniotic fluid at 26 weeks gestation and then increase markedly to term.[11] This increase appears to parallel that of the phospholipids, although the phospholipid to apoprotein ratio is not constant.[12] It is also possible that the relative amounts of the different apoproteins vary during late gestation.

There are dramatic changes in the amount and composition of fetal S during late gestation, and these are subsequently reflected in amniotic fluid. The changes appear to be qualitatively similar across species. In the rabbit, there is a 400% increase in the absolute amount of lavagable phospholipid between day 27 and term (day 31), which includes a tenfold increase in saturated and unsaturated PC,[12] but little change in Sph. There is a resultant 12-fold increase in the PC/Sph ratio. In terms of relative percentages, PC increases dramatically over the final 5 d of gestation, while Sph falls from 38 to 7%. There are slight decreases in the proportions of PI + PS and in PE, but little, if any, change in the percentage of PG.[12] PS formation appears to precede that of PI and PG in fetal lung. In fact, a reciprocal relationship exists between PG and PI in both the alveolar wash and lb during this last part of gestation. The PG/PI ratio and the total amount of acidic phospholipids increases.[13] Little is known of the neutral lipids, although the relative cholesterol concentration in fetal lung tissue is only 25% of that in the mature lung.[7] In the monkey, the appearance of DPPC correlates closely with the appearance of surface-active material.[6] Much less is known of the relative changes in the human, where most of our information comes from changes reflected in the amniotic fluid. These include an increase in PC and in the PG/PI ratio, a fall in the percentage of Sph and PS, and little change in PG and PE between 31 weeks and term. Although the presence of appreciable amounts of PG appears to indicate fetal lung maturity, PG per se does not appear to be essential, and can be replaced by PI without compromising the ability to lower surface tension.

In rabbits and rats, the proportion of 2-palmitate and 2-palmitoleoyl in PC increases greatly, while that of 2-oleoyl decreases. In PE there is an increase in docosahexaenoate, but no other changes.[14] Less PG is saturated than is PC, but the amount increases toward term. In tracheal and pharyngeal aspirates from premature infants recovering from RDS, there is an increase in the percent of palmitic and palmitoleic and a decrease in stearic, oleic, linoleic, and arachidonic acid in PC over the first 14 d after delivery.[15] In conclusion, as term approaches there is a shift to saturated fatty acids, particularly palmitate.

Source of Precursors

Little is known of pool sizes of phospholipid precursors in the lung. The fetal lung can readily utilize glucose for the glycerol backbone, fatty acid synthesis, and possibly myo-inositol synthesis. In some species, glycogen may be an important source of glucose. Certainly in the rabbit fetus, glycogen stores decrease prior to the increase in PC synthesis.[3]

Fatty acids could be derived from free fatty acids, although plasma levels are low.[6] The fetal lung is rich in β-lipoprotein lipase, and the fatty acids could also be derived from circulating triglycerides; however, the β-lipoprotein levels are also low in fetal plasma.[6] Probably the most important source in the fetal lung is *de novo* synthesis using glucose or lactate as precursors.[6] *De novo* synthesis is inhibited by palmitate, which may explain why this source becomes of less importance as plasma fatty acids increase in the neonate.[6] Base-exchange reactions are minimal, but may be important in the synthesis of PS.[16] Choline is probably derived from the diet, and its incorporation into lung phospholipids increases dramatically during the last part of gestation.[3] Much of this work has been done using lung minces or slices under optimal enzyme conditions. In fact, the conditions may not be optimal *in situ*, so that the *in vitro* findings may not reflect the true situation. Certainly, conditions are hypoxic in the fetal lung.

CONTROL OF TURNOVER OF SURFACTANT

The amount of S in lb and alveolus depends on the rates of synthesis, secretion, and removal. A study of the kinetics of the process is complicated, because (1) the different pools of S almost certainly do not act as well-mixed compartments; (2) it is impossible to administer a pulse of radiolabeled precursor; and (3) in the neonatal lung, much of the alveolar S is reutilized. Recent reports show that turnover is very rapid. Using an approach derived from Zilversmit et al., Jacobs[17] has calculated that the turnover time of S-phospholipids in 3-d-old rabbits is less than 4 h. However, it is very difficult to apply these figures to the fetal lung, where S is certainly released, but the majority may be removed via the lung fluid. Furthermore, if the turnover of S in the neonatal lung is controlled by a closed-loop negative feedback system as some propose, and the controlled variable is some property of the hypophase, then the loop will be open in the fluid-filled fetal lung. The extruded lb, rather than forming tubular myelin, will be removed via the lung fluid and appear in the amniotic fluid.[18] In the neonatal lung, release is stimulated by ventilation.[19] In the adult rat lung, the release is directly to the tidal volume[20] and may well be proportional to distortion of the type II cells and the level of cyclic AMP.[21] Possibly, then, fetal breathing releases S. This seems unlikely, for repeated lavage of the adult rat lung to about 80% of total lung capacity with saline at 37°C results in an amount of S recovered comparable to that recovered if lavaged at 2 or 22°C (unpublished results); in the latter cases release would be inhibited. Possibly there is less distortion in the absence of the gas-liquid interface; certainly, septal pleats are not evident in fluid-filled lungs.[22] The importance of distortion will not be determined until morphometric measurements are made on the type II cells during various degrees of inflation both in the fluid- and the gas-filled lung. The only direct way to test the effect of distortion will be to plate isolated type II cells onto a flexible backing and test the effect of flexing. In conclusion, S release in the fetal lung is unlikely to be due to the very small tidal volumes during fetal breathing (refer to the section titled Role of the Autonomic Nervous System).

The Rate-limiting Step in Surfactant Synthesis

There is now good evidence that cholinephosphate cytidylyltransferase (CCT) catalyzes the rate-limiting step in *de novo* PC synthesis in the fetal lung.[23] CCT increases either during late fetal development or after birth, and this parallels choline incorporation into PC.[3] There are similar developmental increases in activity of choline phosphotransferase and phosphatidate phosphatase, but it is very unlikely that these enzymes are rate limiting. CCT is present in microsomal and cytosolic forms. However, although earlier evidence suggested that the latter represented an inactive form and required translocation to the microsomal fraction for activity, it now appears that in the lung, the cytosolic form is directly activated by stimuli including lipids, fatty acids, glucocorticoids, thyroid hormones, and estrogen.[4,24] Possibly, activation of phospholipases in the fetal type II cell changes the level of free fatty acids, leading to the activation of CCT.[25]

Role of the Autonomic Nervous System

β-Sympathomimetics release S and stimulate synthesis in isolated type II cells; both actions can be blocked by propranolol.[26,27] Although type II cells have β-receptors which increase in number near term,[28] there is no evidence of a sympathetic innervation. Circulating catecholamines increase during late gestation and increase further during labor.[3] However, the ventilation-induced release of S in the adult isolated perfused lung is not inhibited by propranolol,[20] suggesting that β-adrenoreceptors are not involved in this response. Likewise, propranolol did not alter release in the neonatal rat lung.[29] Therefore, it is possible that there are either two pools of S in a type II cell, or two types of type II cells, one of which possesses β-receptors.[30]

The role of the parasympathetic system is even less clear. Reports that pilocarpine releases S must be viewed with caution, as the doses used were huge (150 to 250 mg/kg),[31,32] and symptoms of pilocarpine poisoning were reported. Likewise, sectioning the vagus nerve in the adult animal has such widespread effects, including those on pattern of respiration, that it is impossible to ascribe any subsequent change in S to variation in parasympathetic input. Furthermore, cholinergic receptors are not found on isolated type II cells, and muscarinic agents had no effect on either synthesis or release of S.[26,27] Reports that atropine inhibits release of S in neonatal lung must also be evaluated in terms of the possible effect of the large doses used on pattern of respiration, and hence only indirectly on release.[19] In some cases when radioactivity alone is used as an index of release, it is difficult to distinguish between an increase in synthesis or release. Finally, it is possible that pilocarpine is acting via the adrenal medulla and catecholamines.[32]

The release of S in response to increased tidal volume was not altered by indomethacin, cyproheptadine, or tetrodotoxin,[20] suggesting that, at least in the isolated perfused rat lung, prostaglandins, histamine, serotonin, or an intrapulmonary neural reflex are not involved. However, there is some evidence that prostaglandins are involved in the adult rabbit,[33] and recently, that a product of the lipoxygenase pathway releases S in isolated type II cells.[34]

In summary, there is general agreement that β-agonists stimulate synthesis and release of S. However, the amount released appears small compared to that released by increasing tidal volume; hence the large amount of S released at birth is probably via a different mechanism. The apparent slow, steady release of S in the fluid-filled fetal lung may well be mediated by β-receptors and circulating epinephrine. The roles of the parasympathetic system and prostaglandins are more contentious.

Very little is known of the fetal development of the S release mechanism. Even in very prematurely delivered animals, air breathing releases S, leaving little in tissue stores. At this stage there is no evidence that RDS is associated with an immature release mechanism. The depleted stores may reflect inadequate synthesis or insufficient reuptake and reutilization of alveolar S. No information is available regarding these systems. Recently it was reported that the β-adrenergic control of S release does not appear in the rabbit lung until 90% of gestation.[35] However, these results are difficult to interpret because the lungs were serially lavaged at room temperature, using saline containing 1 mM propranolol. The link between rate of release and synthesis is unclear. The fact that in isolated type II cells β-agonists appear to stimulate both independently[26] does not mean that in the intact lung, with normal cell-to-cell contact, release and synthesis are independent.

Effect of Hormones

The ability of estrogen to increase the activity of CCT may be mediated via an effect on phospholipids, rather than by an actual increase in enzyme; the amount of cytosolic lipids does increase after birth. There was no evidence that the activation was associated with translocation to the microsomal fraction. In fact, there is no convincing proof that estrogens have a physiological role in fetal S synthesis.

Glucocorticoids stimulate S synthesis in the fetus, which results in an increase in the PC to Sph ratio in amniotic fluid. The plasma levels of cortisol and cortisone increase in parallel with the increasing incorporation of choline into DPPC, suggesting a physiological role, although the concentration of free steroid may not increase. Interestingly, the incorporation of choline into lung phospholipids also parallels the ability of the fetal rabbit lung to convert inactive cortisone to cortisol,[17] so the lung itself may titrate the amount of cortisol present. Glucocorticoids may act both on specific receptors in type II cells and via the release of a peptide from fibroblasts (fibroblast pneumocyte factor, FPF), which in turn acts on the type II cell. FPF stimulates CCT. In cultured fetal type II cells, FPF stimulates the formation of DPPC and PG, but had no apparent effect on other phospholipids.[37] This property of cortisol to stimulate DPPC synthesis, via release of FPF, is blocked by dihydrotestosterone,[38] which may explain why female fetuses produce more DPPC than males, and ultimately why the incidence of RDS is greater in the latter. Whether or not FPF mediates other effects of glucocorticoids on fetal lung development is not known.

There is an increase in thyroid hormones in fetal plasma during the final trimester, and this is accompanied by an increase in thyroid hormone receptors in lung.[3] In cultured fetal rabbit lung, triiodothyronine stimulates PC synthesis via an action on nuclear receptors, inducing RNA and protein synthesis. Moreover, the effect is potentiated by dexamethasone. Thyroid hormones have also been reported to increase the sensitivity of type II cells to FPF. However, is is still not established that thyroid hormones have a physiological role in fetal S synthesis.[3]

Lung maturation is delayed in fetuses of diabetic mothers. Insulin appears to antagonize the action of glucocorticoids and may also slow the mobilization of glycogen. There is also a report that maternal diabetes mellitus delays development of enzymes involved in lung phospholipid synthesis.[3]

Precursor Control of Synthesis

An elevated concentration of precursor could affect the rate of synthesis if the step is rate limiting and is under first-order kinetics. It is also possible that it may induce an enzyme. There is some evidence that CDP-choline can stimulate PC synthesis.[4]

CONCLUSIONS

Work over the past 15 years has greatly increased our understanding of S homeostasis in both the fetal and the mature lung. The chemical nature of S is largely understood, and the type II cell has been identified as the site of synthesis, storage, and release. The rate-limiting step in synthesis of PC has been identified and the actions of many hormones on the enzyme involved, CCT, have been described. The rate of turnover is markedly faster than previously appreciated. Although we appreciate the importance of the different components and electrolytes in determining the physicochemical properties of S, our knowledge of the interactions between lb, tubular myelin, and monomolecular forms of S *in vivo* is hampered by our ignorance of the composition of the hypophase. There are many gaps in our knowledge of control of S-homeostasis and whether or not a negative feedback loop exists. The relationship between synthesis and release is unclear. Although conflicting reports exist regarding the control of release, some of these can be explained in terms of both the huge doses of drugs used and the serially lavaging of lungs at room temperature long after the death of the animal.

REFERENCES

1. **Gil, J.**, Histological preservation and ultrastructure of alveolar surfactant, *Annu. Rev. Physiol.*, 47, 753, 1985.
2. **King, R. J.**, Pulmonary surfactant, *J. Appl. Physiol.*, 53, 1, 1982.
3. **Rooney, S. A.**, The surfactant system and lung phospholipid biochemistry, *Am. Rev. Respir. Dis.*, 131, 439, 1985.
4. **Farrell, P. M.**, General features of phospholipid metabolism in the developing lung, in *Lung Development: Biological and Chemical Perspectives*, Vol. 1, Farrell, P. M., Ed., Academic Press, New York, 1982, 223.
5. **Perelman, R. H., Farrell, P. M., Engle, M. J., and Kennedy, J. W.**, Developmental aspects of lung lipids, *Annu. Rev. Physiol.*, 47, 803, 1985.
6. **Possmayer, F.**, Biochemistry of pulmonary surfactant during fetal development and in the perinatal period, in *Pulmonary Surfactant*, Robertson, B., Van Golde, L. M. G., and Batenburg, J. J., Eds., Elsevier Science, Amsterdam, 1984, 295.
7. **Sanders, R. L.**, The composition of pulmonary surfactant in lung development, in *Lung Development: Biological and Chemical Perspectives*, Vol. 1, Farrell, P. M., Ed., Academic Press, New York, 1982, 223.
8. **King, R. J.**, Composition and metabolism of the apolipoproteins of pulmonary surfactant, *Annu. Rev. Physiol.*, 47, 775, 1985.
9. **Chevalier, G. and Collet, A. J.**, In vivo incorporation of choline-^{3}H, leucine-^{3}H and galactose-^{3}H in alveolar type II pneumocytes in relation to surfactant synthesis: a quantitative radioautoradiographic study in mouse by electron microscope, *Anat. Rec.*, 174, 289, 1972.
10. **Mason, R J.**, Importance of acyl dihydroxyacetone phosphate pathway in the synthesis of phosphatidylglycerol and phosphatidylcholine in alveolar type II cells, *J. Biol. Chem.*, 253, 3367, 1978.
11. **Gikas, E. G., King, R. J., Mescher, E. J., Platzker, A. C. G., and Kitterman, J. A.**, Radioimmunoassay of pulmonary surface-active material in the tracheal fluid of the fetal lamb, *Am. Rev. Respir. Dis.*, 115, 587, 1977.
12. **Rooney, S. A., Wai-Lee, T. S., Gobran, L., and Motoyama, E. K.**, Phospholipid content, composition and biosynthesis during fetal lung development, *Biochim. Biophys. Acta*, 431, 447, 1976.
13. **Hallman, M. and Gluck, L.**, Formation of acidic phospholipids in rabbit lung during perinatal development, *Pediatr. Res.*, 14, 1250, 1980.
14. **Soodsma, J. F., Mims, L. C., and Harrow, R. D.**, The analysis of the molecular species of fetal rabbit lung phosphatidylcholine by consecutive chromatographic techniques, *Biochim. Biophys. Acta*, 424, 159, 1976.
15. **Shelley, S. A., Kovacevic, M., Paciga, J. E., and Balis, J. U.**, Sequential changes of surfactant phosphatidylcholine in hyaline membrane disease of the newborn, *N. Engl. J. Med.*, 300, 112, 1979.
16. **Van Golde, L. M. G.**, Synthesis of surfactant lipids in the adult lung, *Annu. Rev. Physiol.*, 47, 765, 1985.
17. **Jacobs, H. C.**, Surfactant kinetics, *Semin. Perinatol.*, 8, 258, 1984.
18. **Duck-Chong, C. G.**, Lamellar body content of amniotic fluid: a possible index of fetal lung maturity, *Ann. Clin. Biochem.*, 16, 191, 1979.
19. **Lawson, E. E., Birdwell, R. L., Huang, P. S., and Taeusch, H. W.**, Augmentation of pulmonary surfactant secretion by lung expansion at birth, *Pediatr. Res.*, 13, 611, 1979.
20. **Nicholas, T. E. and Barr, H. A.**, Control of release of surfactant phospholipids in the isolated perfused rat lung, *J. Appl. Physiol.*, 51, 90, 1981.
21. **Nicholas, T. E. and Barr, H. A.**, The release of surfactant in rat lung by brief periods of hyperventilation, *Respir. Physiol.*, 52, 69, 1983.
22. **Gil, J., Bachofen, H., Gehr, P., and Weibel, E. R.**, Alveolar volume-surface area relationship in air- and saline-filled lungs fixed by vascular perfusion, *J. Appl. Physiol.*, 47, 990, 1979.
23. **Post, M., Batenburg, J. T., Schurrmans, E. A., and Van Golde, L. M. G.**, The rate-limiting step in the biosynthesis of phosphatidylcholine by alveolar type II cells from adult rat lung, *Biochim. Biophys. Acta*, 712, 390, 1982.
24. **Chu, A. J. and Rooney, S. A.**, Developmental differences in activation of cholinephosphate cytidyltransferase by lipids in rabbit lung cytosol, *Biochim. Biophys. Acta*, 835, 132, 1985.
25. **Aeberhard, E. E., Barrett, C. T., Kaplan, S. A., and Scott, M. L.**, Regulation of phospholipid synthesis by intracellular phospholipases in fetal rabbit type II pneumocytes, *Biochim. Biophys. Acta*, 833, 473, 1985.
26. **Mettler, N. R., Gray, M. E., Schuffman, S., and LeQuire, U. S.**, β-Adrenergic induced synthesis and secretion of phosphatidylcholine by isolated pulmonary alveolar type II cells, *Lab. Invest.*, 45, 575, 1981.
27. **Dobbs, L. G. and Mason, R. J.**, Pulmonary alveolar type II cells isolated from rats: release of phosphatidylcholine in response to β-adrenergic stimulation, *J. Clin. Invest.*, 63, 378, 1979.

28. **Cheng, J. B., Goldfien, A., Ballard, P. L., and Roberts, J. M.**, Glucocorticoids increase pulmonary β-adrenergic receptors in fetal lung, *Endocrinology,* 107, 1646, 1980.
29. **Nijjar, M. S.**, Ventilation-induced release of phosphatidylcholine from neonatal-rat lungs in vitro, *Biochem. J.,* 221, 577, 1984.
30. **Nicholas, T. E., Power, J. H. T., and Barr, H. A.**, Surfactant homeostasis in the rat lung during swimming exercise, *J. Appl. Physiol.,* 53, 1521, 1982.
31. **Goldenberg, V. E., Buckingham, S., and Sommers, S. C.**, Pilocarpine stimulation of granular pneumocyte secretion, *Lab. Invest.,* 20, 147, 1969.
32. **Corbet, A. J. S., Flax, P., and Rudolph, A. J.**, Reduced surface tension in lungs of fetal rabbits injected with pilocarpine, *J. Appl. Physiol.,* 41, 7, 1976.
33. **Clements, J. A., Oyarzun, M. J., and Baritussio, A.**, Secretion and clearance of lung sufactant: a brief review, *Prog. Respir. Res.,* 15, 20, 1981.
34. **Gilfillan, A. M. and Rooney, S. A.**, Arachidonic acid metabolites stimulate phosphatidylcholine secretion in primary cultures of type II pneumocytes, *Biochim. Biophys. Acta,* 833, 336, 1985.
35. **Corbet, A. J., Koline, H. W., Perreault, T., Frink, J. A., and Rudolph, A. J.**, Development of β-adrenergic control of phospholipid secretion in rabbit lung, *J. Appl. Physiol.,* 58, 2011, 1985.
36. **Nicholas, T. E., Johnson, R. G., Lugg, M. A., and Kim, P. A.**, Pulmonary phospholipid biosynthesis and the ability of the fetal rabbit lung to reduce cortisol during the final 10 days of gestation, *Life Sci.,* 22, 1517, 1984.
37. **Post, M. and Smith, B. T.**, Effect of fibroblast-pneumocyte factor on the synthesis of surfactant phospholipids in type II cells from fetal rat lung, *Biochim. Biophys. Acta,* 793, 297, 1985.
38. **Torday, J. S.**, Dihydrotestosterone inhibits fibroblast-pneumocyte factor-mediated synthesis of saturated phosphatidylcholine by fetal rat lung cells, *Biochim. Biophys. Acta,* 835, 23, 1985.

DEVELOPMENT IN LUNG MECHANICS

Claude Gaultier

INTRODUCTION

In human full-term newborns, lung development is incomplete at birth. However, in contrast to newborn mammals, such as rats,[1] the full-term newborn is already in the alveolar phase with a mean number of alveoli of 50×10^6.[2] The major change after birth concerns multiplication of alveoli which is, in principle, complete by the age of 2.[3] Changes in lung function accompany this developmental process. Unfortunately, these changes are not currently very well understood, and this is due mainly to technical reasons. Methodology rapidly runs up against limits in the study of neonates. Results of studies on lung mechanics done many years ago have to be reviewed in light of recent data. The present review briefly summarizes the current knowledge on development of lung mechanics in humans. Readers are referred to other reviews for a fuller account of this complex and multifaceted topic.[4,5]

CHEST WALL MECHANICS

Lung mechanics depend dramatically upon their environment, i.e., the chest wall. Rib cage configuration changes with growth.[6-9] In newborns, the rib cage is circular and the ribs horizontal. This mechanical arrangement seems inefficient. In newborns, the ribs are already "raised", and this may be one of the reasons why the motion of the rib cage contributes little to tidal volume. Moreover, the angle of insertion of the diaphragm is almost horizontal, which induces decreased efficiency of diaphragmatic contraction. This mechanical disadvantage is worsened in the supine posture, which is the usual nursing position of babies in most modern societies. With age, the shape of the thorax becomes ovoid and the ribs oblique. Concurrently, there is progressive mineralization of the ribs. These changes are of major importance in improving the stiffness of the rib cage.

High chest wall compliance (C_W) is an inherent characteristic of newborn mammals.[10] In human newborns, most current information is derived from measurements during anesthesia and muscle paralysis,[11-13] which are eminently nonphysiological. Further, the measurement of C_W is a technically challenging enterprise. It is therefore not surprising that reported values of C_W differ greatly among authors. Gerhardt and Bancalari[14] reported of C_W of 4.2 ml/cm of H_2O per kilogram, while Reynolds and Esten[12] reported a value of 6 ml/cm of H_2O per kilogram. In addition to methodological problems, part of this difference is almost certainly due to difficulties in normalizing for growth. Normalizing by body weight in infants is potentially misleading. Although in animals C_W decreases rapidly with growth,[10] a similar decrease in C_W in humans during infancy cannot be affirmed because of the lack of sufficient measurements. The uncertainties about the way in which C_W changes with age are regrettable, since chest wall compliance has a major influence on both the maintenance of the end-expiratory lung volume and on the stabilization of ventilation.

The pliable rib cage of newborns and infants is easily deformed by the effect of diaphragmatic contraction whenever the stabilizing effect of the intercostal muscles is inhibited, such as during rapid eye movement (REM) sleep.[15] Paradoxical inward motion of the rib cage (PIRC), i.e., rib cage distortion, occurs throughout all REM sleep in newborns.[16] In the distortion periods, the end-expiratory lung volume decreases.[17] and the efficiency of the diaphragm in generating pressure diminishes.[18] In infants, periods of rib cage distortion during REM sleep shorten with age.[19] However, the age at which PIRC motion disappears entirely is unknown.

FUNCTIONAL RESIDUAL CAPACITY

The functional residual capacity (FRC) is defined as the static-passive balance of forces between the lung and the chest wall. At birth, the recoil pressure at passive FRC is close to atmospheric because of the highly compliant chest wall.[20] As a result of this, the FRC predicted from the passive balance of the inward recoil of the lung and the outward recoil of the chest wall would be about 15% of the total lung capacity (TLC). Such a small FRC seems incompatible with stability of the terminal airways or adequate gas exchange. From the measured FRC[21] in infants and estimated TLC,[21] the dynamic FRC/TLC ratio appears to be about 50%. Thus, there are compelling reasons to believe that the dynamic end-expiratory lung volume is substantially above the passively determined FRC. Indeed, during apnea the end-expiratory lung volume, i.e., the dynamic FRC, falls to a substantially lower level, which is presumably the passive FRC. In newborns, expiration terminates at substantial flow rates, suggesting an active interruption of relaxed respiration.[22] How this active end-expiratory position is maintained is still a matter of debate. It has been suggested that the newborn may have a relatively long time constant with respect to expiratory time; and there might not be enough time to expire down to passive FRC. However, recent measurements of the time constant do not favor this hypothesis. Two other mechanisms have been proposed that could elevate the end-expiratory lung volume: laryngeal braking[23] and postinspiratory diaphragmatic activity.[24,25] Harding et al.[23] have shown in the lamb that adductor muscles of the larynx act as an expiratory brake in the awake state and in non-REM sleep. In the immediate postnatal period, glottic closure appears to be a very important mechanism for establishing FRC. Respiration at this time is quite irregular and is often interrupted abruptly during expiration.[22] On the other hand, it has been suggested that postinspiratory diaphragmatic activity holds the lungs at a higher active end-expiratory lung volume in the awake state and in non-REM sleep.[24] In any case, both mechanisms are switched off in REM sleep, which is the predominant behavioral state of the newborn. Accordingly, Henderson-Smart and Read[17] have shown that the end-expiratory lung volume measured by plethysmography decreased by approximately 30% in newborns going from non-REM to REM sleep. The age at which the FRC is no longer determined dynamically, but rather by the static passive balance of forces, is not known.

Up to now, most studies of FRC measurements in newborns and infants have not taken into account the difference between REM and non-REM sleep during testing. Furthermore, the methods used to measure "*in vivo*" FRC have been recently criticized. FRC can be measured by the helium dilution technique or by plethysmography. When using the helium dilution technique, it has been shown that mixtures with high F_IO_2 have to be eliminated since they induce lower FRC.[26] On the other hand, the plethysmographic technique also has recently been criticized.[27,28] When occlusion is performed at the end of expiration, FRC is lower than when occlusion is performed at the end of inspiration because of inequalities of regional pleural pressure. A further problem is that normalization of FRC with age is still being debated. As suggested by Mead,[29] the best normalization factor for FRC would be dry lung weight, which is impossible to measure *in vivo*. Most *in vivo* studies normalize FRC by body weight or length, and such normalization is potentially misleading.

All these methodological problems explain the dispersion of previous published values of FRC in full-term newborns:[5] FRC per kilogram of body weight varies from 25.7 to 38.6 ml.[30,31] Furthermore, comparison of FRC measured by the helium dilution technique and by plethysmography shows greater values with plethysmography; this suggests the presence of trapped gases in newborns. However, such comparison is presumably erroneous because of methodological errors in both methods as well as the absence of assessment of sleep stages in the studies. In spite of all the methodological problems, it is interesting to observe that, in preterms of different gestational age[32] and in growing infants,[21] it is possible to find a significant relationship between increase in FRC and increase in alveolar number. This result suggests that, despite other influencing factors, FRC essentially depends on alveolar multiplication. Thus, decrease in FRC in pathological conditions may be interpreted as a consequence of impairment of lung growth.[5]

ELASTIC PROPERTIES OF THE LUNG

Little is known about the development of lung connective tissue. Emery[33] showed expansion of the elastic network with growth in peripheral pulmonary parenchyma. More recently, Keely et al.[34] reported that true elastin, expressed as a percentage of dry lung weight, increases after full-term birth to reach adult proportions at about 6 months of postnatal age.

Functional studies of the elastic properties of the lung concern postmortem static pressure-volume (P/V) curves and *in vivo* measurements of lung elastic recoil as well as lung compliance. Unfortunately, postmortem studies have been done at different distending pressures. Stigol et al.,[35] using a maximal pressure of 20 cm of H_2O, did not find any change in the overall shape of the P/V curves during infancy, whereas Fagan,[36] using a 30-cm of H_2O maximal pressure, observed a sharp drop in the proportion of volume retained at low pressure and then a shift to the right of the P/V curves during the first months of life. In addition, Fagan[20] found lung elastic recoil in infancy to be much lower than that observed in children and adults.

In vivo measurements of lung elastic recoil and lung compliance depend on the reliability of esophageal pressure. Firstly, the reliability of esophageal pressure is related to the characteristics of the balloon and the catheter used.[37] Secondly, the influence of chest wall distortion of pleural pressure which varies with changes in chest wall distortion.[38] Thus, all previous studies of esophageal pressure have to be reviewed in light of these methodological problems. *In vivo* measurements of lung elastic recoil suggest low values at FRC, ranging from -0.7 to -2.65 cm of H_2O.[39-41] Measurement of lung compliance suggests that static lung compliance is higher than dynamic lung compliance (C_L dyn)[42] and that C_L dyn is frequency dependent.[42] Repeated measurements of C_L dyn during the first days of life show an increase in C_L dyn, which may be explained by absorption of lung fluid.[43] Estimation of C_W/C_L ratio shows a higher value in newborns (i.e., 18 ml/cm of H_2O over 4 ml/cm of H_2O) than in children, indicating that the lung may be deformed when chest wall distortion is present.

Because of the problems of reliability of esophageal pressure measurement, other techniques have recently been proposed to assess C_L. Since C_W is high in newborns, the compliance of the total respiratory system (C_{RS}) should be a correct assessment of C_L according to the following equation: $1/C_{RS} = 1/C_W + 1/C_L$. C_{RS} can be measured by the technique of the weighted spirometer[44] and by the occlusion technique performed at the end of inspiration[40] or, as proposed by Mortola et al.,[45] during expiration. C_{RS} increases during the first days of life.[45] C_{RS} in full-term newborns after day 1 ranges from 1 to 1.13 m·/cm H_2O/kg.[44,46,47] Passive C_{RS} measured by the occlusion technique is higher than active C_{RS}, suggesting that the active work of breathing is higher than the passive one, at least during the first days of life.[45] In sick, intubated babies, measurement of passive and active total respiratory system mechanics is of interest, although the methodological problems have not yet been fully resolved.

AIRWAY DYNAMICS

We summarize, here, present knowledge concerning only intrathoracic and not upper airway dynamics. From morphometric studies, it is known that conducting airways down to the terminal bronchioles are present at the sixteenth week of gestation, whereas respiratory bronchioles and alveolar ducts are present only at birth[48] or even appear afterwards.[2] During growth, airways increase in caliber. However, different results have been found concerning the proportional growth of central and peripheral airways. For Hislop et al.,[48] the length and diameter of each branch appear to grow proportionately and to maintain a constant relationship to the size of the whole airways, whereas Hogg et al.[49] reported that peripheral airways are disproportionately narrow, suggesting that they grow at delayed rate under 5 years of age.

Airway resistance and/or airway conductance (Gaw) have been assessed by plethysmographic measurement from birth to adolescence. Specific airway conductance (Gaw/FRC plethysmographic) is primarily a measurement of the central airways and is higher in infancy than in childhood.[50] Thus, growth of the lung is nonisotropic. Airways are present and relatively large in the newborn. Growth of lung volume occurs after birth and results in a dysanaptic process. For the same reasons, there is a fall with age in size-corrected flows and in specific upstream conductance.[41]

In unsedated infants, expiratory resistance is higher than inspiratory resistance.[51] This difference was not observed when infants were sedated, probably because of a marked reduction in thyroarytenoid muscle activity.[52] Lung tissue resistance appears to decrease with age.[53]

For the peripheral airways, conflicting results have been reported concerning dynamics. Hogg et al.[49] measured peripheral conductance per gram of lung tissue as a function of age and found a sharp rise in peripheral conductance around the age of 5 years. This finding was in agreement with their pathological discoveries reported above. Other functional studies, looking at maximal flows at FRC partial flow-volume curves (with rapid compression of the chest), did not support the finding that peripheral airways are disproportionately smaller in infants than in adults (when compared with central airways).[41] These problems have yet to be resolved.

Recent studies on dynamics of breathing in newborns, using the flow-volume curves of relaxed expiration after release of airway occlusion, have given values of the passive time constant of the total respiratory system equal to $0.208 \text{ s} + 0.089$ at 1 to 5 d of life.[45] This is close to that observed in adults. The active time constant at 1 to 5 d is lower than the passive one.[45]

MAXIMAL LUNG VOLUMES AND RESPIRATORY MUSCLES

Crying vital capacity (CVC) has been measured in preterm and full-term newborns and infants. Results show that CVC increases with postgestational[54] and postnatal age.[21] These changes are presumably related to improvement in respiratory muscle efficiency. Indeed, maximal transdiaphragmatic pressure, during airway occlusion at the time of crying, increases with postconceptional age. This suggests an improvement[55] in diaphragmatic strength with growth, not only because of increased muscle mass, but also because of changes in structure and compliance of the chest wall.

The diaphragmatic work of breathing has recently been measured in infants. Under conditions of a small rib cage distortion associated with non-REM sleep, it is equal to $56 + 36$ gr cm per breath, compared to $153 + 31$ gr cm per breath when there is maximal distortion of the chest wall in REM sleep.[56] The dramatic increase in diaphragmatic work may be associated with EMG diaphragmatic fatigue.[15] Fatigue is favored by the composition of diaphragmatic fibers, which are poor in resistive-fatigue fibers in preterms and newborns and only reach adult values at 1 year of age.[57] In neonatal disorders, the threshold of diaphragmatic fatigue is reached rapidly when resistive or elastic loads increase. This explains why newborns and infants progress so rapidly to respiratory failure.[58]

DISTRIBUTION OF VENTILATION AND GAS EXCHANGE

Arterial pressure of O_2 (PO_2) is around 70 mmHg on the first day of life.[59] The alveolo-arterial difference in PO_2 is about 30 mmHg during air breathing and 120 mmHg during O_2 breathing. Anatomic right-left shunt and low ventilation-perfusion ratio account almost equally for the total venous admixture. During the first months of life, awake capillary arterialized PO_2 increases dramatically,[60] suggesting a decrease in the inequalities of ventilation-perfusion ratios. In the neonatal period,[61] during chest wall distortion in REM sleep, transcutaneous PO_2 presumably falls because of the increasing inequalities associated with lung deformation. In the growing infant, no further decrease in transcutaneous PO_2 is observed when the chest wall is distorted.[19] This suggests improvement in ventilation distribution.

CONCLUSION

Despite two decades of considerable research, other studies are needed in order to fully describe the complexity of the development of lung mechanics in newborns during the first year of life, i.e., the period of active lung growth.[3] We know that newborns and infants risk respiratory failure when even minimal injury occurs. Furthermore, lung injury can induce abnormal lung development in terms of structure and function.[5] Abnormal lung development later leads to respiratory insufficiency in children and adults. However, further studies of lung function follow-up in infants with lung injury during the active period of growth are necessary to fully describe the dependence of lung function later in life on the early individual "lung growth story".

REFERENCES

1. **Burri, P. H., Dbaly, J., and Weibel, E. R.**, The postnatal growth of the rat lung. I. Morphometry, *Anat. Rec.*, 178, 711, 1974.
2. **Langston, C., Kida, K., Reed, M., and Thurlbeck, W.**, Human lung growth in late gestation and the neonate, *Am. Rev. Respir. Dis.*, 129, 607, 1984.
3. **Thurlbeck, W. M.**, Postnatal human lung growth, *Thorax*, 37, 564, 1982.
4. **Polgar, G. and Weng, T. R.**, The functional development of the respiratory system, *Am. Rev. Respir. Dis.*, 120, 620, 1979.
5. **Gaultier, C. and Girard, F.**, Normal and pathological lung growth structure-function relationships, *Bull. Eur. Physiopathol. Respir.*, 16, 791, 1980.
6. **Takahashi, E. and Atsumi, H.**, Age differences in thoracic form as indicated by the Thoracic Index, *Hum. Biol.*, 27, 65, 1955.
7. **Howatt, W. F. and de Muth, G. R.**, Growth of lung function III. Configuration of the chest, *Pediatrics*, 35, 177, 1965.
8. **Openshaw, P., Edwards, S., and Helms, P.**, Changes in rib cage geometry during childhood, *Thorax*, 39, 634, 1984.
9. **Bryan, A. C. and Gaultier, C.**, Chest wall mechanics in the newborn, in *The Thorax*, Part B, Roussos, C. and Macklem, P., Eds., Marcel Dekker, New York, 1986, 871.
10. **Agostoni, E.**, Volume-pressure relationships to the thorax and lung in the newborn, *J. Appl. Physiol.*, 14, 909, 1959.
11. **Richards, C. C. and Bachman, L.**, Lung and chest wall compliance of apneic paralyzed infants, *J. Clin. Invest.*, 40, 273, 1961.
12. **Reynolds, R. N. and Etsten, B. E.**, Mechanics of respiration in apneic anesthetized infants, *Anesthesiology*, 27, 13, 1966.
13. **Lunn, J. N.**, Measurement of compliance in apneic anesthetized infants, *Anesthesia*, 23, 175, 1968.
14. **Gerhardt, T. and Bancalari, E.**, Chest wall compliance in full-term and premature infants, *Acta Pediatr. Scand.*, 69, 359, 1980.

15. **Muller, N. L., Gulston, G., Cade, D., Whitton, J., Froese, A. B., Bryan, M. H., and Bryan, A. C.**, Effect of chest wall distortion on occlusion pressure and the preterm diaphragm, *J. Appl. Physiol.*, 55, 359, 1983.
16. **Curzi-Dascalova, L.**, Thoraco-abdominal respiratory correlations in infants: constancy and variability in different sleep states, *Early Hum. Dev.*, 2, 25, 1978.
17. **Henderson-Smart, D. J. and Read, D. J. C.**, Reduced lung volume during behavioral active sleep in the newborn, *J. Appl. Physiol.*, 46, 1081, 1979.
18. **Lesouef, P. N., Lopes, J. M., England, S. J., Bryan, M. H., and Bryan, A. C.**, Effect of chest wall distortion on occlusion pressure and the preterm diaphragm, *J. Appl. Physiol.*, 55, 359, 1983.
19. **Gaultier, C. L., Praud, J. P., Canet, E., Delaperche, M. F., and D'Allest, A. M.**, Paradoxical inward rib cage motion during REM sleep in infants and young children, submitted.
20. **Fagan, D.**, Functional development of the human lung, in *The Anatomy of the Developing Lung*, Emery, J., Ed., Levenham, Suffolk, 1969, 191.
21. **Gaultier, C., Boule, M., Allaire, Y., Clement, A., and Girard, F.**, Growth of lung volumes during the first three years of life, *Bull. Eur. Physiopathol. Respir.*, 15, 1103, 1979.
22. **Fisher, J. T., Mortola, J. P., Smith, J. B., Fox, G. S., and Weeks, S.**, Respiration in newborns, development of the control of breathing, *Am. Rev. Respir. Dis.*, 125, 650, 1982.
23. **Harding, R. D., Johnson, P., and McClelland, M. E.**, Respiratory function of the larynx in the developing sheep and the influence of sleep state, *Respir. Physiol.*, 40, 165, 1980.
24. **Muller, N. L., Volgyesi, G., Becker, L., Bryan, M. H., and Bryan, A. C.**, Diaphragmatic muscle tone, *J. Appl. Physiol.*, 47, 279, 1979.
25. **Griffiths, G. B., Noworaj, A., and Mortola, J. P.**, End-expiratory level and breathing pattern in the newborn, *J. Appl. Physiol.*, 55, 243, 1983.
26. **Geubelle, F., Francotte, N., Beyer, M., Louis, I., and Logvinoff, M. M.**, Functional residual capacity and thoracic gas volume in normoxic and hyperoxic newborns, *Acta Paediatr. Belg.*, 30, 221, 1977.
27. **Beardsmore, C. S., Stocks, J., and Silverman, M.**, Problems in measurement of thoracic gas volume in infancy, *J. Appl. Physiol.*, 52, 995, 1982.
28. **Helms, P.**, Problems with plethysmographic estimation of lung volume in infants and young children, *J. Appl. Physiol.*, 53, 698, 1982.
29. **Mead, J.**, Mechanical properties of lung, *Physiol. Rev.*, 41, 281, 1961.
30. **Geubelle, F., Karlberg, P., Koch, G., Lind, J., and Wallberg, G.**, L'aeration du poumon chez le nouveaune, *Biol. Neonate*, 1, 169, 1959.
31. **Doershuk, C. F., Fisher, B. J., and Mattews, L. W.**, Specific airway resistance from the perinatal period into adulthood, *Am. Rev. Respir. Dis.*, 109, 452, 1974.
32. **Lacourt, G. and Polgar, G.**, Development of pulmonary function in late gestation, *Acta Paediatr. Scand.*, 63, 81, 1974.
33. **Emery, J. L.**, Connective tissue and lymphatics, in *The Anatomy of the Developing Lung*, Emery, I., Ed., Levenhaum, Suffolk, 1969, 203.
34. **Kelly, F. W., Fagan, D. G., and Webster, S. I.**, Quantity and character of elastin in developing lung parenchymal tissues in normal infants and in infants with respiratory distress syndrome, *J. Lab. Clin. Med.*, 90, 981, 1977.
35. **Stigol, L. C., Vawter, G. F., and Mead, J.**, Studies on elastic recoil of the lung in a pediatric population, *Am. Rev. Respir. Dis.*, 105, 552, 1972.
36. **Fagan, D. G.**, Post-mortem studies of the semistatic volume-pressure characteristics of the infant's lungs, *Thorax*, 31, 534, 1976.
37. **Asher, M. I., Coates, A. L., Collinge, J. M., and Milic-Emili, J.**, Measurement of pleural pressures in neonates, *J. Appl. Physiol.*, 52, 491, 1982.
38. **Lesouef, P. N., Lopes, J. M., England, S. J., Bryan, M. H., and Bryan, A. C.**, Influence of chest wall on esophageal pressure, *J. Appl Physiol.*, 55, 353, 1983.
39. **Senterre, J. and Geubelle, F.**, Measurement of endo-oesophageal pressure in the newborn, *Biol. Neonate*, 16, 47, 1970.
40. **Helms, P., Beardsmore, C. S., and Stocks, J.**, Absolute intra-esophageal pressure at functional residual capacity in infants, *J. Appl. Physiol.*, 51, 270, 1981.
41. **Taussig, L. M., Landau, L. I., Godfrey, S., and Arad, I.**, Determinants of forced expiratory flows in newborn infants, *J. Appl. Physiol.*, 53, 1220, 1982.
42. **Olinsky, A., Bryan, A. C., and Bryan, M. H.**, A simple method of measuring total respiratory system compliance in newborn infants, *S. Afr. Med. J.*, 50, 128, 1976.
43. **Drorbaugh, J. E., Segal, S., Sutherland, J. M., Oppe, T. E., Cherry, M. A., and Smith, C. A.**, Compliance of lung during first week of life, *Am. J. Dis. Child.*, 104, 63, 1963.
44. **Tepper, R. S., Pagtakhan, R. D., and Tuassig, L. M.**, Noninvasive determination of total respiratory system compliance in infants by the weighted-spirometer method, *Am. Rev. Respir. Dis.*, 130, 461, 1984.

45. **Mortola, J. P., Fischer, J. T., Smith, B., Fox, G., and Weeks, S.,** Dynamics of breathing in infants, *J. Appl. Physiol.,* 52, 1209, 1982.
46. **Mortola, J. P., Milic-Emili, J., Noworaj, A., Smith, B., Fox, G., and Weeks, S.,** Muscle pressure and flow during expiration in infants, *Am. Rev. Respir. Dis.,* 129, 49, 1984.
47. **Midgal, M., Gaultier, C., Dreizzen, E., Dehan, M., and Chambille, B.,** *J. Physiol. (Paris),* in press.
48. **Hislop, A., Muir, D. C., Jacobsen, J., Simon, G., and Reid, L.,** Postnatal growth and function of the pre-acinar airways, *Thorax,* 27, 265, 1972.
49. **Hogg, J. C., Williams, J., Richardson, J. B., Macklem, P. T., and Thurbeck, W. T.,** Age as a factor in the distribution of lower-airway conductance and in the pathologic anatomy of obstructive lung disease, *N. Engl. J. Med.,* 282, 1283, 1970.
50. **Stocks, J. and Godfrey, S.,** Specific airway conductance in relation to postconceptional age during infancy, *J. Appl. Physiol.,* 43, 144, 1977.
51. **Wohl, M. E., Stigol, L. C., and Mead, J.,** Resistance of total respiratory system in healthy infants and in infants with bronchiolitis, *Pediatrics,* 43, 495, 1969.
52. **England, S. J., Lesouef, P. N., Bryan, M. H., and Bryan, A. C.,** The role of upper airway in airway resistance in infants, *Am. Rev. Respir. Dis.,* 131, A225, 1985.
53. **Polgar, G. and String, S. T.,** The viscous resistance of lung tissues in newborn infants, *J. Pediatr.,* 69, 787, 1965.
54. **Krauss, A. N., Klain, D. B., Dahms, B. B., and Auld, P. A.,** Vital capacity in premature infants, *Am. Rev. Respir. Dis.,* 108, 1361, 1973.
55. **Scott, C. B., Nickerson, B. G., Sargent, C. W., Platzker, A. C. G., Watburton, D., and Keens, T. G.,** Developmental pattern of maximal transdiaphragmatic pressure in infants during crying, *Pediatr. Res.,* 17, 707, 1983.
56. **Guslits, B., Howard, S., Bryan, M. H., England, S. J., and Bryan, A. C.,** Diaphragmatic work of breathing, *Fed. Proc., Fed. Am. Soc. Exp. Biol.,* 45, 1023, 1986.
57. **Keens, T. G., Bryan, A. C., Levison, H., and Ianuzzo, C. D.,** Developmental pattern of muscle fiber types in human ventilatory muscles, *J. Appl. Physiol.,* 44, 909, 1978.
58. **Milic-Emili, J.,** Respiratory muscle fatigue and its implications in RDS, in *Pulmonary Surfactant System,* Losmi, E. V. and Scarpelli, E. M., Eds., Elsevier, Amsterdam, 1983, 135.
59. **Koch, G.,** Alveolar ventilation, diffusing capacity and the a-aPO_2 difference in the newborn infant, *Respir. Physiol.,* 4, 168, 1968.
60. **Gaultier, C., Boule, M., Allaire, Y., Clement, A., Buvry, A., and Girard, F.,** Determination of capillary oxygen tension in infants and children, *Bull. Eur. Physiopathol. Respir.,* 14, 287, 1978.
61. **Martin, R. J., Okke, A., and Rudin, D.,** Arterial oxygen tension during active and quiet sleep in the normal neonate, *J. Pediatr.,* 94, 271, 1979.

SUDDEN INFANT DEATH SYNDROME (SIDS)

Toke Hoppenbrouwers and Joan Hodgman

INTRODUCTION

Incidence and Definition

SIDS has occurred since biblical times and historically has been attributed to a large number of causes, including smothering, infanticide, child abuse, and infections.[1,2] It has been documented in virtually every developed country, but little is known about its incidence in Third World countries.[3-22] Only during the last 20 years has it been included in the *Handbook of Nomenclature* as a specific disease entity with unknown etiology. SIDS is defined as the sudden death of an infant which is unexpected by medical history and in which a thorough postmortem examination fails to demonstrate an adequate cause of death.[23] An estimated 6500 babies die each year of SIDS in the U.S. alone, and approximately 80% die during a time they are presumed to be asleep, hence the designation crib death or cot death.[24,25]

Reported incidences vary from a low of 0.3 to 0.5 per 1000 live births in Israel and Scandinavia[17,26] to a high of 3.0 to 4.2 per 1000 live births in Ontario, Canada and Belfast, Ireland.[14,16] The rate reported in the U.S. varies between 1.5 to 3.0 per 1000 live births, with higher and lower incidences in certain segments of the population. For instance, black and Eskimo infants have a two to four times increased risk of dying from SIDS.[27,28] For more extensive discussions we refer the reader to the excellent articles by Valdes Dapena,[29] Beckwith,[1] and Peterson.[30]

Epidemiology

While the etiology of SIDS remains a matter of speculation, a number of epidemiological risk factors have been established. Some of these are uniquely related to SIDS, such as age and season of death. As can be seen in Figure 1, SIDS occurs most frequently between 2 and 3 months of age, in contrast to death prior to 1 month from all other causes.[27,31,32] Similarly, SIDS occurs more frequently in the fall and winter months compared to the summer, again a different distribution than infant deaths from other causes (Figure 2), although death from respiratory infections follows a similar pattern.[33] This relationship is preserved in New Zealand and Australia, in the Southern Hemisphere, where winter and summer are reversed in time.[34]

Other factors are related to infant mortality in general (Table 1). Deaths from all causes, including SIDS, are more common in infants born to women of low socioeconomic class.[25,35-37] Few studies have successfully separated class from race, but there is some evidence that the extremely high incidence of SIDS in black infants is reduced, once socioeconomic class and birthweight are equated.[44,45] Smoking during pregnancy is another general factor associated with an increased risk for SIDS in the offspring,[37,41-43] together with such maternal factors as teenage pregnancy and short interpregnancy intervals.[25,35,38,39,41] While breast feeding has been considered to provide protection from SIDS, this has not been conclusively established.[37,46] An upper respiratory infection frequently precedes SIDS and has in the past led to pathological diagnoses, such as interstitial pneumonia, which cannot entirely explain the death.[1]

IDENTIFICATION OF RISK INFANTS

Most deaths from SIDS occur in healthy term-born infants with no history that would suggest an increased risk. However, certain groups have been identified who are more

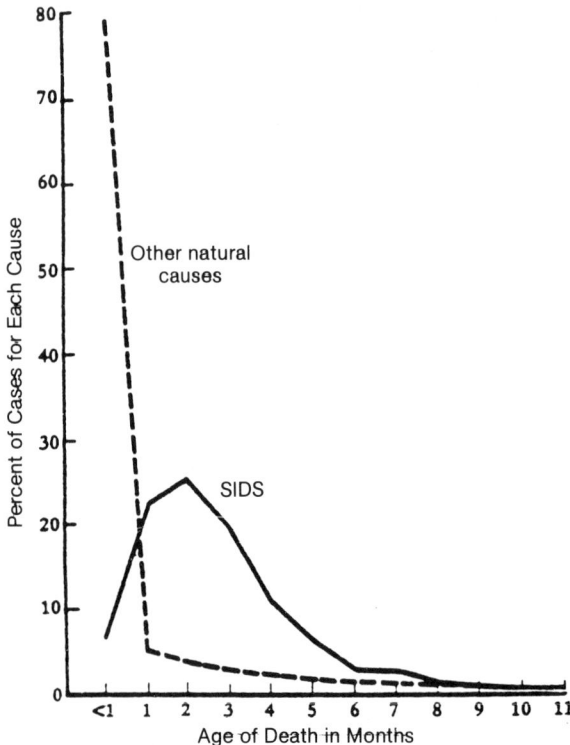

FIGURE 1. Sudden infant death syndrome (age at death) (1976 to 1978): research update (fall 1980). Maternal and infant health. (Courtesy of Dr. J. Grether, California State Department of Health Services, San Francisco.)

vulnerable. Among infants born to a family with a previous SIDS, the rate of SIDS has been reported to be four to five times that of the general population,[16,47,48] but when maternal age and parity is controlled, this may be reduced to about twice that of the general population.[49] Although this increase is highly significant statistically, the actual rate of $1/2$ to 1% makes the risk for the individual subsequent sibling small.

The second identifiable risk group is the premature infant where risk for SIDS appears to increase with decreasing gestational age.[45,50,51] In 178 babies weighing less than 1500 g discharged from our high-risk nurseries between 1979 and 1981, 2 died of SIDS, for a rate of 1.1%.

An infant who has experienced a life-threatening apnea for which no explanation can be found on diagnostic study used to be designated a near miss for SIDS or an aborted SIDS.[52] A recent consensus report proposed the term *apparent life-threatening event* (ALTE) for this clinical syndrome. Of these events, 50% have an identifiable etiology, but the cause of the remainder is unknown. The diagnosis of apnea of infancy (AOI) is reserved for those infants for whom no specific cause of ALTE or apnea can be identified.[53] These infants are at increased risk of subsequently dying of SIDS, with a reported death rate varying from 1.4 to 2.5% in large series.[54,55] Infants who required vigorous positive pressure resuscitation have been reported at greatest risk in one series,[56] but that has not been the experience of others.[57,58] One of a twin birth is at increased risk for SIDS, and the risk for the twin sibling is enhanced, once one has died.[44] Finally, mothers who abused substances during pregnancy, especially cocaine and "crack", deliver infants who seem to be at an approximately fivefold increased risk for SIDS.[59,60]

Although the risk of SIDS is increased in these groups of infants, probabilities are low,

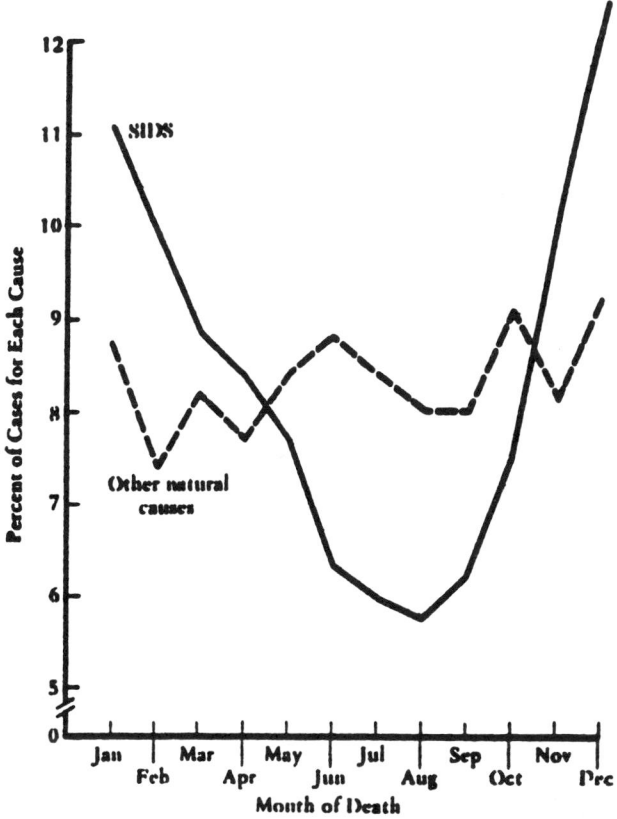

FIGURE 2. Sudden infant death syndrome (month of death) (1976 to 1978): research update (fall 1980). Maternal and infant health. (Courtesy of Dr. J. Grether, California State Department of Health Services, San Francisco.)

Table 1
FACTORS RELATED TO SIDS AND HIGH
INFANT MORTALITY AND MORBIDITY

	Ref.
Low birthweight/preterm birth	16,25,35
Males	30
Low socioeconomic class	25,35-37
Teenage mother	25,35,38,39
Few prenatal visits	14,35,37,40
Short intervals between pregnancies	41
Smoking during pregnancy	37,41-43

and at the present, the individual infant within the group who will actually die of SIDS cannot be identified ahead of time. The additional risk factors listed in Table 1 have not been useful in increasing predictive power.

HYPOTHESES ABOUT ETIOLOGY

Abnormal Physiologic Patterns and Risk for SIDS

In spite of heightened public interest and intensive research, the etiology of SIDS is still unknown. There is a consensus that, although an identifiable entity, the etiology may not be identical in all infants. For instance, it was found recently that 1 to 2% of infant deaths, indistinguishable from SIDS, can now be attributed to botulism.[61] A number of investigators stress that an interaction of several variables probably leads to vulnerability and death.[15,24,62,63] New information has emerged indicating that the infant who will die of SIDS may not be quite as healthy as previously believed, and that SIDS may have a medical history, as do most other diseases.[29,37]

Study of SIDS is hampered by our inability to predict accurately which infant will subsequently succumb. Most functional studies have been carried out on the three statistical risk groups of subsequent siblings, premature infants, and infants with AOI where the probability of dying is low. In the last few years, physiological data from three studies of babies who acutally died of SIDS, totaling approximately 100 infants, have been reported in the literature.[55,64-66] The studies of risk groups suffer from both the presumptive relationship with SIDS and the variety in the findings, while the studies of infants who subsequently died are limited by the small numbers and the lack of clear clues.

When infants with AOI were compared with appropriate controls, few significant differences were found in cardiorespiratory behavior. The most striking finding has been the variation of results among investigators. This is not surprising in light of the range of severity of an event and the lack of appropriate controls, especially in the earlier studies. For example, and increase in periodic breathing was reported by Kelly and Shannon,[67] a decrease was reported by Navelet et al.,[68] and no difference from normals was reported by Guillemenault et al.[69] and Hodgman et al.[57] A similar lack of consistency exists in the findings of apnea recorded during diagnostic study after the initial apneic episode.[70-74] Studies of response to hypercarbic stimuli have also produced conflicting results. A decrease in ventilatory response to CO_2 was reported by Shannon et al.,[75] not confirmed by either Brady et al.[76] or Ariagno et al.,[77] while Haddad et al.[78] found an increased CO_2 response in infants with AOI. Responses to a mild chronic hypoxic challenge have differed from normal, but again, the specific response has varied. Brady et al.[76] found a decrease in respiratory rate, and increase in periodic breathing, and longer apnea, while Ariagno et al.[77] and McCulloch et al.[79] did not confirm these findings, but revealed a decreased arousal in response to hypoxia.

In spite of this variation in reported findings, some agreement has emerged. One of the most consistent findings has been an increase in respiratory rates in risk infants. We first observed this in subsequent siblings (Figure 3),[80] and several reports have since described this finding in the same risk group,[81] in infants with AOI and other risk infants,[71,82,83] and in some who actually died of SIDS.[65,82] Heart rates have been found increased in risk groups and in SIDS in those few sleep studies where they have been reported.[84-86]

Sources of variation among investigations include differences in monitoring techniques. Length of monitoring sessions will influence the results, with long sessions accentuating numbers of apnea.[87] Time of night will make a difference, especially as the infant develops circadian rhythms in cardiorespiratory behavior.[88] The observation of rapid change during the first 3 months of life in normal infants with a relative stability after that time (vide infra) demands accurate age matching of controls. Controls for preterm infants present a particular problem. Although the practice of using postconceptional rather than postnatal ages for comparisons decreases some of the variance due to different gestational ages at birth, the preterm infant does not necessarily develop along the identical path of the full-term infant. This becomes of importance in SIDS research because preterm infants either comprise or are overrepresented in two of the groups at increased risk.

The ability to identify the subsequent sibling during gestation has made it possible to study this group before delivery. Increased variability and an increase in bradycardia was found in the fetal heart rate at 32 and 36 weeks of gestation.[89] This is of particular interest in light of the increasing data suggesting that SIDS infants are abnormal from before birth. In many respects, subsequent siblings cannot be distinguished from healthy controls. Subtle

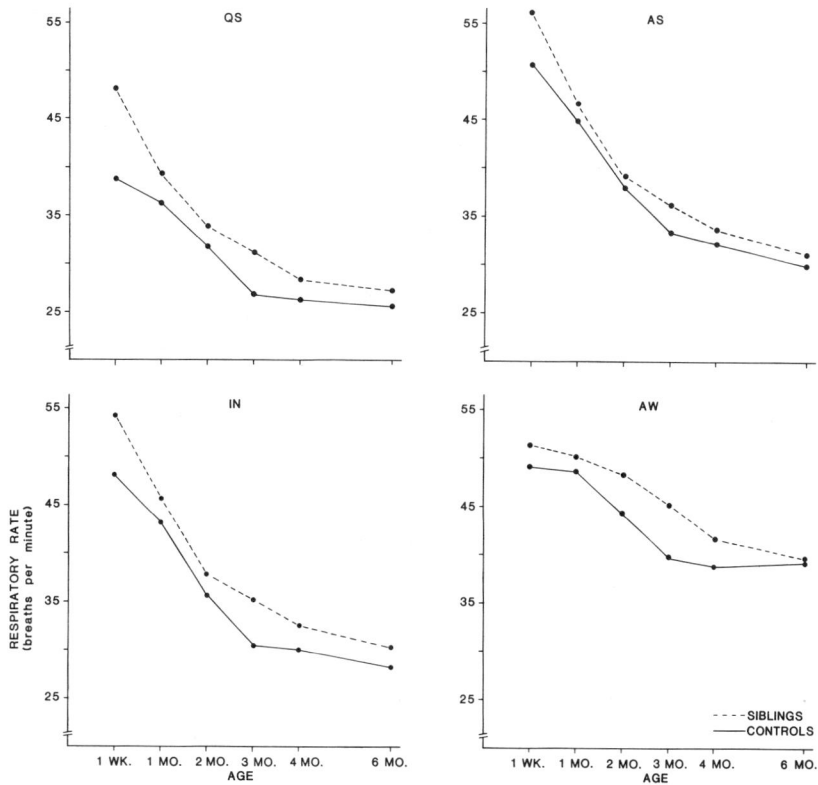

FIGURE 3. Median respiratory rate (ordinate) in breaths per minute as a function of age (abscissa) in various states. Group means are based on 20 infants. At 3 months of age subsequent siblings breathed faster in all states. (From Hoppenbrouwers, T., Hodgman, J. E., McGinty, D., Harper, R. M., and Sterman, M. B., *Pediatrics*, 66, 205, 1980. Reproduced by permission of *Pediatrics*, © 1980.)

differences, however, such as higher respiratory and heart rates, were present as mentioned above.[80,81,84,90]

Maternal anxiety during pregnancy and early infancy is understandably increased with subsequent siblings. One of the manifestations of this anxiety is reflected in the earlier feeding of solids to the subsequent sibling, resulting in more rapid early weight gain when compared to control infants.[91] Increased heart rates have been associated with caloric intake in the young infant.[92] The difference in respiration, however, was already present during the first week of life not only in subsequent siblings, but also in infants who later died of SIDS.[65,90] Whether these subtle findings are due to environmental stimuli rather than to intrinsic abnormalities is unclear; the former is suggested by our inability to replicate the original cardiorespiratory findings in a second carefully matched group of siblings and controls.[93]

Little information is available regarding cardiorespiratory behavior of the preterm baby during infancy beyond the neonatal period. The relationship, if any, between apnea in the nursery and SIDS following discharge has just begun to be systematically explored, and the available evidence suggests that the infant with apnea in the nursery is not necessarily at increased risk for SIDS.[37,55] Apnea of prematurity occurring after the first week is relatively uncommon, appearing in 1% of the premature population. In our large center with over 100 premature infants delivered monthly, we find approximately 1 case of unexplained apnea per month present beyond the first week. There is some suggestion that these particular infants are at increased risk for later AOI.[94]

Putative Role of Hypoxia, Apnea, and Arousal Failure

Naeye[95] firt reported that approximately 60% of SIDS showed evidence of an abnormal increase of smooth muscle in the small pulmonary arteries, possibly indicating chronic hypoxemia prior to death. He has since described a number of other markers of chronic hypoxemia in SIDS, including increased weight of the wall of the right ventricle, increased retention of periadrenal brown fat cells, increased hepatic erythropoiesis, and increased chromaffin tissue in the adrenal medulla.[96-98] A number of groups have confirmed the presence of one or more of these markers,[99-103] but some findings could not be replicated.[104,105] On balance, there seem to be at least three markers that have withstood efforts to disprove their validity.[106] Two other findings have been added recently that await replication: both bone marrow erythropoesis and fetal hemoglobin were found elevated in infants who died of SIDS, a finding which could reflect mild hypoxia.[107,108]

The hypothesis that SIDS is preceded by mild hypoxia is supported by the findings of increased respiratory and heart rates in risk infants which can be seen as compensatory responses to mild hypoxia.[65,80-86] Increased respiratory rates in quiet sleep were seen in infants with documented hypoxemia due to congenital heart disease,[109] while kittens monitored in mildly reduced ambient oxygen also exhibited increased respiratory rates.[110]

Unexplained central, mixed, or obstructive apnea, especially during sleep, was first proposed as the most likely mechanism for hypoxemia.[111] Two lines of evidence shed doubt on the primary role of these apnea as a cause of hypoxemia and SIDS. First, severe apnea preceding death were only seen in a very small proportion of SIDS.[16,37] Second, pauses in breathing are an integral part of normal respiration during sleep in the young infant,[112] and these pauses do not cause significant drops in blood oxygen tension[73] as measured by transcutaneous measures. The recent use of pulse oximetry measuring oxygen desaturation reveals very brief episodes of desaturation during apnea of periodic breathing that may have gone unnoticed with transcutaneous measurements.[113] The clinical significance of these events has yet to be determined.

The role of obstructive apnea in the etiology of SIDS has been of particular interest, but is far from established. Obstructive apnea is not easy to measure because obstructions are typically accompanied by movements when polygraphic measurements provide the least meaningful data. Microphones, nasal and buccal thermistors, and strain gauges have been used in combination and still have not provided accurate counts. Esophageal tracings provide more accurate information, but are difficult to obtain routinely. Estimates of the prevalence of obstructive and mixed apnea vary tremendously between laboratories, with some observing obstructions rarely and others commonly.[114-116] Our own data indicate that significant obstructions are uncommon, and many brief obstructions are typically accompanied by movements and *not* bradycardia.[117] In a few risk infants, anatomically narrowed airways have been documented.[118] A *functional* narrowing of the airways during sleep, however, causing hypoxemia and death, has been postulated, but never proven.

We have entertained the hypothesis that ambient pollution, such as that engendered by maternal smoking, could contribute to the mild hypoxemia. We studied the relationship between ambient pollution and SIDS in Los Angeles County with the help of time-series analyses[32] and demonstrated a temporal relationship between the seasonal increase of CO, SO_2, lead, NO_2, and hydrocarbons in the winter and SIDS. Peaks in these pollutants preceded the seasonal increase in SIDS by 7 weeks (Figure 4), a finding replicated by others.[119] CO and NO_2, in particular, are prime candidates for causing mild tissue hypoxemia. In the body, CO can be measured by levels of carboxyhemoglobin. A study by Radford and Drizd,[120] based on 27,000 individuals between 6 months and 74 years, showed a remarkable association between SIDS risk factors and high mean concentrations of carboxyhemoglobin. These included elevated levels in smokers, children living in metropolitan areas, children of poor parents, and black children. They also reported a seasonal increase with a peak in the winter months when the SIDS rate is elevated.

Part B: Cardiovascular and Respiratory Development 187

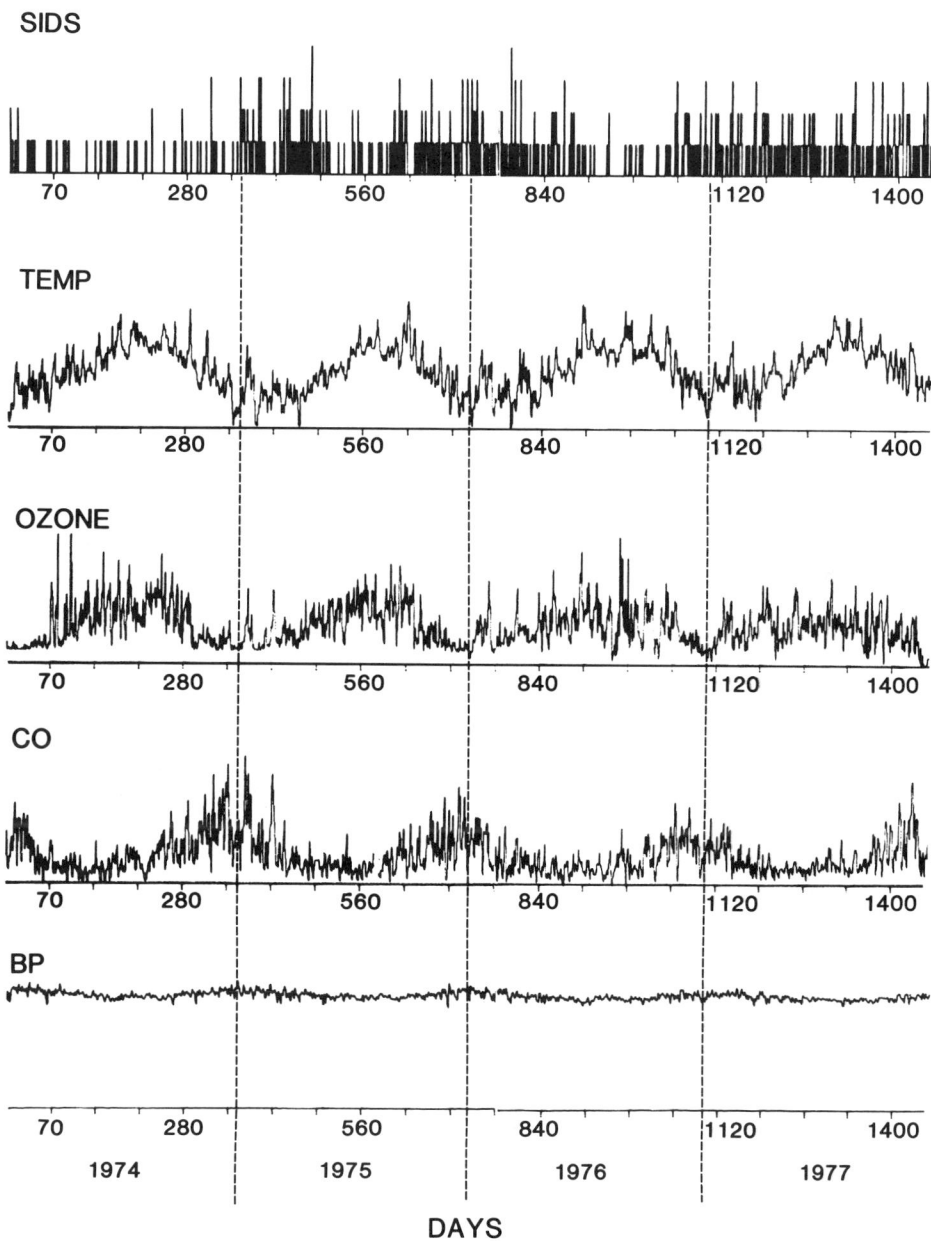

FIGURE 4. Daily incidence of SIDS in Los Angeles County between January 1974 and December 1977: abscissa — each increment represents 70 d, totaling 1460 d; ordinate — number of SIDS per day ranges between 0 and 4. Notice that low temperature during the winter coincided with an increase in SIDS, carbon monoxide (CO), barometric pressure (BP), and a decrease in ozone. The maxima and minima of CO and SIDS were out of phase. Fast Fourier analysis revealed that the lag time between CO and SIDS is 7 weeks. (From Hoppenbrouwers, T., Calub, M., Arakawa, K., and Hodgman, J. E., *Am. J. Epidemiol.*, 113, 623, 1981. With permission.)

Several lines of evidence point to the potential significance of an arousal failure in SIDS, a hypothesis which is currently being pursued. Firstly, laboratory sleep studies in subsequent siblings of SIDS showed that these infants had a propensity, once asleep, to remain asleep, whereas controls tended to awaken.[121] In addition, subsequent siblings had a decreased tendency to enter short waking periods at 2 and 3 months of age (Figure 5).[122] The studies

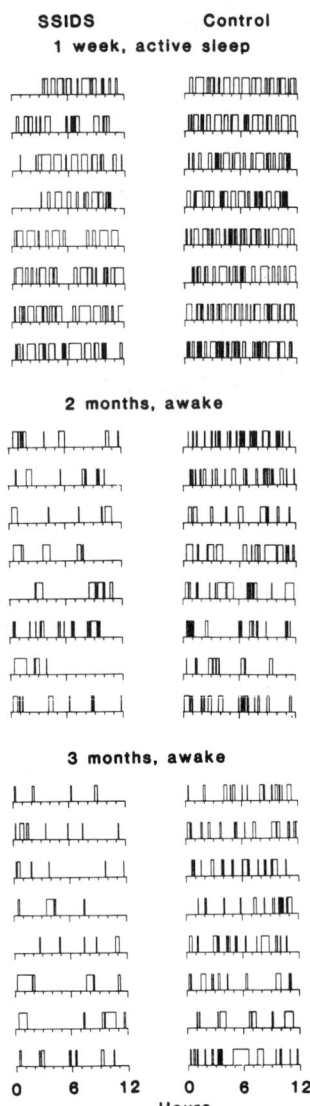

FIGURE 5. Plots of binary state sequences from selected subsequent siblings (SSIDS) and control infants. The presence of a state is indicated by a rise from baseline. Note the reduced incidence of short awakenings in subsequent siblings at 2 and 3 months of age. (From Harper, R. M., Leake, B., Hoffman, H., Walker, D. O., Hoppenbrouwers, T., Hodgman, J. E., and Sterman, M. B., *Science*, 213, 1030, 1981. With permission from AAAS, © 1981.)

of sleep in infants with AOI are less clear cut.[57,69,94,123,124] Some studies suggest that these infants have a reduction in quiet sleep (QS), an increase in active sleep (AS), and a reduction in awakenings. While these patterns appear fairly consistent, the timing of these seemingly aberrant patterns varies from laboratory to laboratory and is almost certainly influenced by methodology and other normal environmental factors. Secondly, functional changes in the arousal system were suggested by Anderson-Huntington and Rosenblith,[125] who reported that infants who subsequently died of SIDS showed a less strong defensive reaction to partial nasal occlusion than control infants, a finding replicated in subsequent siblings.[126] Several groups have examined the arousal response to decreased O_2 and increased CO_2.[77,79,127] The findings are most consistent for decreased O_2; a higher arousal threshold was found in infants

Table 2
QUIET SLEEP MOTILITY DENSITY IN CONTROLS AND SIBLINGS

	1 Month	2 Months	3 Months
Controls	4.9	5.2	4.7
Subsequent siblings	1.8[a]	3.0[b]	4.1

Note: Number of minutes with motility per 100 min of sleep.

[a] $p < 0.001$.
[b] $p < 0.002$.

with AOI, subsequent siblings, and infants born to substance-abusing mothers.[128,129] In one study, arousal to sensory stimuli was also examined and found elevated in risk infants.[130]

It is generally accepted that arousal is regulated by one or more reciprocal excitatory and inhibitory systems, with the brainstem reticular formation (RF) acting as an excitatory center which initiates and maintains levels of arousal.[131] Movements during sleep can stimulate the RF. We examined somatic activity during QS in our subsequent siblings and found that these infants exhibited significantly less movement at 1 and 2 months of age than control infants (Table 2).[132] Guillemenault and Coons[123] reported a similar sleep-related reduction in motility in near-miss infants.

Pathological findings in the brainstem have been reported by several groups. Naeye[98] found that approximately 50% of SIDS exhibited an abnormal proliferation of astroglial fibers in the lateral RF of the brainstem, a finding replicated by others,[133,134] but not found uniquely related to SIDS by Summers and Parker.[135] Quattrochi et al.,[136] described abnormalities in reticular dendritic spine development, a maturational deficit in the brainstem.

The presence of functional deficits in the brainstem have not been proven, but are suggested by the findings of Korobkin and Guillemenault,[137] who carried out semiquantitative neurological assessments in infants with AOI and found shoulder hypotonia to be the most common phenomenon. In the early 1970s, Salk et al.[138] reported a significant decrease in habituation in one infant who died of SIDS. Interestingly, at least one theory postulates that habituation is organized in the brainstem.[139]

Past preoccupation with apnea as the major cause of SIDS has obscured the fact that the vast majority of SIDS are healthy term infants without evidence of apnea preceding death.[16,37,55] Any theory of SIDS must explain the cause of death in these typical infants. Although two highly respected pathologists recently argued about the validity of the findings, suggesting that mild hypoxemia preceded death from SIDS,[140] the cumulative evidence indicates this remains a plausible hypothesis. The potential role of an arousal failure in the etiology of SIDS is currently under study and may involve an increased arousal threshold coupled with hypoventilation and hypoxemia.

Developmental Patterns and Consequences for SIDS

The mechanism of SIDS is probably complex, involving interactions at several levels of the neuraxis, between the organism and the environment and spanning both fetal life and infancy. Any theory about SIDS needs to explain why the risk is highest between 1 and 3 months of age and why the newborn tends to be spared.

Recent observations have demonstrated dramatic changes in physiological and behavioral development during the first 3 months of life in normal infants (Figure 6).[141-144] While respiratory rates during sleep decline in a monotonic fashion, heart rates and respiratory sinus arrhythmia exhibit an inverse nonmonotonic trend. Heart rates increased up to 1 month of age and then began to decrease, while respiratory sinus arrhythmia followed an opposite

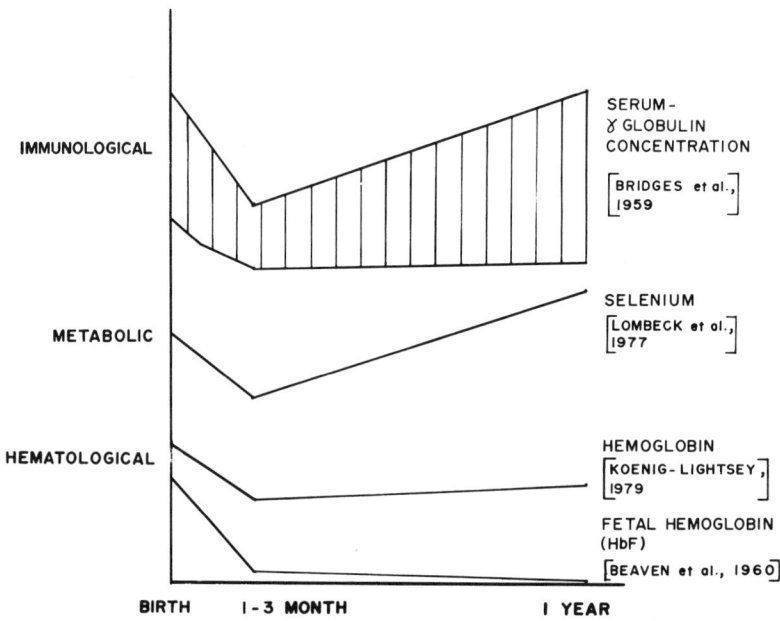

FIGURE 6. Schematic representation of the development of various physiologic functions. Note the nonmonotonic trends.

trend (Figure 7).[145-149] Crying was most prevalent between 1 and 2 months of age, at the time when active sleep percentages were still declining and quiet sleep percentages rose (Figure 8).[150-153] This is also the time when manifestations of a circadian influence first appear.[88] In the immature animal, nonlinear patterns have also been documented. For instance, altitude hypoxia caused a higher degree of central nervous system (CNS) acidity in 12-d-old rats than in those either older or younger.[154] Sasaki[155] studied the development of laryngeal function in beagle pups during the first year of life. He found transient laryngeal hyperexcitability in 2- and 3-month-old pups, compared to either younger or older ones, partially due to a reduction of inhibitory influences from the brainstem. Therefore, this interval between 1 and 3 months of age, when the SIDS rate is at its peak, is a time of integration in the CNS, when numerous functions change, some nonlinearly. The described variables are almost certainly only a small portion of those involved in this CNS reorganization. A recent article, for instance, provides normative data for neurotransmitter metabolites in the cerebrospinal fluid of normal infants. Again, abrupt monotonic and nonmonotonic changes were observed during this same time period.[156]

Vulnerability to deleterious influences tends to increase during periods of rapid change. This rule applies, for instance, in the first trimester during organogenesis when fetal morphology changes rapidly and is easily compromised by toxic agents.[157] A system should also be vulnerable when its physiologic and functional characteristics change dramatically as between 1 and 3 months of age.

The immature organism is known to have a diminished capacity to oxidize and conjugate drugs and foreign compounds.[158] Studies on the metabolism of toxic heavy metals such as lead revealed that absorption is greatly enhanced in the young animal; the distribution is different from that of adults in that the brain accumulates much larger quantities in the immature organism. Spontaneous excretion of heavy metals is reduced in the young. Although toxicity is lower for the young organism, this advantage is overwhelmed by three factors: higher absorption rate, lower excretion, and unfavorable distribution of toxic materials.[159,160] Therefore, not only is the immature organism vulnerable to exogenous and

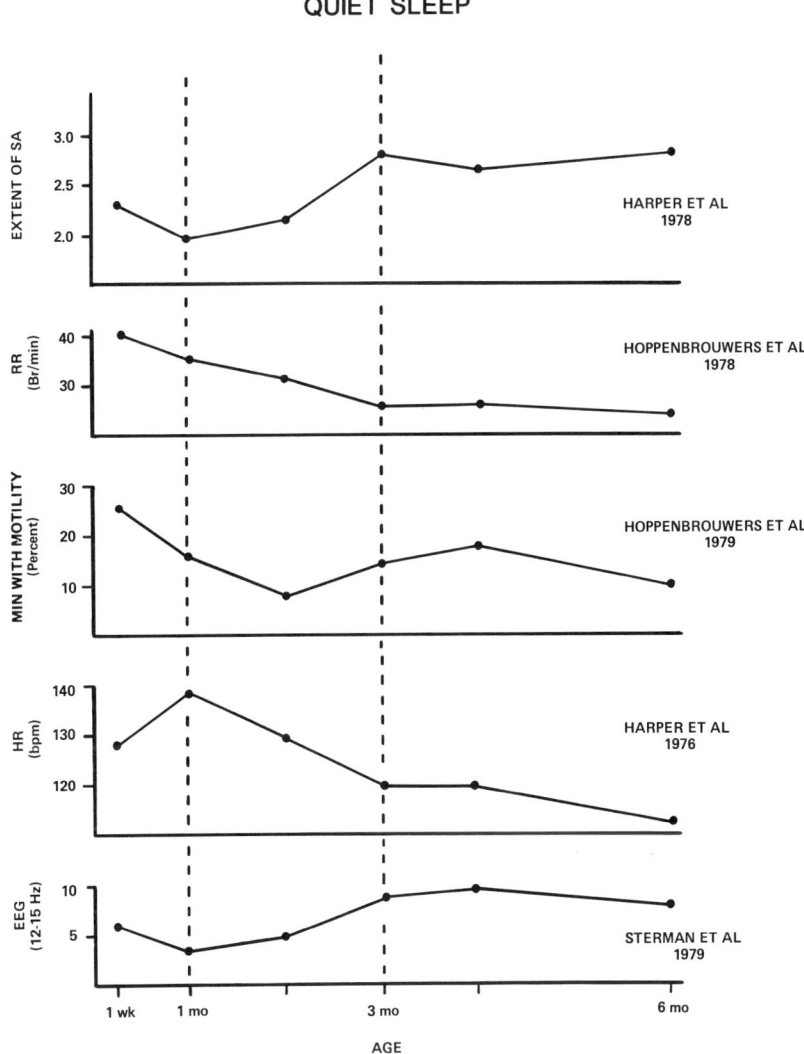

FIGURE 7. Schematic representation of the ontogenesis of a number of physiologic variables, measured during quiet sleep (QS): SA, respiratory sinus arrhythmia; RR, respiratory rate; HR, heart rate; Min, minutes. (From Hoppenbrouwers, T. and Hodgman, J. E., *Neuropaediatrie*, 13, 36, 1982. With permission.)

endogenous insults because of the many ongoing functional changes, but it is also vulnerable because of its inability to rid itself of these influences as promptly as the adult.

CNS Maturation in Infants at Risk for SIDS

When we compared the physiologic development of subsequent siblings and healthy, age-matched controls, we expected that the controls would be relatively more mature. It was a surprise, therefore, when in several respects the subsequent siblings appeared ahead of the control infants. In the early development of the EEG, the subsequent siblings showed an increment in power of 12- to 15-Hz activity, which represents the sleep spindle, before the control infants (Figure 9).[161] Circadian influences upon heart and respiratory rates, with daytime peaks and nighttime troughs, tend to emerge about the same time as the sleep spindle emerges, between 2 and 3 months of age. Compared to controls, manifestations of a circadian influence appeared sooner in subsequent siblings (Figure 10).[162]

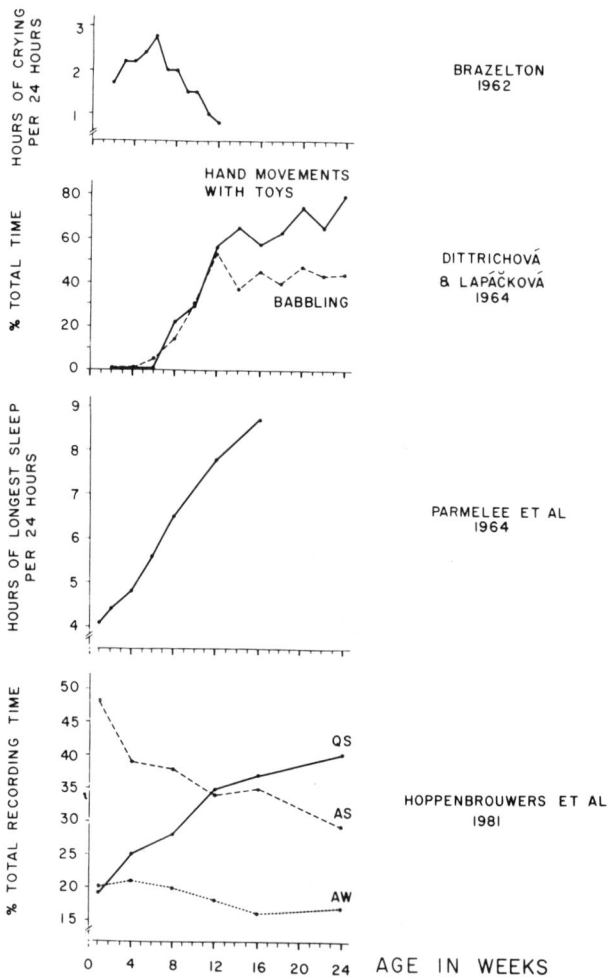

FIGURE 8. Development of various behaviors during the first 6 months of life in infants. Percent time crying is transiently increased between 1 and 2 months, while babbling, quiet sleep (QS), and longest sleep time undergo a sharp rise. (From Hoppenbrouwers, T. and Hodgman, J. E., *Neuropaediatrie*, 13, 36, 1982. With permission.)

A number of reports suggest a developmental model in which a physiologic deficit can bring about a transient acceleration in maturation. One or more abnormal stimuli, such as undernutrition or mild hypoxia, may upset the normal developmental schedule and transiently cause an "overshoot" in some functional systems. This may be initially adaptive, but ultimately cause a depletion of the infant's reserve, whereupon a sudden worsening of the infant's condition may be expected. Examples of this can be found in the lung where premature infants who have suffered mild intrauterine asphyxia show evidence of accelerated maturity.[163,164] Also small-for-gestational-age preterms, who probably have experienced mild hypoxia *in utero*, exhibited at comparable ages faster brainstem conduction times than appropriate for gestational-age preterms.[165] This phenomenon is not restricted to humans. In rats, Hollenberg et al.[166] described how a hypoxic stress applied early in the neonatal period augmented the rate of division and the ultimate number of cardiac muscle cells.

It should be noted that the consequences of strong stimuli can be entirely different from those of chronic, mild stimuli. It is much more likely to encounter a compensatory response

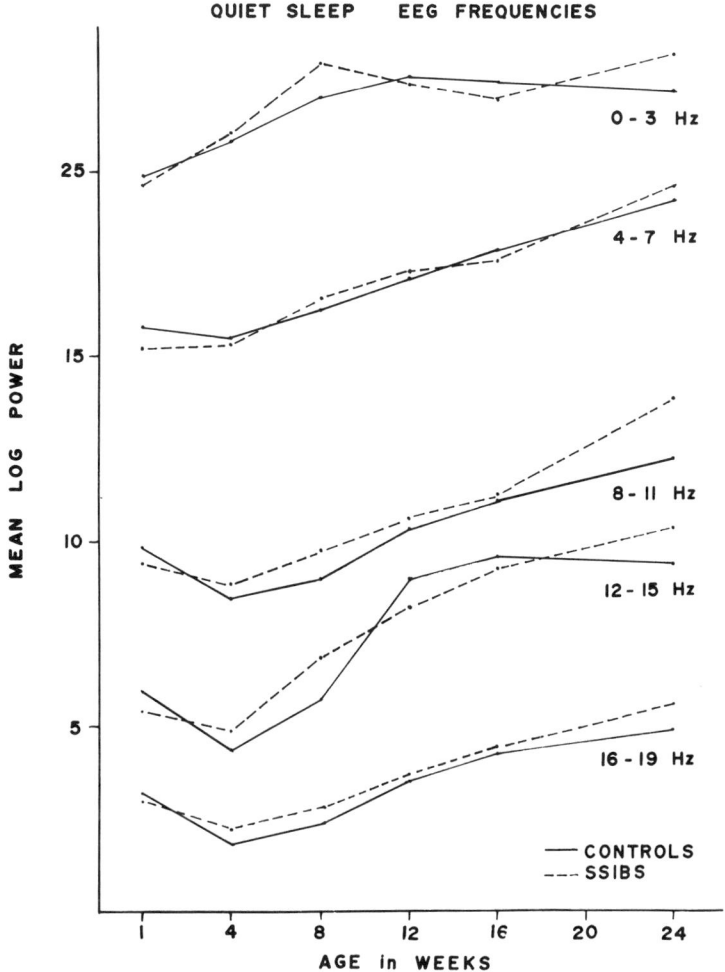

FIGURE 9. Mean log EEG spectral power during quiet sleep (QS) in subsequent siblings (SSIBS) and control infants. Note the elevated power in 12 to 15 Hz at 8 weeks in SSIBS compared to controls.

in the latter, which, by virtue of its mild nature, may go unrecognized. For instance, a severe oxygen deficit almost invariably causes insomnia and restlessness;[167] a mild oxygen deficit, however, may cause drowsiness and sleep in adults and babies.[168,169]

Accelerated morphologic or functional maturation will not be manifest simultaneously at all levels of the neuraxis.[170] The developing nervous system exhibits regional differences in metabolic rate and blood flow, which are not constant across the entire developmental period, even in the same structures. For instance, Himwich and Fazekas[171] demonstrated that medullar metabolic rate was twice that of the adult in the 1-week-old puppy. The metabolic rate of the cortex was only half of that in the medulla right after birth, but increased dramatically so that in adulthood it was twice that of the medulla.

It is tempting to postulate that the early appearance of the sleep spindle and circadian rhythms reflect an early maturation of the forebrain. The EEG sleep spindle requires the maturation of thalamocortical circuits, while circadian rhythms are believed to require the maturity of a hypothalamic pacemaker in the basal forebrain.[172,173] The gradual development of the forebrain tends to be associated with the emergence of inhibitory influences, including a rise in quiet sleep. In this light it is interesting to note that subsequent siblings exhibited less movement during quiet sleep, which could well reflect this enhanced inhibition.

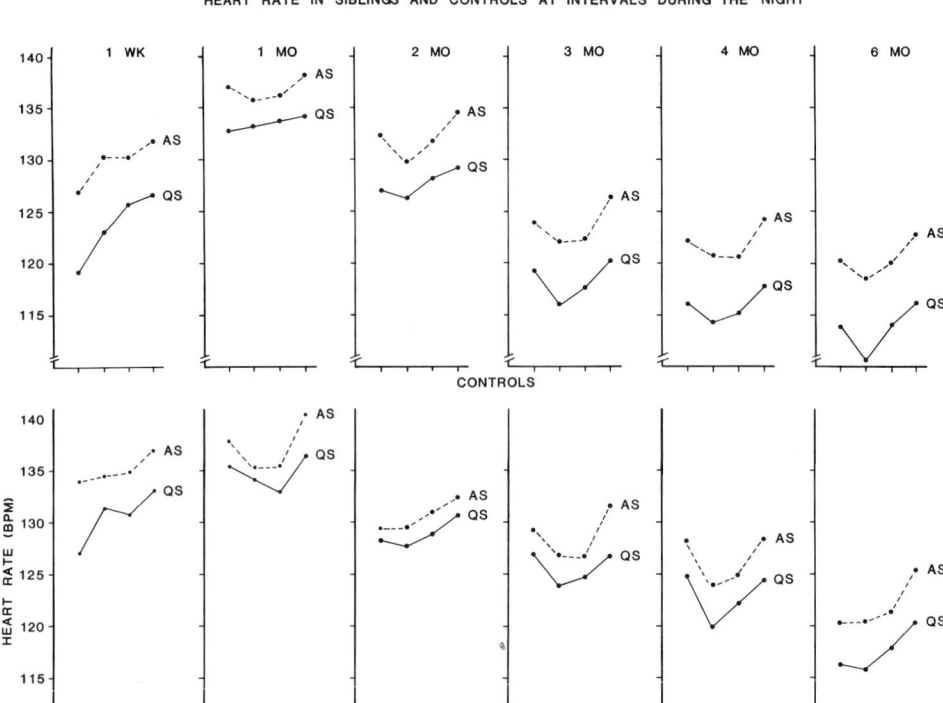

FIGURE 10. Heart rate (HR) in beats per minute as a function of state, age, and interval of the night: top, controls; bottom, subsequent siblings. Note a transient increase in HR at 1 and 2 months of age in both groups. In controls, a circadian influence on HR was seen in active sleep (AS) prior to quiet sleep (QS). In subsequent siblings, this influence appears already at 1 month in QS.

Arousal requires the integrity of the brainstem reticular activating system. With increasing age, descending cortical influences will alter arousal thresholds. If this time table is disturbed and risk infants exhibit descending cortical inhibitory influences sooner, their arousal threshold should be prematurely elevated. This is indeed what we found in the subsequent sibling. They tended to show fewer spontaneous arousals throughout the night, which can be seen as an adaptive mechanism initially, because oxygen consumption is reduced during sleep.[171] In the long run, however, this strategy may stimulate the development of neuronal circuits favoring sleep rather than arousal, and therefore jeopardize the infant's ability to awaken spontaneously when blood oxygen decreases to dangerously low levels. At the moment, there are research strategies, such as nuclear magnetic resonance spectroscopy or positron emission tomography, which can elucidate the neural development of selective brain regions, including aspects of synaptic functioning in sleeping infants. Such studies would begin to establish whether or not we should seek the etiology of SIDS in subtle functional CNS abnormalities. Alternatively, computer guided pathological studies based on carefully selected control series and SIDS, as are now being carried out by Kinney and co-workers,[170] might reveal CNS abnormalities that have thus far gone unnoticed.

Basis for Interest in Neurotransmitters

The relevance of a discussion of neurotransmitters in the context of SIDS is predicated on two considerations: (1) some neurotransmitters have been found altered in SIDS, and (2) a number of investigators have studied the effect of mild hypoxia on neurotransmitters and reported some intriguing findings.

Denoroy et al.[175] found diminished activity of two enzymes involved in the synthesis of catecholamines in brainstem sites in SIDS. Ozand and Tildon[176] described a diminished activity of dopamine beta hydroxylase, an enzyme involved in noradrenergic neurotransmission. They also found increased activity of tyrosine hydroxylase, the rate-limiting enzyme for dopamine, in SIDS infants compared to controls. In the former study, changes were found in the brainstem; in the latter, brainstem activity was unaltered, while the hypothalamus, caudate nucleus, and putamen showed altered activity. Finally, Bergstrom et al.[177] observed higher concentrations of substance P in the medulla of SIDS victims. While this is a promising avenue of inquiry, more controlled studies need to be performed before the true significance of these data can be determined.

Hedner et al.[178] systematically examined the differences in prolonged and short exposures to mildly and severely reduced ambient oxygen levels in the young rat. He reported monoamines in different brain regions in 1-, 4-, 14-, and 28-d-old rat pups. In a mild hypoxic environment (12% O_2) the rat pups showed rapid breathing and decreased spontaneous motility. Activities of the enzymes tyrosine and tryptophan hydroxylase, rate limiting in the synthesis of dopamine (DA), norepinephrine (NE), and serotonin (5 HT), initially decreased with $^1/_2$-h exposures, but then increased significantly during 6 h of exposure. This phenomenon was observed only in the youngest animals (1 to 4 d). 5-Hydroxytryptophan (5-HTP), the immediate precursor of 5 HT, followed the same pattern. DA levels remained low in all but the 1-d-old animals in response to prolonged mild hypoxia. Therefore, hypoxic levels that were found not to change cerebral energy production[179] caused significant changes in monoamine synthesis. In the youngest animals this level of hypoxia gave rise to an unexpected increase in rate-limiting enzymes, in particular those of 5 HT. Oxygen deficits are expected to affect neurotransmitters by inhibiting the oxygen-dependent, rate-limiting enzymes such as tyrosine and tryptophan hydroxylase, which in turn decrease the levels of NE, DA, and 5 HT. At some ages, such a homeostatic adjustment does not take place; *increases* in the activity of the enzymes and levels of the neurotransmitters are observed, rather than decreases. These changes must have physiologic and behavioral sequelae, although the nature, duration, and reversibility of these are unknown.

Vaccari et al.[180] reported similar findings in rat pups who were bred and raised at high altitude (3800 m, 12.8% O_2) for 60 d. Under these conditions, increases in monoamine enzymes were much more common than decreases, and they were seen more frequently in synthetic enzymes than in catabolic ones. These findings show that the immature organism responds to moderate hypoxia by an increase rather than a decrease in synthetic metabolism. In all studies these phenomena were transient, being limited to the first few weeks of life. A similar effect in acetylcholine was reported by Timiras and Woolley[181] in rats studied at high altitude. Acetylcholinesterase activity, an estimate of ACH, was decreased in animals older than 24 d, but enhanced in the neocortex, brainstem, hypothalamus, and cerebellum in the 12-d-old animals. Miller et al.[182] exposed pregnant rats to mild ambient hypoxia (15% O_2) and measured synaptic development in the nucleus tractus solitarius in infant rats at 1, 5, 10, 15, and 30 d. Exposed rats could not be differentiated from controls except by number of vesicles per bouton, which were increased in the oxygen-deprived animals. Together, these studies indicate that mild hypoxia can induce what appear to be signs of accelerated functional activity in the CNS.

Conclusions

Between 1972 and 1982, the etiology of SIDS has been vigorously pursued. While groups of infants at increased risk for SIDS have been identified, the mechanism of SIDS has thus far eluded us. The final mechanism of death is still uncertain, although the evidence strongly favors a respiratory rather than a cardiac failure. Because of the incidence and distribution of SIDS, a straight Mendellian genetic origin is deemed unlikely, but the absence of a polygenic inheritance has not been established unequivocally.[30] Infants who die of SIDS have had in general a benign history, although compared with carefully matched controls, they have experienced a slightly less favorable intra- and extrauterine environment. The peak incidence of SIDS occurs between 1 and 3 months of age, a time now recognized as one of tremendous functional changes including those related to circadian and ultradian cycles. This maturational interval can yield functional instability and excessive stability, both of which may render every infant transiently more vulnerable to unusual stimuli such as hypoxia.[183] One promising hypothesis postulates that during that maturational interval, infants who have experienced deficits in utero and the immediate postnatal period will be more prone to a failure to arouse when hypoventilation or hypoxia would call for such an arousal. Newer computer-driven research strategies can now be enlisted to examine the involvement of subtle CNS abnormalities in infants at risk and infants who died of SIDS.

MANAGEMENT

Indication for Laboratory and Home Monitoring

The majority of studies using cardiorespiratory monitoring have been performed in the infant with ALTE. The diagnosis in all cases has been based on the history of an apneic event considered life threatening by the caretakers. Most of these have occurred at home, but some have been observed in the hospital. Additional criteria include the presence of pallor or cyanosis, neuromuscular change, usually hypotonia, but occasionally stiffening, and the need for vigorous resuscitation by the caretakers. In all cases where no cause for the apnea could be uncovered by clinical study, the apnea has been designated AOI.[53]

A work-up for an infant with ALTE should rule out known causes of apnea, such as seizures and sepsis; it should also include overnight cardiorespiratory monitoring, as abnormalities are more likely to occur during the early morning hours. Excessive central apnea and cardiac abnormalities can be readily detected on standard impedance and electrocardiogram (ECG) monitors. Detection of obstructive apnea, however, requiring at least a nasal thermistor, but preferably also a buccal thermistor or an esophageal catheter, is more difficult. Even with these precautions, it is difficult to clearly differentiate obstructive apnea from normal behaviors. For example, in the premature it was shown that movements of the infant which are normal and perhaps advantageous cannot be separated from obstructive apnea with chest motion, but no air flow.[184] Although sleep has not been reported grossly abnormal in risk infants, proper interpretation of cardiorespiratory variables requires that sleep state be known.

Detailed study of responses to inhaled gases do not seem indicated in the usual clinical setting. Some studies have suggested that about 1% of the ALTE are accompanied by either an obstruction or a cardiac arrhythmia.[185,186] One fifth of the infants with ALTE in our own study had some peculiarity such as excessive apnea, or fixed or abnormally variable heart rate, but no two infants showed the same findings.[53] Since the majority of infants will have normal findings, this overnight recording will be disappointing most of the time except in a negative sense. The significance of the abnormal findings in our study is unknown and care should be exercised in attributing clinical importance to them, as similar findings occasionally also occur in normal infants. Whether or not treatment is indicated for the rare infant with apnea beyond the norm is unknown. Theophilyne and caffeine have been shown to improve respiratory function, but their usefulness in treating risk infants has not been established by a controlled study.[187,188]

Even less information is available concerning the proper care of infants with cardiac findings of too little or too much variability. Much is made of the association of "brady-

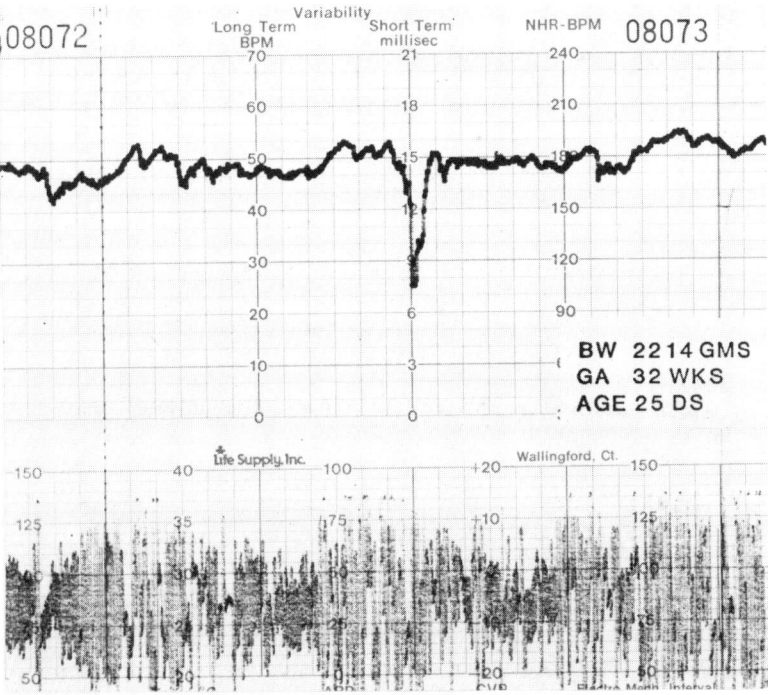

FIGURE 11. V-shaped cardiac deceleration, without apnea in healthy premature infant, born at 32 weeks of gestation.

cardia'' with apnea as a risk factor; however, the phenomenon is really cardiac deceleration rather than a persistent slowing of the heart rate (Figure 11). Most of these drops in heart rate are a normal response to many exogenous and endogenous stimuli, including apnea, and imply normal responsiveness of the autonomic nervous system. Hypoxia does not appear to cause the decelerations, because they occur simultaneously with the onset of apnea and prior to decreases in O_2 levels.[74] It is reasonable to equate most of these decelerations with the V-shaped dips in fetal heart rate tracings, known to be benign.[189] For more details we refer the reader to a nosology to be published by the Association of Professional Sleep Societies.[190]

Cardiorespiratory monitoring of the asymptomatic subsequent sibling for clinical diagnosis is not warranted. It is not helpful to monitor these infants in the nursery in an effort to identify those few who will have future difficulties, because early findings have not predicted later behavior in risk infants.[37,55,191] Our own studies of the premature infant with persistent apnea in the nursery indicate that cardiorespiratory behavior, blood gas levels, and sleep are remarkably similar to that of controls on follow-up at 1 and 3 months of age (Table 3), provided the infants are matched for corrected chronological ages.[94] Nonetheless, based on a small sample, premature infants with persistant or late apnea in the nursery seem to be at higher risk for subsequent ALTE.

Long-term monitoring at home has been recommended for infants perceived to be at risk for SIDS in an effort to interrupt life-threatening events. The popularity of the treatment is presumably based on the seriousness of the condition for which it is recommended, in spite of the fact that no studies have demonstrated its effectiveness.[192] Although the risk for SIDS in identifiable risk groups is small, it is risk of death and frightening to both parents and physicians.

The selection of infants for home monitoring has been based primarily on epidemiology.

Table 3
TRANSCUTANEOUS GASES AND AUTONOMIC VARIABLES IN APNEIC AND CONTROL INFANTS DURING QUIET SLEEP

	RR MD	RR IQ	HR MD	HR IQ	O_2 MN	CO_2 MN	SkT MN
40 Weeks							
Apneic	43.7	6.4	150.8	3.7	68.3	38.9	35.1
	(10.9)	(2.8)	(8.9)	(1.2)	(9.0)	(3.7)	(1.3)
Control	38.8	7.3	150.5	3.8	64.0	42.4	35.9
	(3.7)	(3.6)	(6.7)	(1.2)	(8.7)	(3.4)	(1.3)
1 Month							
Apneic	39.7	5.9	139.1	5.9	73.0	41.4	35.5
	(8.5)	(2.0)	(7.7)	(2.6)	(15.7)	(4.7)	(1.0)
Control	38.0	5.1	142.9	4.8	69.4	38.4	36.3
	(3.2)	(1.1)	(7.5)	(1.7)	(14.3)	(4.5)	(1.0)
3 Months							
Apneic	32.5	5.6	131.5	6.3	79.6	37.7	35.1
	(7.3)	(1.8)	(10.8)	(2.5)	(8.8)	(2.2)	(1.1)
Control	29.5	5.8	126.0	6.4	77.0	39.4	34.9
	(3.3)	(1.9)	(14.5)	(2.7)	(8.3)	(3.1)	(0.8)

Note: RR = Respiratory rate; HR = heart rate; SkT = skin temperature; MD = median; IQ = interquartile range; MN = mean.

The addition of criteria from the infant's performance during cardiorespiratory monitoring has been recommended in spite of the fact that most investigators agree that the infant at risk of dying cannot be identified by cardiorespiratory parameters. The use of pneumogram criteria developed on term infants to identify which premature infant to monitor seems particularly inappropriate. In studies of healthy preterm infants, we have found repetitive apnea from 20- to 30-s duration in 4.5% of the 66 normal infants studied.[193] Periodic breathing, used to justify monitoring, occurred in 85% of these healthy infants. In addition, periodic breathing was present during 20% or more of an 8-h nighttime recording in 15% of the group. Cardiac decelerations were also common, being recorded in 68% of the infants. Recent reports throw further doubt on the wisdom of using pneumogram criteria to identify risk, as there appears little correlation between apnea or cardiac irregularity in the neonatal period and subsequent death from SIDS.[37,55]

Recurrence rates of ALTE have varied between 43 and 67%.[57,58,114,194] Interestingly, the rates are not different between series where home monitoring was regularly employed and those where it was not. Which infant will experience a recurrence cannot be accurately predicted, although multiple recurrent apnea have been associated with a need for positive pressure resuscitation.[56] Experience with home monitors has been reported by a number of investigators.[56,58,194-196] The available data do not suggest a marked effect from monitoring. Death from SIDS has occurred in monitored infants,[56,58,114,196] and the mortality rate of 3.8% for monitored infants is not different from the 2.5% reported before monitoring was used.[54]

Evaluation of the psychological effect of prolonged monitoring has been reported by two groups, both of which enthusiastically espouse home monitoring.[197,198] The reports document responses varying from relief of anxiety to severe stress on the family constellation. A more recent follow-up study of families who monitored their infants at home showed distressing interruptions of family life, such as a high incidence of divorce.[199] In the most recent follow-up study of monitored infants, Kahn et al.[200] found no differences in intellectual function at 10 years of age.

Infants with ALTE or AOI are at increased risk to die of SIDS and present a difficult problem for the physician. In spite of the lack of confirmation that long-term home monitoring is beneficial, this treatment continues to be recommended as no effective alternative is available. It must be kept in mind, however, that only a smaller number of infants who die of SIDS exhibit apneic episodes prior to death. Furthermore, parents should be advised that monitoring cannot guarantee prevention of SIDS. Monitoring of subsequent siblings may be justified on the basis of parental anxiety, again in the absence of effective alternative treatments.

Assistance to the Family

Because SIDS is unique in being both unexpected and medically unexplained, the families of these infants have an especially difficult time coping with the death. Frequently, parents express the feeling that only another parent who has experienced the same loss can understand their predicament. Early referrals to parent support groups, such as the National Foundation for SIDS or the Guild for Infant Survival, have been beneficial in helping the parents cope with their loss.

Any professional who deals with these families must be aware of their special needs. Mourning and grief take in excess of a year in many parents. Only one third report undergoing a considerably shorter mourning period.[201] Grieving follows expected patterns with the exception that the initial shock is followed immediately by an intense need for information about SIDS.[202] It is important the families be provided with accurate information that can help to alleviate their guilt and confusion. Because there is no known cause for SIDS, the families engage in a painful scrutiny of their caretaking. Under these circumstances it is also easy to see how blame can be attributed, either subtly or bluntly, to the parents.

Questions about another child to replace the one that is lost may be raised early by well-meaning friends and relatives. This suggestion is frequently resented by the parents. Eventually, parents will address the issue of subsequent children when they are ready. At this time they can benefit from a candid discussion of the risk of SIDS and the choices available in the management of subsequent siblings. For more guidance on assisting parents who have lost an infant to SIDS, we refer the reader to other valuable sources of information.[203-205]

ACKNOWLEDGMENT

The studies from our laboratory referred to in this chapter were carried out between 1972 and 1985. They were supported by funds from the National Institutes of Health, Child Health and Human Development, Contract #NO1-HD-2-2777 and HD4-2810 and Grant #HD 13689. We also wish to acknowledge the financial support of the Los Angeles Chapter of the National Foundation for SIDS, The Orange County Chapter of the Guild for Infant Survival, and the Arthur Zimtbaum Foundation of New York. The studies reported here were done in collaboration with Drs. M. B. Sterman, R. M. Harper, D. J. McGinty, M. Durand, and L. A. Cabal.

REFERENCES

1. **Beckwith, J. B.**, The sudden infant death syndrome, in *Current Problems in Pediatrics*, Vol. 3, Gluck, L., Cone, T., Falkner, F., and Green, M., Eds., Year Book Medical Publishers, Chicago, 1975, 1.
2. **Franciosi, R. A.**, Evolution of SIDS diagnosis, *Minn. Med.*, 2, 411, 1983.
3. **Bowden, K. M. and French, E. L.**, Unexpected death in infants and young children: second series, *Med. J. Austr.*, 38, 926, 1951.
4. **Adelson, L. and Kinney, E. R.**, Sudden and unexpected death in infancy and childhood, *Pediatrics*, 17, 663, 1956.

5. **Jacobsen, T. and Voigt, J.**, Sudden and unexpected infant death. II. Result of medicolegal autopsies of 356 infants aged 0—2 years, *Acta Med. Leg. Soc.*, 9, 133, 1956.
6. **Del Mundo, F. and Cordero, L. S.**, A review on sudden deaths in children, *J. Philipp. Med. Assoc.*, 34, 155, 1958.
7. **Hagiwara, J.**, A statistical study on sudden death in infants, *Jpn. J. Leg. Med.*, 13, 222, 1959.
8. **Houstek, J., Benesova, D., and Holy, J.**, Sudden death among children of the Prague district during 1956—1958, *Cesk. Pediatr.*, 14, 590, 1959.
9. **Maresch, W.**, On the etiology of sudden death in infants, *Z. Kinderheilkd.*, 84, 565, 1960.
10. **Fontaine, G.**, The sudden death of the infant, *Sem. Med. Prof. Med. Soc.*, 38, 624, 1962.
11. **Makarov, I. V. and Seliverstova, A. I.**, On sudden death in children, *Kazan. Med. Zh.*, 1, 38, 1962.
12. **Muller, G.**, Sudden death in infancy, in *Pathologic Anatomy and Dynamics*, Georg Thieme Verlag, Stuttgart, 1963, 145.
13. **Carpenter, R. G. and Shaddick, C. W.**, A study of cot deaths in England and Wales in 1960, *Br. J. Prev. Soc. Med.*, 19, 1,1965.
14. **Steele, R. and Langworth, J. T.**, The relationship of antenatal and postnatal factors to sudden unexpected death in infancy, *Can. Med. Assoc. J.*, 94, 1165, 1966.
15. **Cameron, A. H. and Asher, P.**, Cot deaths in Birmingham 1958-61, *Med. Sci. Law*, 5, 187, 1967.
16. **Frogatt, P., Lynas, M. A., and MacKenzie, G.**, Epidemiology of sudden unexpected death in infants ("cot death") in Northern Ireland, *Br. J. Prev. Soc. Med.*, 25, 119, 1971.
17. **Bloch, A.**, Sudden infant death syndrome in the Ashkelon district: a 10-year survey, *Isr. J. Med. Sci.*, 9, 452, 1973.
18. **Baak, J. P. A. and Huber, J.**, Incidence of SIDS in The Netherlands, in *SIDS 1974, Proc. Francis E. Camps Int. Symp. on Sudden and Unexpected Deaths in Infancy*, Robinson, R. R., Ed., The Canadian Foundation for the Study of Infant Deaths, Toronto, 1974, 157.
19. **Tonkin, S.**, Epidemiology of SIDS in Auckland, New Zealand, in *SIDS 1974, Proceedings of the Francis E. Camps International Symposium on Sudden and Unexpected Deaths in Infancy*, Robinson, R. R., Ed., The Canadian Foundation for the Study of Infant Deaths, Toronto, 1974, 169.
20. **Fohlin, L.**, SIDS occurrence in Stockholm 1968—1972, in *SIDS 1974, Proc. Francis E. Camps Int. Sym. on Sudden and Unexpected Deaths in Infancy*, Robinson, R. R., Ed., The Canadian Foundation for the Study of Infant Deaths, Toronto, 1974, 147.
21. **Kahn, A.**, Unexpected and unexplained infant death. Epidemiologic and comparative study of Belgian and immigrant populations, *Arch. Fr. Pediatr.*, 36(7), 712, 1979 (in French).
22. **Mason, J. K., Harkness, R. A., Elton, R. A., and Bartholomew, S.**, Cot deaths in Edinburgh: infant feeding and socioeconomic factors, *J. Epidem. Commun. Health*, 34, 35, 1980.
23. **Beckwith, J. B.**, Discussion of terminology and definition of sudden infant death syndrome, in *Proc. 2nd Int. Conf. on Causes of Sudden Deaths in Infants*, Bergman, A. G., Beckwith, J. B., and Ray, C. G., Eds., University of Washington Press, Seattle, 1970.
24. **Valdes Dapena, M.**, Sudden and unexpected death in infants: the scope of our ignorance, *Pediatr. Clin. North Am.*, 10, 693, 1963.
25. **Bergman, A. B., Ray, C. G., Pomeroy, M. A., Wahl, P. W., and Beckwith, J. B.**, Studies of the sudden infant death syndrome in King County, Washington. III. Epidemiology, *Pediatrics*, 49, 860, 1972.
26. **Jorgenson, T., Biering-Sorensen, F., and Hilden, J.**, Sudden infant death in Copenhagen 1956—1971, *Acta Pediatr. Scand.*, 68, 11, 1979.
27. **Grether, J. K.**, Sudden Infant Death Syndrome: A Research Update, Fall 1980. Maternal and Infant Health Section, California State Department of Health Services, San Francisco, 1980.
28. **Kaplan, D. W., Bauman, A. E., and Krous, H. F.**, Epidemiology of sudden infant death syndrome in American Indians, *Pediatrics*, 74, 1041, 1984.
29. **Valdes Dapena, M. A.**, Sudden infant death syndrome: a review of the medical literature 1974—1979, *Pediatrics*, 66, 597, 1980.
30. **Peterson, D.**, The epidemiology of sudden infant death syndrome, in *Sudden Infant Death Syndrome, Medical Aspects and Psychological Management*, Culbertson, J. L., Krous, H. F., and Bendell, R. D., Eds., The John Hopkins University Press, Baltimore, 1988, 3.
31. **Hoppenbrouwers, T. and Hodgman, J. E.**, Sudden infant death syndrome (SIDS), *Public Health Rev.*, 11, 363, 1983.
32. **Hoppenbrouwers, T., Calub, M., Arakawa, K., and Hodgman, J. E.**, Seasonal relationship of sudden infant death syndrome and environmental pollutants, *Am. J. Epidemiol.*, 113, 623, 1981.
33. **Foy, H. M. and Ray, C. G.**, Epidemiology of sudden infant death syndrome and lower respiratory tract disease in young children: a comparison, *Am. J. Epidemiol.*, 98, 69, 1973.
34. **Beal, S. M.**, Some epidemiological factors about sudden infant death syndrome (SIDS) in South Australia, in *Sudden Infant Death Syndrome*, Tildon, J. T., Roeder, L. M., and Steinschneider, A., Eds., Academic Press, New York, 1983, 15.
35. **Kraus, J. and Borhani, N. O.**, Post-neonatal sudden unexplained death in California: a cohort study, *Am. J. Epidemiol.*, 95, 497, 1972.

36. **Valdes-Dapena, M. A., Birle, L. J., McGovery, J. A., McGillen, J. F., and Colwell, F. H.,** Sudden infant death in infancy: a statistical analysis of certain socioeconomic factors, *J. Pediatr.,* 73, 387, 1968.
37. **Hoffman, H. J., Damus, K., Hillman, L., and Krongrad, E.,** Risk factors for SIDS: results of the National Institute of Child Health and Human Development SIDS Cooperative Epidemiological Study, *Ann. N.Y. Acad. Science,* 533, 13, 1988.
38. **Peterson, D. R., vanBelle, G., and Chinn, N. M.,** Sudden infant death syndrome and maternal age: etiologic implications, *JAMA,* 247, 2250, 1982.
39. **Babson, S. G. and Clarke, N. G.,** Relationship between infant death and maternal age, *J. Pediatr.,* 103, 391, 1983.
40. **Hoppenbrouwers, T.,** Reassessment of so-called predictive factors for SIDS, paper presented at Research Perspectives in the Sudden Infant Death Syndrome Conference for Research Workers, NICHD, Alexandria, VA, 1977.
41. **Lewak, N., Van den Berg, B., and Beckwith, J. B.,** Sudden infant death syndrome risk factors, *Clin. Pediatr.,* 18, 404, 1979.
42. **Bergman, A.,** Relationship of passive cigarette smoking to sudden infant death syndrome, *J. Pediatr.,* 58, 665, 1976.
43. **Peterson, D. R.,** The sudden infant death syndrome — reassessment of growth retardation in relation to maternal smoking and the hypoxia hypothesis, *Am. J. Epidemiol.,* 113, 583, 1981.
44. **Kraus, J. F.,** Methodologic considerations in the search for risk factors unique to sudden infant death syndrome, in *Sudden Infant Death Syndrome,* Tildon, J. T., Roeder, L. M., and Steinschneider, A., Eds., Academic Press, New York, 1983, 43.
45. **Black, L., David, R. J., Brouillette, R. T., and Hunt, C. E.,** Effects of birth weight and ethnicity on incidence of sudden infant death syndrome, *J. Pediatr.,* 108, 209, 1986.
46. **Carpenter, R. G., Gardner, A., Jepson, M., Taylor, E. M., Salvin, A., Sunderland, R., Emery, J. L., Pursall, E., and Roe, J.,** Prevention of unexpected infant death, *Lancet,* 723, 1983.
47. **Frogatt, P.,** Epidemiologic aspects of Northern Ireland study, *Proc. Second Int. Conf. on Causes of Sudden Death in Infants,* University of Washington Press, Seattle, 1970, 32.
48. **Peterson, D. R., Chin, N. M., and Fisher, L. D.,** The sudden infant death syndrome: repetitions in families, *J. Pediatr.,* 97, 265, 1980.
49. **Peterson, D. R., Sobotta, E. E., and Daling, J. R.,** Infant mortality among subsequent siblings of infants who died of sudden infant death syndrome, *J. Pediatr.,* 108, 911, 1986.
50. **Kulkarni, P., Hall, R. T., Rhodes, P. G., and Sheehan, M. B.,** Postneonatal infant mortality in infants admitted to a neonatal intensive care unit, *Pediatrics,* 62, 178, 1978.
51. **Teberg, A. J., Wu, P. Y. K., Hodgman, J. E., Mich, C., Garfinkle, J., Azen, S., and Wingert, W. A.,** Infants with birth weight under 1500 g: physical, neurological, and developmental outcome, *Crit. Care Med.,* 10, 10, 1982.
52. **Gunteroth, W. G.,** Sudden infant death syndrome (crib death), *Am. Heart J.,* 93, 784, 1977.
53. Infantile Apnea and Home Monitoring Consensus Report, U.S. Department of Health and Human Services, NIH Publ. #87-2905, Washington, D.C., 1987.
54. **Bergman, A. B., Beckwith, J. B., and Ray, C. G.,** The apnea monitor business, *Pediatrics,* 56, 1, 1975.
55. **Southall, D. P., Richards, J. M., Rhoden, K. J., Alexander, J. R., Shinebourne, E. A., Arrowsmith, W. A., Cree, J. E., Fleming, P. J., Goncalves, A., and Orme, R. L'E.,** Prolonged apnea and cardiac arrhythmias in infants discharged from neonatal intensive care units: failure to predict an increased risk for sudden infant death syndrome, *Pediatrics,* 70, 844, 1983.
56. **Kelly, D. H., Shannon, D. C., and O'Connell, K.,** Care of infants with near-miss sudden infant death syndrome, *Pediatrics,* 61, 511, 1978.
57. **Hodgman, J. E., Hoppenbrouwers, T., Geidel, S., Hadeed, A., Sterman, M. B., Harper, R. M., and McGinty, D. J.,** Respiratory behavior in near-miss sudden infant death syndrome, *Pediatrics,* 69, 785, 1982.
58. **Duffy, P. and Bryant, M. H.,** Home apnea monitoring in "near-miss" sudden infant death syndrome (SIDS) and in siblings of SIDS victims, *Pediatrics,* 70, 69, 1982.
59. **Rosen, T. S. and Johnson, H. L.,** Drug addicted mothers, their infants and SIDS, *Ann. N.Y. Acad. Science,* 533, 89, 1988.
60. **Chasnoff, I. C., Burns, K. A., and Burns, W. J.,** Cocaine use in pregnancy: perinatal morbidity and mortality, *Neurotoxicol. Teratol.,* 9, 291, 1987.
61. **Arnon, S. S., Midura, T. F., Clay, S. A., Wood, R. M., and Chin, J.,** Infant botulism: epidemiological, clinical and laboratory aspects, *JAMA,* 237, 1946, 1977.
62. **McGinty, D. J. and Harper, T. M.,** Sleep physiology and SIDS: animal and human studies, in *SIDS 1974, Proc. Francis E. Camps Int. Symp. on Sudden and Unexpected Deaths in Infants,* Robinson, R. R., Ed., The Canadian Foundation for the Study of Sudden Infant Deaths, Toronto, 1974, 201.
63. **Hoppenbrouwers, T. and Hodgman, J. E.,** Sudden infant death syndrome (SIDS): an integration of ontogenetic pathologic, physiologic and epidemiologic factors, *Neuropaediatrie,* 13, 36, 1982.

64. **Hoppenbrouwers, T., Zanini, B., and Hodgman, J. E.,** Intrapartum fetal hear rate and sudden infant death syndrome, *Am. J. Obstet. Gynecol.,* 133, 217, 1979.
65. **Shannon, D. C., Kelly, D. H., Akselrod, S., and Kilborn, K. M.,** Increased respiratory frequency and variability in high risk babies who die of sudden infant death syndrome, *Pediatr. Res.,* 22, 158, 1987.
66. **Kahn, A., Blum, D., Rebuffat, E., Sottiaux, M., Levitt, J., Bochner, A., Alexander, M., Grosswasser, M. D., and Muller, M. F.,** Polysomnographic studies of infants who subsequently died of sudden infant death syndrome, *Pediatrics,* 82, 721, 1988.
67. **Kelly, D. H. and Shannon, D. C.,** Periodic breathing in infants with near-miss sudden infant death syndrome, *Pediatrics,* 61, 511, 1979.
68. **Navelet, Y., Benoit, O., and Lacombe, J.,** Respiration et sommeil du nourrison, *Rev. Electroencephalogr. Neurophysiol. Clin.,* 9, 258, 1979.
69. **Guillemenault, C., Ariagno, R., Korobkin, R., Nagel, L., Baldwin, R., Coons, S., and Owen, M.,** Mixed and obstructive sleep apnea and near miss for sudden infant death syndrome. II. Comparison of near miss and normal control infants by age, *Pediatrics,* 64, 882, 1979.
70. **Monod, N., Curzi-Dascalova, L., Guidasci, S., and Valenzuela, S.,** Pauses respiratoires et sommeil chez le nouveau-ne et le nourrisson. *Rev. Electroencephalogr. Neurophysiol. Clin.,* 6, 105, 1976.
71. **Nogues, B. and Sampson-Dollfus, D.,** Etude comparative de la respiration pendant le sommeil chez des bebes temoins et des bebes a risque de morte subite (enfants de 2 mois a 1 an), *Waking Sleeping,* 3, 263, 1979.
72. **Steinschneider, A., Weinstein, S. L., and Diamond, E.,** The sudden infant death syndrome and apnea/obstruction during neonatal sleep and feeding, *Pediatrics,* 70, 858, 1982.
73. **Kahn, A., Blum, D., Engelman, E., and Waterschoot, P.,** Effects of central apneas on transcutaneous PO_2 in control subjects, siblings of victims of sudden infant death syndrome and near miss infants, *Pediatrics,* 69, 413, 1982.
74. **Kahn, A., Blum, D., Waterschoot, P., Engelman, E., and Smets, P.,** Effects of obstructive sleep apneas on transcutaneous oxygen pressure in control infants, siblings of sudden infant death syndrome victims, and near miss infants: comparison with the effects of central sleep apneas, *Pediatrics,* 70, 852, 1982.
75. **Shannon, D. C., Kelly, D. H., and O'Connell, K.,** Abnormal regulation of ventilation in infants at risk for sudden-infant-death syndrome, *N. Engl. J. Med.,* 297, 747, 1977.
76. **Brady, J. P., Ariagno, R. L., Watts, J. L., Goldman, S. L., and Dumpit, F. M.,** Apnea, hypoxemia, and aborted sudden infant death syndrome, *Pediatrics,* 62, 686, 1978.
77. **Ariagno, R., Nagel, L., and Guillemenault, C.,** Waking and ventilatory responses during sleep in infants near-miss for sudden infant death syndrome, *Sleep,* 3, 351, 1980.
78. **Haddad, G. G., Leistner, H. L., Lai, T. L., and Mellins, R. B.,** Ventilation and ventilatory pattern during sleep in aborted sudden infant death, *Pediatr. Res.,* 15, 879, 1981.
79. **McCulloch, K., Brouillette, R. T., Guzzetta, A. J., and Hunt, C. E.,** Arousal responses in near-miss sudden infant death syndrome and in normal infants, *J. Pediatr.,* 101, 911, 1982.
80. **Hoppenbrouwers, T., Hodgman, J. E., Harper, R. M., McGinty, D. J., and Sterman, M. B.,** Incidence of apnea in infants at high and low risk for sudden infant death syndrome (SIDS), *Pediatr. Res.,* 10, 425, 1976.
81. **Carse, E. A., Henderson-Smart, D. J., Johnson, P., Whyte, P., and Wilkinson, A. R.,** Transcutaneous oxygen tension measurements during sleep in the newborn baby and infants, *Arch. Dis. Child.,* 55, 134, 1980.
82. **Thoman, E. B., Miano, V. N., and Freese, M.,** The role of respiratory instability in the sudden infant death syndrome, *Dev. Med. Child. Neurol.,* 19, 729, 1977.
83. **Franks, C. J., Watson, J. B. G., Brown, B. H., and Foster, E. F.,** Respiratory patterns and risk of sudden unexpected death in infancy, *Arch. Dis. Child.,* 55, 595, 1980.
84. **Harper, R. M., Leake, B., Hoppenbrouwers, T., Sterman, M. B., McGinty, D. J., and Hodgman, J. E.,** Polygraphic studies of normal infants and infants at risk for the sudden infant death syndrome: heart rate and variability as a function of state, *Pediatr. Res.,* 12 778, 1978.
85. **Leistner, H. L., Haddad, G. G., Epstein, R. A., Lai, T. L., Epstein, M. A. F., and Mellins, R. B.,** Heart rate and heart rate variability during sleep in aborted sudden infant death syndrome, *J. Pediatr.,* 97, 51, 1980.
86. **Wilson, A. J., Stevens, B., Franks, C. I., and Southall, D. P.,** Analysis of long-term cardiorespiratory recordings from infants who subsequently suffered SIDS, *Ann. N.Y. Acad. Sci.,* 533, 390, 1988.
87. **Hoppenbrouwers, T., Hodgman, J. E., Arakawa, K., Harper, R., and Sterman, M. B.,** Respiration during the first six months of life in normal infants. III. Computer identification of breathing pauses, *Pediatr. Res.,* 14, 1230, 1980.
88. **Hoppenbrouwers, T., Jensen, D., Hodgman, J., Harper, R., and Sterman, M.,** Respiration during the first six months of life in normal infants. III. The emergence of a circadian pattern, *Neuropaediatrie,* 10, 264, 1979.
89. **Hoppenbrouwers, T., Hodgman, J. E., Harper, R. M., and Sterman, M. B.,** Fetal heart rates in siblings of infants with sudden infant death syndrome, *Obstet. Gynecol.,* 58, 319, 1981.

90. **Hoppenbrouwers, T., Hodgman, J. E., McGinty, D., Harper, R. M., and Sterman, M. B.,** Sudden infant death syndrome: sleep apnea and respiration in subsequent siblings, *Pediatrics,* 66, 205, 1980.
91. **Hodgman, J. E. and Hoppenbrouwers, T.,** The infant with unexplained apnea; near miss for SIDS, in *Ontogeny of Sleep and Cardiopulmonary Regulation: Factors Related to Risk for the Sudden Infant Death Syndrome,* Section III, Part B, Sterman, M. B., Hodgman, J. E., Stark, C. R., and Hoffman, H. J., Eds., NICHD, Washington, D.C., in press.
92. **Chessex, P., Reichman, B. L., Verellen, G. J. E., Putet, G., Smith, J. M., Heim, T., and Swyer, P. R.,** Relation between heart rate and energy expenditure in the newborn, *Pediatr. Res.,* 15, 1077, 1981.
93. **Hoppenbrouwers, T., Hodgman, J. E., Arakawa, K., Durand, M., and Cabal, L.,** Cardiorespiratory behavior and transcutaneous gases during sleep. III. Subsequent siblings, in press.
94. **Hoppenbrouwers, T., Hodgman, J. E., Arakawa, K., Cabal, L. A., and Durand, M.,** Cardiorespiratory behavior and transcutaneous gases during sleep. I. A follow-up study of premature infants with apnea, in press.
95. **Naeye, R. L.,** Hypoxemia and the sudden infant death syndrome, *N. Engl. J. Med.,* 289, 1167, 1973.
96. **Naeye, R. L.,** Hypoxia and the sudden infant death syndrome, *Science,* 186, 837, 1974.
97. **Naeye, R. L., Walen, P., and Ryser, M.,** Cardiac and other abnormalities in the sudden infant death syndrome, *Am. J. Pathol.,* 82, 1, 1976.
98. **Naeye, R. L.,** Brain stem and adrenal abnormalities in the sudden-infant-death syndrome, *Am. J. Clin. Pathol.,* 66, 526, 1976.
99. **Weiler, G. and DeHaardt, J.,** Morphometrical investigations into alterations of the wall thickness of small pulmonary arteries after birth and in cases of sudden infant death syndrome (SIDS), *Forensic Sci. Int.,* 21, 33, 1983.
100. **Emery, J. L. and Dinsdale, F.,** Structure of periadrenal brown fat in childhood in both expected and cot death, *Arch. Dis. Child.,* 53, 154, 1978.
101. **Valdes-Dapena, M., Gillane, M. M., and Catherman, R.,** Brown fat retention in sudden infant death syndrome, *Arch. Pathol. Lab. Med.,* 100, 547, 1976.
102. **Valdes-Dapena, M., Gillane, M. M., and Doss, D.,** Extramedullary hematopoiesis in the liver in the sudden infant death syndrome, *Arch. Pathol. Lab. Med.,* 103, 513, 1979.
103. **Williams, D., Vawter, G., and Reid, L.,** Increased muscularity of the pulmonary circulation in victims of sudden infant death syndrome, *Pediatrics,* 63, 18, 1979.
104. **Kendeel, S. R. and Ferris, J. A. J.,** Apparent hypoxic changes in pulmonary arterioles and small arteries in infancy, *J. Clin. Pathol.,* 30, 481, 1977.
105. **Valdes-Dapena, M. A., Gillane, M. M., and Catherman, R.,** The question of right vetricular hypertrophy in the sudden infant death syndrome, *Arch. Pathol. Lab. Med.,* 104, 184, 1980.
106. **Valdes-Dapena, M.,** A pathologist's perspective on possible mechanisms in SIDS, *Ann. N.Y. Acad. Sci.,* 533, 31, 1988.
107. **Giulian, G. G., Gilbert, E. F., and Moss, R. L.,** Elevated fetal hemoglobin levels in sudden infant death syndrome, *N. Engl. J. Med.,* 316, 1122, 1987.
108. **Naeye, R. L., Olsson, J. M., and Combs, J. W.,** New brain stem and bone marrow abnormalities in victims of sudden infant death syndrome, *J. Perinatol.,* 9, 180, 1989.
109. **Bruck, K., Adams, F. H., and Bruck, M.,** Temperature regulation in infants with chronic hypoxemia, *Pediatrics,* 30, 350, 1962.
110. **Baker, T. and McGinty, D. J.,** Reversal of cardiopulmonary failure during AS in hypoxic kittens. Implications for sudden infant death, *Science,* 198, 419, 1977.
111. **Steinschneider, A.,** Prolonged apnea and the sudden infant death syndrome: clinical and laboratory observations, *Pediatrics,* 50, 646, 1972.
112. **Hoppenbrouwers, T., Hodgman, J. E., Harper, R. M., Hofmann, E., Sterman, M. B., and McGinty, D. J.,** Polygraphic studies of normal infants during the first six months of life. III. Incidence of apnea and periodic breathing, *Pediatrics,* 60, 418, 1977.
113. **Bryan, C.,** Respiratory control, in (EDS), Sudden Infant Death Syndrome, Risk factors and Basic Mechanisms, Harper, R. M. and Hoffman, H. J., Eds., PMA Publishing, New York, 1988, 249.
114. **Rosen, C. L., Frost, J. D., and Harrison, G. M.,** Infant apnea: polygraphic studies and follow-up monitoring, *Pediatrics,* 71, 731, 1983.
115. **Monod, N., Curzi-Dascolova, L., Guidasci, S., and Valenzuela, S.,** Pauses respiratoires et sommeil chez le nouveau-ne et le nourrisson, *E.E.G. Neurophysiol.,* 6, 105, 1976.
116. **Orr, W. C., Stahl, M. L., and Duke, J. C.,** Sleep-related breathing patterns in infantile apnea, in *Sudden Infant Death Syndrome, Risk Factors and Basic Mechanisms,* Harper, R. M. and Hoffman, H. J., Eds., PMA Publishing, New York, 1988, 467.
117. **Hoppenbrouwers, T., Hodgman, J. E., Pollock, L., and Cabal, L.,** Obstructive sleep apnea in infants at low and increased risk for SIDS, *Clin. Res.,* 38, 195A, 1990.
118. **Guillemenault, C., Heldt, G., Powell, N., and Riley, R.,** Small upper airway in near-miss sudden infant death syndrome infants and their families, *Lancet,* 1, 402, 1986.

119. **Erickson, M. M. and Hillman, L. S.,** The relationship of airborne lead to sudden infant death syndrome, Abstract, International Res. Conf. on SIDS, Baltimore, 1982.
120. **Radford, E. P. and Drizd, T. A.,** Blood carbon monoxide levels in persons 3—74 years of age: United States, 1976—80, Advance Data from Vital and Health Statistics, No. 76, DHHS Publ. No. 82-1250, Public Health Service, Hyattsville, MD, 1982.
121. **Hoppenbrouwers, T., Hodgman, J. E., Harper, R. M., and Sterman, M. B.,** Falling asleep, waking up and transitions from one sleep state to another in subsequent siblings (SS) and control (CT) infants, *Sleep Res.*, 9, 103, 1980.
122. **Harper, R. M., Leake, B., Hoffman, H., II, Walter, D. O., Hoppenbrouwers, T., Hodgman, J. E., and Sterman, M. B.,** Periodicity of sleep states is altered in infants at risk for the sudden infant death syndrome, *Science*, 213, 1030, 1981.
123. **Guillemenault, C. and Coons, S.,** Sleep states and maturation of sleep: a comparative study between full-term normal controls and near-miss SIDS infants, in *Sudden Infant Death Syndrome*, Tildon, J. T., Roeder, L. M., and Steinschneider, A., Eds., Academic Press, New York, 1983, 401.
124. **Haddad, G. G., Walsh, E. M., Leistner, H. L., Grodin, W. K., and Mellins, R. B.,** Abnormal maturation of sleep states in infants with aborted sudden infant death syndrome, *Pediatr. Res.*, 75, 1055, 1981.
125. **Anderson-Huntington, R. B. and Rosenblith, J. F.,** Central nervous system damage as a possible component of unexpected deaths in infancy, *Dev. Med. Child. Neurol.*, 18, 480, 1976.
126. **Newman, N. M., Frost, J. K., Bury, L., Jordan, K., and Phillips, K.,** Responses to partial nasal obstruction in sleeping infants, *Aust. Paediatr. J.*, 22, 111, 1986.
127. **Van der Hal, A. L., Rodriguez, A. M., Sargent, C. W., Platzker, A. C., and Keens, T. G.,** Hypoxic and hypercapnic arousal responses and prediction of subsequent apnea in apnea of infancy, *Pediatrics*, 75, 848, 1985.
128. **Rodriguez, A. M., Warburton, D., and Keens, T. G.,** Elevated catecholamine levels and abnormal hypoxic arousal in apnea of infancy, *Pediatrics*, 79, 269, 1987.
129. **Davidson-Ward, S. L., Bautisa, D. B., and Keens, T. G.,** Abnormal hypoxic arousal responses in subsequent siblings of SIDS victims, *Clin. Res.*, 36, 240A, 1988.
130. **Davidson-Ward, S. L.,** personal communication.
131. **Dell, P.,** Reticular homeostasis and critical ractivity, *Prog. Brain Res.*, 1, 82, 1973.
132. **Hoppenbrouwers, T., Jensen, D., Hodgman, J. E., Harper, R. M., and Sterman, M. B.,** Body movements during quiet sleep (QS) in subsequent siblings of SIDS, *Clin. Res.*, 30, 136A, 1982.
133. **Takashima, S., Armstrong, D., Becker, L., and Bryan, C.,** Cerebral hypofusion in the sudden infant death syndrome? Brainstem gliosis and vasculature, *Ann. Neurol.*, 4, 257, 1978.
134. **Kinney, H. C., Burger, P. C., Harrell, F. E., and Hudson, R. P.,** 'Reactive gliosis' in the Medulla Oblongata of victims of the sudden infant death syndrome, *Pediatrics*, 72, 181, 1983.
135. **Summers, G. C., and Parker, J. C.,** The brain stem in sudden infant death syndrome: a postmortem survey, *Am. J. Foresinc Med. Pathol.*, 2, 121, 1981.
136. **Quattrochi, J. J., Baba, N., Liss, L., and Adrion, W.,** Sudden infant death syndrome: preliminary study of reticular dendrite spines in infants with SIDS, *Brain Res.*, 181, 245, 1980.
137. **Korobkin, R. and Guillemenault, C.,** Neurologic abnormalities in near miss for sudden infant death syndrome infants, *Pediatrics*, 64, 309, 1979.
138. **Salk, L., Grellong, B. A., and Deitrich, J.,** Sudden infant death: abnormal cardiac habituation and poor autonomic control, *N. Engl. J. Med.*, 291, 219, 1974.
139. **Groves, P. M. and Lynch, G. S.,** Mechanisms of habituation in the brainstem, *Psychol. Rev.*, 79, 237, 1972.
140. **Beckwith, J. B.,** Chronic hypoxemia in the sudden infant death syndrome: a critical review of the data base, in *Sudden Infant Death Syndrome*, Tildon, J. T., Roeder, L. M., and Steinschneider, A., Eds., Academic Press, New York, 1983, 145.
141. **Bridges, R., Condie, R. M., Zak, S., and Good, R.,** The morphologic basis of antibody formation development during the neonatal period, *J. Lab. Clin. Med.*, 53, 331, 1959.
142. **Lombeck, L., Kasperek, K., Harbisch, H. D., Feinendegen, L. E., and Bremer, H. J.,** The selenium state of healthy children, *Eur. J. Pediatr.*, 125, 81, 1977.
143. **Koenig, H. M. and Lightsey, A. L.,** "Growth curves" for hemoglobin and red cell size during infancy, *Clin. Res.*, 27, 108a, 1979.
144. **Beaver, G. H., Ellis, M. J., and White, J. C.,** Studies on human fetal hemoglobin. II. Fetal hemoglobin levels in healthy children and adults and in certain hematological disorders, *Br. J. Haematol.*, 6, 201, 1960.
145. **Harper, R. M., Walter, D. O., Leake, B., Hoffman, H. J., Sieck, G. C., Sterman, M. B., Hoppenbrouwers, T., and Hodgman, J. E.,** Development of sinus arrhythmia during sleeping and waking states in normal infants, *Sleep*, 1, 33, 1978.
146. **Hoppenbrouwers, T., Harper, R. M., Hodgman, J. E., Sterman, M. B., and McGinty, D. J.,** Polygraphic studies of normal infants during the first six months of life. II. Respiratory rate and variability as a function of state, *Pediatr. Res.*, 12, 120, 1978.

147. **Hoppenbrouwers, T., Hodgman, J. E., Harper, R. M., and Sterman, M. B.,** Motility patterns as a function of age and time of night, *Sleep Res.,* 8, 124, 1979.
148. **Harper, R. M., Hoppenbrouwers, T., Sterman, M. B., McGinty, D. J., and Hodgman, J. E.,** Polygraphic studies of normal infants during the first six months of life. I. Heart rate and variability as a function of state, *Pediatr. Res.,* 10, 945, 1976.
149. **Sterman, M. B., Harper, R. M., Havens, B., Hoppenbrouwers, T., McGinty, D. J., and Hodgman, J. E.,** Quantitative analysis of infant EEG development during quiet sleep, *Electroencephalogr. Clin. Neurophysiol.,* 43, 371, 1977.
150. **Brazelton, T. B.,** Crying in infancy, *Pediatrics,* 29, 579, 1962.
151. **Dittrichova, J. and Lapackova, V.,** Development of the waking state in young infants, *Child Dev.,* 35, 365, 1964.
152. **Parmelee, A. H., Wenner, W. H., and Schultz, H. R.,** Infant sleep patterns from birth to 16 weeks of age, *J. Pediatr.,* 65, 576, 1964.
153. **Hoppenbrouwers, T., Hodgman, J. E., Arakawa, K., Geidel, S. A., and Skiman, M. B.,** Sleep and waking states in infancy; normative studies, *Sleep,* 11, 387, 1988.
154. **Jilik, L., Trojan, S., and Travnickova, E.,** Lactic acid and glycogen changes in the rat brain due to aerogenic (altitude) hypoxia during ontogenesis, *Physiol. Bohemoslov.,* 15, 532, 1966.
155. **Sasaki, C. T.,** Development of laryngeal function: etiologic significance in the sudden infant death syndrome, *Laryngoscope,* 89, 1964, 1979.
156. **Langlais, P. J., Walsh, F. X., Bird, E. D., and Levy, H. L.,** Cerebrospinal fluid neurotransmitter metabolites in neurologically normal infants and children, *Pediatrics,* 75, 580, 1985.
157. **Dobbing, J. and Sands, J.,** Quantitative growth and development of human brain, *Arch. Dis. Child.,* 48, 757, 1973.
158. **Kappas, A. and Alvares, A. P.,** How the liver metabolizes foreign substances, *Sci. Am.,* 232, 22, 1975.
159. **Jugo, S.,** Metabolism of toxic heavy metals in growing organisms: a review, *Environ. Res.,* 13, 36, 1977.
160. **Kostial, K., Kello, D., Jugo, S., Rabar, I., and Maljkovic, T.,** Influence of age on metal metabolism and toxicity, *Environ. Health Perspect.,* 25, 81, 1978,.
161. **Sterman, M. B., McGinty, D. J., Harper, R. M., Hoppenbrouwers, T., and Hodgman, J. E.,** Developmental comparison of sleep EEG power spectral patterns in infants at low and high risk for sudden death, *Electroencephalogr. Clin. Neurophysiol.,* 53, 166, 1982.
162. **Hoppenbrouwers, T., Jensen, D. K., Hodgman, J. E., Harper, R. M., and Sterman, M. B.,** The emergence of a circadian pattern in respiratory rates: comparison between control infants and subsequent siblings of SIDS, *Pediatr. Res.,* 14, 345, 1980.
163. **Gould, J. B., Gluck, L., and Kulovitch, M. V.,** The relationship between accelerated pulmonary maturity and accelerated neurological maturity in certain chronically stressed pregnancies, *Am. J. Obstet. Gynecol.,* 127, 181, 1977.
164. **Gluck, L., Kulovich, M. V., and Hallman, M.,** Effects of chronic intrauterine asphyxia in organ maturation, in *Intrauterine Asphyxia and the Developing Fetal Brain,* Gluck, L., Ed., Year Book Medical Publishers, Chicago, 1977, 93.
165. **Henderson-Smart, D. J., Pettigrew, A. G., and Campbell, D. J.,** Prenatal stress, brain stem neural maturation, and apnea in preterm infants, in *Sudden Infant Death Syndrome,* Tildon, J. T., Roeder, L. M., and Steinschneider, A., Eds., Academic Press, New York, 1983, 293.
166. **Hollenberg, M., Honbor, N., and Samorodin, A. J.,** Effects of hypoxia on cardiac growth in neonatal rat, *Am. J. Physiol.,* 231, 1445, 1976.
167. **Reite, M., Jackson, D., Cahoon, R. L., and Weil, J. V.,** Sleep physiology at high altitude, *Electroencephalogr. Clin. Neurophysiol.,* 38, 463, 1975.
168. **Petajan, J. H.,** Influence of altitude on the electroencephalogram. Sleep induction at moderate altitude, Technical Report, Contract No. DADA-17-68-C-8018. U.S. Army Medical Research Development Command, Washington, D.C., 1971.
169. **James, L. S. and Rowe, R. D.** The pattern of response of pulmonary and systematic arterial pressures in newborn and older infants to short periods of hypoxia, *J. Pediatr.,* 51, 5, 1957.
170. **Kinney, H. C., Brody, B. A., Kloman, A. S., and Gilles, F. H.,** Sequence of central nervous system myelination in human infancy. II. Patterns of myelination in autopsied infants, *J. Neuropathol. Exp. Neurol.,* 47, 217, 1988.
171. **Himwich, H. E. and Fazekas, J. F.,** Comparative studies of the metabolism of the brain of infant and adult dogs, *Am. J. Physiol.,* 132, 454, 1941.
172. **Sterman, M. B. and Clemente, C. D.,** Basal forebrain structures and sleep, *Acta Neurol. Latinoam.,* 14, 228, 1968.
173. **Moore, R. Y. and Eichler, V. B.,** Loss of a circadian adrenal corticosterone rhythm following suprachiasmatic lesions in the rat, *Brain Res.,* 42, 201, 1972.
174. **Stothers, J. K. and Warner, R. M.,** Oxygen consumption and sleep state in the newborn, *J. Physiol.,* 269, 57, 1977.

175. **Denoroy, L., Kopp, N., Gay, N., Bertrand, E., Pujol, J. F., and Gilley, R.**, Activités des enzymes de synthèse des catécholamines dans des régions du tronc cérébral au cours de la mort subite du nourrisson, *C.R. Acad. Sci. Ser. D Paris*, 291, 245, 1981.
176. **Ozand, P. T. and Tildon, J. T.**, Alterations of catecholamine enzymes in several brain regions of victims of sudden infant death syndrome, *Life Sci.*, 32, 1765, 1983.
177. **Bergstrom, L., Lagercranz, H., and Terenius, L.**, Post-mortem analyses of neuropeptides in brains from sudden infant death victims, *Brain Res.*, 323, 279, 1984.
178. **Hedner, T., Lundborg, P., and Engel, J.**, Effect of hypoxia on monoamine synthesis in brains of developing rats, *Biol. Neonate*, 32, 229, 1977.
179. **MacMillan, V. and Siesjo, B. K.**, Brain energy metabolism in hypoxemia, *Scand. J. Clin. Lab. Invest.*, 30, 127, 1972.
180. **Vaccari, A., Brotman, S., Cimino, J., and Timiras, P. S.**, Adaptive changes induced by high altitude in the development of brain monoamine enzymes, *Neurochem. Res.*, 3, 295, 1978.
181. **Timiras, P. S. and Woolley, D. E.**, Functional and morphologic development of brain and other organs of rats at high altitude, *Fed. Proc., Fed. Am. Soc. Exp. Biol.*, 25, 1312, 1966.
182. **Miller, A. J., McKoon, M., Pinneau, M., and Silverstein, R.**, Postnatal Synaptic Development of the Nucleus Solitarius After Chronic Prenatal Hypoxia, paper presented at the International Research Conference on SIDS, Alexandria, VA, 1982.
183. **Hoppenbrouwers, T.**, Sudden Infant Death Syndrome (SIDS) and Sleep, *Proc. 11th Annu. Conf. of the IEEE/Engineering in Medicine and Biology Society*, 11(1), 402, 1989.
184. **van Someren, V. and Stothers, J. K.**, A critical dissection of obstructive apnea in the human infant, *Pediatrics*, 71, 721, 1983.
185. **Guillemenault, C., Ariagno, R., Coons, S., Winkle, R., Korobkin, R., Baldwin, R., and Souquet, M.**, Near miss sudden infant death syndrome in eight infants with apnea related cardiac arrhythmias, *Pediatrics*, 76, 236, 1985.
186. **Guillemenault, C., Souquet, M., and Ariagno, R. L.**, Five cases of near-miss sudden infant death syndrome and development of obstructive sleep apnea syndrome, *Pediatrics*, 73, 71, 1984.
187. **Hunt, C. E., Brouillette, R. T., and Hanson, D.**, Theophylline improves pneumogram abnormalities in infants at risk for sudden infant death syndrome, *J. Pediatr.*, 103, 969, 1983.
188. **Aranda, J. V., Turmen, T., Davis, J., Trippenbach, T., Grondin, D., Zinman, R., and Watters, G.**, Effect of caffeine on control of breathing in infantile apnea, *J. Pediatr.*, 103, 975, 1983.
189. **Hon, E. H.**, An atlas of fetal heart rate patterns, Harty Press, New Haven, CT, 1968.
190. International Classification of Sleep Disorders, in press.
191. **Jones, R. W. A., Sharp, C., Rabb, L. R., Lambert, B. R., and Chamberlain, D. A.**, 1028 Neonatal electrocardiograms, *Arch. Dis. Child.*, 54, 427, 1979.
192. **Hodgman, J. E. and Hoppenbrouwers, T.**, Home monitoring for the sudden infant death syndrome: the case against, *Ann. N.Y. Acad. Sci.*, 533, 164, 1988.
193. **Hodgman, J. E., Gonzalez, F., Hoppenbrouwers, T., and Cabal, L.**, Apnea, transient episodes of bradycardia and periodic breathing in preterms, *Am. J. Dis. Child.*, 144, 54, 1990.
194. **Ariagno, R. L., Guillemenault, C., Korobkin, R., Boeddiker, M., and Baldwin, R.**, "Near-miss" for sudden infant death syndrome infants: a clinical problem, *Pediatrics*, 71, 726, 1983.
195. **Kahn, A. and Blum, D.**, Home monitoring of infants considered at risk for the sudden infant death syndrome. Four year's experience (1977—1981), *Eur. J. Pediatr.*, 139, 94, 1982.
196. **Lewak, N.**, Sudden infant death syndrome in a hospitalized infant on an apnea monitor, *Pediatrics*, 56, 296, 1975.
197. **Black, L., Hersher, L., and Steinschneider, A.**, Impact of the apnea monitor on family life, *Pediatrics*, 62, 681, 1978.
198. **Cain, L. P., Kelly, D. H., and Shannon, D. C.**, Parent's perceptions of the psychological and social impact of home monitoring, *Pediatrics*, 66, 37, 1980.
199. **Wasserman, A. L.**, A prospective study of the impact of home monitoring on the family, *Pediatrics*, 74, 323, 1984.
200. **Kahn, A., Sottiaux, M., Appelboom-Fondu, J., Blum, D., Rebuffat, E., and Levitt, J.**, Long-term development of children monitored as infants for an apparent life-threatening event during sleep: a 10-year follow-up study, *Pediatrics*, 83, 668, 1989.
201. **Wortman, C. B. and Silver, R. L.**, Is "processing" a loss necessary for adjustment? A study of parental reactions to death of an infant, submitted.
202. **Smialek, Z.**, Observations on immediate reactions of families to sudden infant death, *Pediatrics*, 62, 160, 1978.
203. **Goldring, J., Limerick, S., and Macfarlac, A.**, *Sudden Infant Death, Patterns, Puzzles and Problems*, University of Washington Press, Seattle, 1985.

204. **Mandell, F.,** The family and sudden infant death syndrome, in *Sudden Infant Death Syndrome, Medical Aspects and Psychological Management,* Culbertson, J. L., Krous, H. F., and Bendell, R. D., Eds., Johns Hopkins University Press, Baltimore, 1988.
205. **Hoppenbrouwers, T. and Hodgman, J. E.,** *SIDS, Facts, Feelings, Research and Resources,* in press.

THE DEVELOPMENT OF RESPIRATORY CAPACITY DURING EXERCISE IN CHILDREN AND ADOLESCENTS

Thomas D. Fahey

INTRODUCTION

The pulmonary system serves as the interface between the environment and the cells of the body for the movement of oxygen and carbon dioxide. In addition, this system plays a vital role in the regulation of pH. At rest, the demands on this system are minimal, with oxygen consumption being approximately 0.3 l·min^{-1} or 3.5 ml·kg^{-1}·min^{-1} (1 MET, or resting metabolic equivalent), and CO_2 expiration being slightly less than this.[1] During maximal exercise, the demands on the system increase tremendously: the metabolic rate may be more than 25 times above rest in a well-trained, male adult endurance athlete and more than 18 times above rest in the well-trained male adolescent endurance athlete. Minute ventilation rates ($\dot{V}e$), which are only 6 to 12 l·min^{-1} at rest, may exceed 125 to 200 l·min^{-1} during maximal exercise in the adult.[2,3] In children, the rates are more variable, depending on factors such as age, height, and maturational level. However, in relation to body size, minute ventilation values are similar to those of adults.[2,4,5]

The movement of O_2 and CO_2 into and out of the bloodstream (respectively) depend on the coordination of the three basic functions of the pulmonary system: ventilation, diffusion, and perfusion. The process of ventilation involves the movement of ambient air from the environment into the alveoli of the lungs. Diffusion is the passive movement of gases across the alveolar membrane, into or out of the pulmonary circulation; and perfusion involves the movement of blood into the pulmonary vasculature.

In the healthy adult, the pulmonary system does not appear to limit maximal oxygen consumption ($\dot{V}O_2$max).[1] Arterial oxygen pressure (PaO_2) and hemoglobin saturation remain at almost resting levels during maximal exercise, and the ventilatory reserve is not exhausted. Children have a lower absolute ventilatory capacity than adults, yet have similar relative ventilatory and maximal oxygen consumption capacities, except at the very highest levels of performance (elite adult endurance athletes have much higher relative aerobic capacities than do elite child endurance athletes).

This chapter will explore the processes of pulmonary ventilation, diffusion, and perfusion in developing children during exercise. Specific topics include normative values of exercise respiratory function during growth, the influence of genetics and endurance training on these processes, and ventilation as a limiting factor of $\dot{V}O_2$max in children and adolescents.

THE CLINICAL SIGNIFICANCE OF PULMONARY EXERCISE DATA

Pulmonary exercise studies are of obvious interest in enhancing the knowledge of the role of the lungs in exercise and oxygen transport capacity. Such studies are also extremely valuable to the pediatric clinician in determining the presence or severity of obstructive airway diseases.[6,7] Patient exercise responses for minute volume, arterial blood gases, diffusion, and perfusion can be compared with values from normal populations in an effort to uncover any underlying pathology. In addition, the challenge of exercise administered in a clinical setting is perhaps the best method of inducing exertional dyspnea and exercise-induced bronchospasm among patients complaining of exercise intolerance and shortness of breath.[8,9]

The clinical usefulness of respiratory exercise data in children and adolescence depends upon their proper interpretation. While the normal values for most measurements of pul-

Table 1
MAXIMAL EXERCISE VENTILATION VOLUME (l · min^{-1})

Study	Sex	4—6	7—8	9—10	11—12	13—14	15—16	17—18	Ref.
Rutenfranz et al.	M	—	—	—	82	99	117	112	5
	F	—	—	—	79	88	81	80	
Andersen et al.	M	—	48	45	57	57	78	—	13
	F	—	44	45	51	54	50	—	
Åstrand	M	40	62	71	75	113	110	110	2
	F	34	57	61	80	88	94	93	
Robinson	M	33	—	53	—	92	—	121	4

Note: Data were sometimes smoothed to fit into comparable age categories.

Table 2
MAXIMAL EXERCISE BREATHING FREQUENCY (b · min^{-1})

Study	Sex	4—6	7—8	9—10	11—12	13—14	15—16	17—18	Ref.
Rutenfranz et al.	M	—	—	—	—	54	50	48	5
	F	—	—	—	—	52	43	43	
Andersen et al.	M	—	57	39	49	35	38	—	13
	F	—	61	50	49	37	42	—	
Åstrand	M	70	67	58	54	53	45	42	2
	F	66	67	61	54	52	51	48	
Robinson	M	62	—	57	—	50	—	46	4

Note: Data were sometimes smoothed to fit into comparable age categories.

monary function are +/− 20 to 30% of the mean, the appropriate normal value must be chosen in light of factors such as height, weight, age, sex, race, and physical fitness.[9-11] The sometimes rapid physical changes that occur during growth may confound the interpretation of respiratory exercise data. In addition, the measurement of ventilatory capacity during exercise in young children may be subject to considerable error.[12] Children often have limited attention spans, have difficulty following instructions, may be intimidated by the instrumentation and clinical or laboratory procedures, and may lack the motivation to perform maximally during an exercise tolerance test.[13] Typical of the difficulty of obtaining valid data on young subjects was a statement by Robinson, "The youngest boys were unwilling to continue work after it ceased to be fun..."[5] It is important that test results (or, for the matter, the results of published studies) be validated by objective measures of maximal exercise performance such as maximal respiratory exchange ratio ($\dot{V}CO_2/\dot{V}O_2$), maximal heart rate, or maximal blood lactate.

VENTILATION

Extensive reports of ventilatory response to exercise have been presented by Robinson,[4] Morse et al.,[14] Åstrand,[15] Andersen et al.,[13] and Rutenfranz et al.[5] Incidental reporting of exercise pulmonary data on children and adolescence also appear in numerous other studies.[16-18] Summaries of minute volume, breathing frequency, and tidal volume achieved during maximal exercise by children of various ages appears in Tables 1 to 3.

Minute ventilation (Ve) increases as a function of exercise intensity in children and adults due to increases in breathing frequency (f_b) and tidal volume (V_t). To achieve the

Table 3
MAXIMAL EXERCISE TIDAL VOLUME (l)

Study	Sex	\multicolumn{7}{c}{Age (years)}	Ref.

Study	Sex	4—6	7—8	9—10	11—12	13—14	15—16	17—18	Ref.
Rutenfranz et al.	M	—	—	—	—	1.9	2.4	2.4	5
	F	—	—	—	—	1.7	1.9	1.8	
Andersen et al.	M	—	0.9	1.2	1.2	1.7	2.1	—	13
	F	—	0.7	0.9	1.1	1.5	1.2	—	
Åstrand	M	0.6	1.1	1.3	1.6	2.5	2.8	2.9	2
	F	0.5	0.9	1.1	1.6	1.9	2.0	2.0	
Robinson	M	0.5	—	0.9	—	1.9	—	2.7	4

Note: Data were sometimes smoothed to fit into comparable age categories.

high minute volumes of exercise, children rely on a higher breathing frequency and lower tidal volume than adults.[4,5,13] Ventilation increases as a linear function of exercise intensity and oxygen consumption ($\dot{V}O_2$) until about 60 to 70% of maximal capacity, when ventilation increases at a faster rate than oxygen consumption.[19-21] This ventilation inflection point (Tvent) occurs at approximately the same relative point in the ventilation-oxygen consumption relationship in children and adults. The possible significance of Tvent will be briefly discussed alter in this chapter.

Ventilation accelerates faster in young children during maximal exercise, reaching 73% of $\dot{V}$emax within 1 min in 6 year olds and 50 to 62% in 11 to 17 year olds.[5] Because the phenomenon of rapid ventilatory acceleration is only evident in very young subjects, it is possible that it is due to measurement error (i.e., lack of subject cooperation resulting in a submaximal effort).

Maximal exercise ventilation volume increases with age until young adulthood. However, $\dot{V}$emax is approximately the same in growing children and adults when the values are expressed per unit body weight, height, or body surface area.[2,14] Morse et al.[14] and Robinson[4] have found that $\dot{V}$emax/body weight (kilograms) is 1.6 to 1.8 in subjects between 6 and 25 years of age. However, Gadhoke and Jones[22] found that $\dot{V}$emax had greater relationship to age than height, weight, body surface area, or vital capacity. Åstrand[2,15] reported typical maximal exercise minute ventilation rates (l·min^{-1}, body temperature, ambient pressure, saturated, BTPS) for boys as 40 l (4 to 6 years old), 75 l (12 to 13 years old), and 122 l (20 years old). With only minor discrepancies, these values are similar to data presented by other investigators.[4,5,18] Variations between studies in findings can probably be attributed to differences between subjects in stature and physical fitness.

The increase in $\dot{V}$emax during growth parallels increases in lung volumes, such as vital capacity, functional residual capacity, and total lung capacity, and volume-dependent ventilatory capacity, such as maximum voluntary ventilation, forced inspiratory volume per second, and forced expiratory volume per second.[23] However, alveolar ventilation accounts for 80 to 85% of the total ventilation during maximal exercise in children, but only 75 to 80% in adults.[24]

A number of factors may lead to confusion when attempting to assess normal values for maximal exercise ventilation from growth alone. $\dot{V}$emax increases abruptly during the adolescent growth spurt in boys, but not in girls.[4] While $\dot{V}$emax is largely determined by vital capacity and the strength and endurance of the respiratory muscles, these factors can be affected by training and are subject to racial and sexual variation.

Mirwald et al.[25] found that the largest increases in maximal oxygen consumption occur during the peak height velocity (i.e., when the child is growing at the greatest rate). This trend was greater in trained than in untrained children. Since $\dot{V}$emax is closely related to $\dot{V}O_2$max, it may be that training during a critical growth period may result in a disproportionate increase in maximal ventilatory capacity than would be predicted from growth in stature alone.

Hsu and co-workers[26,27] have demonstrated differences in ventilatory function between Mexican-American, white, and black subjects, and Bar-Or[28] has reported differences between Arab and Jewish children in vital capacity and FEV_1 (forced expiratory volume per second). While the differences between Arabs and Jews can be attributed to height, other racial differences are not so easily explained. McBride and Wohl[9] have suggested that racial differences in pulmonary function may be due to variations in ratios of sitting to standing height. Thus, the basic relationship between stature and ventilatory function may not be altered, but must be scrutinized when an investigator is faced with apparent differences between populations.

$\dot{V}emax$ is approximately 5 to 10% higher in boys than in girls during childhood, but the gap widens abruptly during the pubertal growth spurt in boys.[4,13] At that time, $\dot{V}emax$ increases to a much greater extent in boys than in girls. By age 20, the values for boys are approximately 20% greater than in girls: 120 $l \cdot min^{-1}$ in males and 90 $l \cdot min^{-1}$ in females. Male values are higher both in absolute and relative to body size in the adult and older adolescent, but are approximately the same relative to body size in prepubertal children.[2] $\dot{V}emax$ is highly related to growth below a height of 160 cm in girls and to at least 170 cm in boys.[4] $\dot{V}emax$ seem to follow the pattern of declining or leveling of female physical performance characteristics at puberty.[29] The explanation of this phenomenon may lie in sex differences in the amount and intensity of physical activity during adolescence.[1]

Ventilatory equivalent, the ventilatory volume required for the consumption of a liter of oxygen ($\dot{V}e/\dot{V}O_2$), is considered to be measure of pulmonary efficiency. While some studies have found no systematic age or sex differences in this measure,[4,11,18] others have found that it is lower in adults than in children.[2,13] Typical values for 8- to 17-year-old boys and girls are 29 to 35 during maximal exercise and 22 to 28 at 60% of $\dot{V}O_2max$.[4,18,30] Ventilatory equivalent changes with training in children, so the divergent findings may reflect differences in aerobic capacity rather than age-related differences.[30]

The ventilatory response of children and adolescents to running and cycling (bicycle ergometer) is approximately the same in relation to oxygen consumption, except for a smaller respiratory exchange ratio ($\dot{V}CO_2/\dot{V}O_2$) during cycling.[31] However, Åstrand[2] has found that the rhythmical pedalling cadence used in laboratory experiments affects $\dot{V}e$ in children, particularly at submaximal exercise intensities.

Submaximal exercise ventilation relative to weight, height, or body surface area tends to be slightly higher in children than in adults.[5] The 11- to 13-year-old male subjects of Koch and Eriksson,[30] exercising at a power output of 500 $kg \cdot min^{-1}$, achieved a $\dot{V}e$ of 24.9 $l \cdot min^{-1}$ and a $\dot{V}A$ of 21.9 $l \cdot min^{-1}$, indicating a high degree of ventilatory efficiency (i.e., 88%). These subjects were able to maintain a PaO_2 of 99 mmHg.

$\dot{V}e$ is similar in boys and girls during submaximal exercise when the exercise intensity is set at 50 to 65% of $\dot{V}O_2max$.[5,13] However, girls have a higher $\dot{V}e$ at any power output relative to weight, even though oxygen consumption is the same.[32]

Breathing Frequency and Tidal Volume

Respiratory frequency during maximum and submaximum exercise tends be higher in children than in adults.[2,4,13,33] Young children tend to rely on increased f_b, while adults tend to rely more on increased V_t to achieve a given level of ventilation.[4,13] Young children (i.e., 6 years old) use only 45% of vital capacity as tidal air during maximal exercise, while older teenagers use over 55%.[5] The increased respiratory rate, which is also evident at rest, has the effect of decreasing arterial PCO_2.[5]

Children exhibit the typical adult increase in tidal volume that occurs during the transition from rest to light exercise. However, when proceding from moderate to heavy exercise, tidal volume increases less in children than in adults. Instead, children rely on a relatively greater increase in breathing frequency. At age 8 years, children increase V_t by approximately 50% when going from rest to maximal exercise. By age 17 years, V_t increases by 150% during this transition.[4,13] Breathing frequency increases linearly with exercise intensity and ventilation volume in children and adults; however, the rate of increase is inversely related to age.

The smaller tidal volume and higher breathing frequency might be expected to lead to a higher dead space:tidal volume ratio (V_d/V_t) and thus a lower alveolar ventilation. However, this effect is reduced by a smaller anatomical dead space in children.[22] The V_d/V_t of approximately 20% found in children during exercise is similar to that found in adults.[22,24] Estimations for maximal exercise range from approximately 20% by Shephard and Bar-Or[24] to 11% by Koch and Erikkson.[30]

Maximal exercise V_t, which is highly related to total lung volume, increases with age until age 13 years in girls and 15 years in boys. After that, the increases are much more gradual and variable until adult values are assumed. There are no sex differences in exercise throughout childhood and adolescence in f_b.[4]

Ventilatory Inflection Point

The ventilation inflection point (Tvent), the point where the relationship between ventilation and oxygen consumption becomes nonlinear, is highly correlated to height.[19] However, when expressed relative to weight, Tvent occurs at approximately the same point in children and adults. Tvent occurs at a lower percentage of VO_2max in girls than in boys, again probably due to their relatively lower levels of physical fitness.

Several investigators have linked Tvent with increases in blood lactate that occur during heavy exercise.[21,35] They have called this point the anaerobic threshold and have hypothesized that it is caused by anaerobiosis in muscle. However, several studies have shown that the lactate and ventilatory inflection points are unrelated, and that the observed correlations between the two are coincidental.[35] Studies using subjects with McArdle's disease (patients lack the enzyme phosphorylase and are unable to produce lactate), subjects who were glycogen depleted, and subjects who were trained during the experiment demonstrated a dissociation between the lactate and ventilatory inflection points.[37-39] While the mechanism of Tvent is unknown, it probably lies in the neural mechanisms that are the primary controllers of ventilation during exercise.[1] The reader is referred to two recent discussions of this topic.[1,35]

DIFFUSION AND PERFUSION

Pulmonary diffusing capacity and capillary blood volume depend upon physical characteristics (i.e., body size) and lung volume.[40,41] The diffusing capacity of the lung (D_L) increases with age, height, weight, body surface area, and pulmonary capillary blood volume (Q_c), but is comparable to adults when expressed in relation to $\dot{V}O_2$max (ml·kg^{-1}·min^{-1}). During mild treadmill exercise, D_L varies from approximately 12 ml·min^{-1}·mmHg^{-1} at age 4 years to 30 ml·min^{-1}·mmHg^{-1} at age 12 years.[42] Koch and Eriksson studied D_{Lco} during near maximal exercise in 11- to 13-year-old boys and found it to be approximately 24 ml·mmHg^{-1}·min^{-1}, which, when expressed relative to body size (i.e., body surface area), is approximately the same as seen in healthy adults.[43,44]

Children show an increase in Q_c with age until adolescence, when the changes are more variable.[43] The ratio of V_c/V_A is lower in children, probably due to the lower relative number of alveoli in children compared with adults. Yet, arterial oxygen pressure (P_aO_2) and hemoglobin saturation during maximal exercise are similar in children and adults.[45,46] This indicates that the diffusion reserve (lung diffusion time — time required for hemoglobin saturation) is probably adequate in children and probably does not limit VO_2max in these subjects at sea level.

The venous admixture does not appear to be a significant factor affecting circulatory oxygenation during exercise in children.[43] The venous admixture is nonoxygenated blood that bypasses the ventilated parts of the lungs through direct venoarterial communication. It includes right-left shunts in the heart, abnormal communications between the aorta and pulmonary artery and between the pulmonary arteries and veins within the lungs, bronchial and Thesbian drainage into the postpulmonary circulation, and perfusion of unventilated alveoli.[43] The venous admixture falls from about 5 to 3% in both children and adults. The increased $PAO_2 - PaO_2$ reflects a decrease in venous PO_2 that is characteristic during maximal exercise.[20]

EXERCISE TRAINING

Exercise training tends to produce a more "adult-like" pulmonary exercise response in children.[30] Maximal exercise ventilation increases with training, while submaximal exercise $\dot{V}e$ decreases. Respiratory rate decreases and tidal volume increases at any intensity of exercise. In addition, exercise tidal volume in the trained child represents a higher percentage of vital capacity than in the untrained child. The changes in ventilatory characteristics may be, in part, due to improved endurance of the ventilatory muscles.[47]

Several studies did not find significant changes in maximal or submaximal ventilation with training.[10,45,48,49] These studies typify some of the problems inherent in training studies, particularly with younger subjects. Daniels and Oldridge[48] failed to use a control group, which made it difficult to separate the effects of training from normal changes that occur during growth. Vaccaro and Clarke,[10] who did not find significant changes in $\dot{V}e$max, FEV_1, and maximal voluntary ventilation following a season of competitive swimming, nevertheless demonstrated large mean increases in these factors. They used a small number of subjects, which makes it difficult to find significant differences, even when differences may exist. Certainly, the researcher is faced with extreme difficulty in first enlisting and then trying to obtain the cooperation of large numbers of young subjects for a prolonged period of time.

Koch and Eriksson[30] found that 16 weeks of physical training in 11- to 13-year-old boys resulted in a 22% increase in $\dot{V}O_2$max, a 22% increasee in alveolar ventilation, and a 49% increase in $\dot{V}e$max (mainly due to a 39% increase in tidal volume). While the ventilatory equivalent increased by 22%, the alveolar ventilatory equivalent remained the same. Training increased the ratio of tidal volume to vital capacity from 0.40 to 0.52. The latter value is a level similar to that found in adults.[24]

The maximal exercise hyperventilation following the training period may be caused by the increased muscle fiber tension due to the higher exercise intensity. This may result in a greater stimulation of the respiratory center by the limb mechanoreceptors and increased activity of the motor cortex.[1,30] So, although ventilatory equivalent increases with training, suggesting a decrease in ventilatory economy, both PAO_2 and PaO_2 increase as well, which demonstrates an increase in pulmonary exercise effectiveness with training.

$\dot{V}e$max is related to lung volume.[2,50] Engstrom et al.[50] found that girls involved in competitive swimming for 3 or more years showed substantially greater values for vital capacity, total lung capacity, and functional residual than sedentary girls or girls who had trained for 1 year or less. Training resulted in an increase in vital capacity that was disproportional to linear growth and the attending growth in lung capacity. Follow-up studies on the "girl swimmers" as adults showed that the increases in vital capacity, total lung capacity, and functional residual capacity remained higher than would be expected by growth alone.[51-53] On the other hand, higher values for $\dot{V}e$max, $\dot{V}O_2$max, blood volume, and hemoglobin demonstrated during the subjects' competitive period were no longer evident 10 years later. It is probable that large lung volumes, which have been evident in the subjects studied in several investigations of trained swimmers, is a prerequisite to becoming a successful swimmer.[52]

Physical training does not appear to induce any change in exercise carbon monoxide diffusion capacity (D_{Lco}) despite large increases in cardiac output and ventilatory capacities (Koch and Ericksson).[43] However, Vaccaro et al.[41] have demonstrated that well-trained child and adolescent swimmers had higher resting D_{Lco} than normal children of the same age and body dimensions. As with the large lung capacities of the "girl swimmers" of Åstrand and co-workers,[54] a greater than average lung diffusing capacity may be a prerequisite for success in swimming.

V_c, which is affected by body size, posture, exercise, and stroke volume, is the most important factor affecting lung diffusing capacity (D_L) is healthy subjects.[43] Exercise training appears to have no effect on pulmonary capillary blood volume beyond that which occurs during normal growth. Koch and Eriksson[43] hypothesized that since continued development and remodeling of alveolar structure with new formation of alveolar spaces occur from birth until the age of 8 years, physical training during those years (early and middle childhood) may have a positive effect on D_L and Q_c. Physical training results in a significant decrease in resting anatomical right-left shunt from 4.6 to 2.9% of cardiac output.

GENETICS AND PULMONARY FUNCTION DURING EXERCISE

The functional capacity of a physiological system is due to the inherent characteristics of the species, individual genetic variation, and environmental influences. The genetic component of physical performance has been shown considerable interest by researchers in recent years.[55-60] The contribution of heredity to a factor is determined by comparing intrapair variances in monozygous and dizygous twins. Genetics have been shown to have a strong influence on many variables associated with physical performance, such as $\dot{V}O_2$max, muscle respiratory capacity, running velocity, strength, power, muscle fiber composition, vital capacity, and body composition.[57,59]

Klissouras[59] and co-workers found no significant intrapair differences between identical and nonidentical twins in $\dot{V}e$max. Although $\dot{V}e$max is highly related to $\dot{V}O_2$max, other factors, such as the percent of slow-twitch muscle fibers, may account for much of the genetic component in aerobic capacity.[57]

PULMONARY FUNCTION AS A LIMITING FACTOR OF $\dot{V}O_2$max IN CHILDREN

In the adult, ventilation is not considered to be a limiting factor of $\dot{V}O_2$max at sea level.[1] Ventilation can increase to a much greater extent during exercise than cardiac output or oxygen consumption. $\dot{V}e$max remains well below maximum voluntary ventilation and little more than 50% of vital capacity is used to achieve the tidal volumes of heavy exercise. VAO_2 remains within 85 to 90% of $\dot{V}e$, and PaO_2 remains the same or drops slightly during maximal exercise.[11] The flat shape of the oxyhemoglobin dissociation curve at higher O_2 partial pressures ensures that hemoglobin saturation does not decrease appreciably.

Gas exchange in the lungs is adequate in children during maximal exercise. Alveolar-arterial PO_2 of approximately 15 mmHg and hemoglobin saturation of 93 to 94% are similar to those observed in healthy adults.[61] In addition, arterial pH is to reduced to the same extent in children as in adults, possibly indicating a less potent Bohr effect during exercise in children.[61]

SUMMARY

Maximal minute ventilation volume increases with age during childhood and adolescence. When $\dot{V}_{emax}$ is expressed per unit height, weight, or body surface area, the values are similar to those of adults. Submaximal ventilation per kilogram of body weight is higher in children than in adults. The ventilation inflection point is similar in both. Children achieve their exercise minute volumes by using a higher breathing frequency and lower tidal volume. Children also use a lower percentage of vital capacity to achieve their exercise tidal volumes than do adults. The higher breathing frequency results in a slightly lower PCO_2 than in adults. Ventilation, diffusion, and perfusion are adequate in children during exercise and do not appear to limit $\dot{V}O_2max$.

Training tends to make the pulmonary responses of children more "adult-like". Submaximal ventilation and breathing frequency decrease, while tidal volume/vital capacity maximal ventilation and tidal volume increase. Training increases vital capacity in swimmers, but running training seems to have no effect other than that which occurs during normal growth. Training has no effect on pulmonary diffusing capacity.

REFERENCES

1. **Brooks, G. A. and Fahey, T. D.**, *Exercise Physiology, Human Bioenergetics and its Applications*, Macmillan, New York, 1984.
2. **Åstrand, P.-O.**, *Experimental Studies of Physical Working Capacity in Relation to Sex and Age*, Munksgaard, Copenhagen, 1952.
3. **Fahey, T. D., Akka, L., and Rolph, R.**, Body composition and $\dot{V}O_2max$ of exceptional weight-trained athletes, *J. Appl. Physiol.*, 39, 559, 1975.
4. **Robinson, S.**, Experimental studies of physical fitness in relation to age, *Arbeitsphysiologie*, 10, 251, 1938.
5. **Rutenfranz, K. L., Andersen, V., Seliger, F., Klimmer, J., Ilmarinen, M., Ruppel, R., and Kylian, H.**, Exercise ventilation during the growth spurt period, comparison between two European countries, *Eur. J. Pediatr.*, 136, 135, 1981.
6. **Thoren, C.**, Exercise testing in children, *Paediatrician*, 7, 100, 1978.
7. **Bar-Or, O.**, Clinical implications of pediatric exercise physiology, *Ann. Clin. Res.*, 34, 97, 1982.
8. **Godfrey, S.**, *Exercise Testing in Children*, W. B. Saunders, London, 1974.
9. **McBride, J. T. and Wohl, M. E. B.**, Pulmonary function tests, *Pediatr. Clin. N. Am.*, 26, 537, 1979.
10. **Vaccaro, P. and Clarke, D. H.**, Cardiorespiratory alterations in 9 to 11 year old children following a season of competitive swimming, *Med. Sci. Sports*, 10, 204, 1978.
11. **Mellerowicz, H. and Smodlaka, V. N.**, *Ergometry, Basics of Medical Exercise Testing*, Urban & Schwarzenberg, Baltimore, 1981.
12. **Wall, M. A.**, Office pulmonary function testing, *Pediatr. Clin. N. Am.*, 31, 773, 1984.
13. **Andersen, K. L., Seliger, V., Rutenfranz, J., and Messel, S.**, Physical performance capacity of children in Norway. III. Respiratory responses to graded exercise loadings — population parameters in a rural community, *Eur. J. Appl. Physiol.*, 33, 265, 1974.
14. **Morse, M., Schultz, F. W., and Cassels, D. E.**, Relation of age to physiological responses of the older boy (10—17 years) to exercise, *J. Appl. Physiol.*, 1, 683, 1949.
15. **Astrand, P.-O.**, Human physical fitness with special reference to sex and age, *Physiol. Rev.*, 36, 307, 1956.
16. **Nagle, F. J., Hagberg, J., and Kamei, S.**, Maximal O_2 uptake of boys and girls — ages 14—17, *Eur. J. Appl. Physiol.*, 36, 75, 1977.
17. **Knuttgen, H.**, Aerobic capacity of adolescents, *J. Appl. Physiol.*, 22, 655, 1967.
18. **Fahey, T. D., Del Valle-Zuris, A., Oehelsen, G., Trieb, M., and Seymour, J.**, Pubertal stage differences in hormonal and hematological responses to maximal exercise in males, *J. Appl. Physiol.*, 46, 823, 1979.
19. **Cooper, D. M. and Weiler-Ravell, D.**, Gas exchange response to exercise in children, *Am. Rev. Respir. Dis.*, 129, 47, 1984.
20. **Jones, N. L.**, Normal values for pulmonary gas exchange during exercise, *Am. Rev. Respir. Dis.*, 129, 44, 1984.

21. **Wasserman, K., Whipp, B. J., Koyal, S. N., and Beaver, W. L.,** Anaerobic threshold and respiratory gas exchange during exercise, *J. Appl. Physiol.*, 35, 236, 1973.
22. **Gadhoke, S. and Jones, N. L.,** The responses to exercise in boys aged 9—15 years, *Clin. Sci.*, 37, 789, 1969.
23. **Koch, G.,** Aerobic power, lung dimensions, ventilatory capacity, and muscle blood flow in 12- to 16-year-old boys with high physical activity, in *Children and Exercise,* Berg, K. and Eriksson, B. O., Eds., University Park Press, Baltimore, 1980, 99.
24. **Shephard, R. J. and Bar-Or, O.,** Alveolar ventilation in near maximum exercise. Data on pre-adolescent children and young adults, *Med. Sci. Sports*, 2, 83, 1970.
25. **Mirwald, R., Bailey, D. A., Cameron, N., and Rasmussen, R. L.,** Longitudinal comparison of aerobic power in active and inactive boys aged 7 to 17 years, *Ann. Hum. Biol.*, 8, 405, 1981.
26. **Hsu, K. H. K., Jenkins, D. E., Hsi, B. P., Bourhofer, E., Hsu, F. C. F., and Jacob, S. C.,** Ventilatory functions of normal children and young adults — Mexican American, white, and black. II. Wright peak flowmeter, *J. Pediatr.*, 95, 192, 1979.
27. **Hsu, K. H. K., Jenkins, D. E., Hsi, B. P., Bourhofer, E., Thompson, V., Tanakawa, N., and Hsieh, G. S. J.,** Ventilatory functions of normal children and young adults — Mexican-American, white, and black, *J. Pediatr.*, 95, 14, 1979.
28. **Bar-Or, O.,** A comparison of responses to exercise and lung functions of Israeli Arabic and Jewish 12 to 14 year-old boys, in *Pediatric Work Physiology,* Bar-Or, O., Ed., Wigate Institute for Physical Education and Sport, Tel-Aviv, 1973, 59.
29. **Shephard, R. J.,** *Physical Activity and Growth,* Year Book Medical Publishers, Chicago, 1982.
30. **Koch, G. and Eriksson, B. O.,** Effect of physical training on pulmonary ventilation and gas exchange during submaximal and maximal work in boys aged 11 to 13 years, *Scand. J. Clin. Lab. Invest.*, 31, 87, 1973.
31. **Anderson, S. D. and Godfrey, S.,** Cardio-respiratory response to treadmill exercise in normal children, *Clin. Sci.*, 40, 433, 1971.
32. **Macek, M. and Vavra, J.,** Cardiopulmonary and metabolic changes during exercise in children 6—14 years old, *J. Appl. Physiol.*, 30, 200, 1971.
33. **Sobolova, V., Seliger, V., Grussova, D., Machovcova, J., and Zelenka, V.,** The influence of age and sports training in swimming on physical fitenss, *Acta Paediatr. Scand. Suppl.*, 217, 63, 1971.
34. **Davis, J. A., Frank, M. H., Whipp, B. J., and Wasserman, K.,** Anaerobic threshold alterations caused by endurance training in middle-aged men, *J. Appl. Physiol.*, 46, 1039, 1979.
35. **Brooks, G. A.,** Anaerobic threshold: review of the concept and directions for future research, *Med. Sci. Sports Exer.*, 17, 22, 1985.
36. **Hagberg, J., Coyle, E. M., Carroll, J. E., Miller, J. M., Martin, W. H., and Brooke, M. H.,** Exercise hyperventilation in patients with McArdles's disease, *J. Appl. Physiol.*, 52, 991,1982.
37. **Hughs, E. F., Turner, S. C., and Brooks, G. A.,** Effects of glycogen depletion and pedaling speed on "anaerobic threshold", *J. Appl. Physiol.*, 52, 1598, 1982.
38. **Segal, S. S. and Brooks, G. A.,** Effects of glycogen depletion and workload on postexercise O_2 consumption and blood lactate, *J. Appl. Physiol.*, 47, 512, 1979.
39. **Gaesser, G. A., Poole, D. C., and Garder, B. P.,** Dissociation between $\dot{V}O_2$max and ventilatory threshold responses to endurance training, *Eur. J. Appl. Physiol.*, 53, 242, 1984.
40. **Baran, D. and Englert, M.,** Characteristics of the pulmonary capillary network in normal children, in *Pediatric Work Physiology,* Bar-Or, O., Ed., Wingate Institute for Physical Education and Sport, Tel-Aviv, 1973, 219.
41. **Vaccaro, P., Zauner, C. W., and Updyke, W. F.,** Resting and exercise respiratory function in well trained child swimmers, *J. Sports Med. Phys. Fit.*, 17, 297, 1977.
42. **Giammona, S. T. and Daly, W. J.,** Pulmonary diffusing capacity in normal children, aged 4 to 13 years, *Am. J. Dis. Child.*, 110, 144, 1965.
43. **Koch, G. and Eriksson, B. O.,** Effect of physical trainig on anatomical R-L shunt at rest and pulmonary diffusing capacity during near-maximal exercise in boys 11—13 years old, *Scand. J. Clin. Lab. Invest.*, 31, 95, 1973.
44. **Koch, G. and Eriksson, B. O.,** Anatomical right-to-left shunt at rest and ventilation, gas exchange and pulmonary diffusing capacity during exercise in 11 to 13 year old boys before physical training, in *Pediatric Work Physiology,* Bar-Or, O., Ed., Wingate Institute for Physical Education and Sport, Tel-Aviv, 1973, 151.
45. **Ekblom, B.,** Effect of physical training in adolescent boys, *J. Appl. Physiol.*, 27, 350, 1969.
46. **Ekblom, B.,** Effect of physical training on oxygen transport system in man, *Acta Physiol. Scand. Suppl.*, 328, 1969.
47. **Bar-Or, O.,** *Pediatric Sports Medicine for the Practitioner: from Physiologic Principles to Clinical Applications,* Springer-Verlag, New York, 1983, 51.
48. **Daniels, J. and Oldridge, N.,** Changes in oxygen consumption of young boys during growth and running training, *Med. Sci. Sports*, 3, 161, 1971.

49. **Daniels, J., Oldridge, N., Nagle, F., and White, B.**, Differences and changes in VO$_2$ among young runners 10 to 18 years of age, *Med. Sci. Sports,* 10, 200, 1978.
50. **Engstrom, I., Eriksson, B. O., Karlberg, P., Saltin, B., and Thoren, C.**, Preliminary report on the development of lung volumes in young girl swimmers, *Acta Paediatr. Scand.*, 217, 73, 1971.
51. **Ericksson, B. O., Engstrom, I., Karlberg, A., Saltin, B., and Thoren, C.**, A physiological analysis of former girl swimmers, *Acta Paediatr. Scand. Suppl.*, 217, 68, 1971.
52. **Ericksson, B. O., Engstrom, I., Karlberg, P., Lundin, A., Saltin, B., and Thoren, C.**, Long-term effect of previous swim training in girls, a 10-year follow-up of the "girl swimmers", *Acta Paediatr. Scand.*, 67, 285, 1978.
53. **Ericksson, B. O., Lundin, A., and Saltin, B.**, Physical training of post-active girl swimmers, in *Pediatric Work Physiology,* Bar-Or, O., Ed., Wingate Institute for Physical Education and Sport, Tel-Aviv, 1973, 217.
54. **Åstrand, P.-O., Engstrom, L., Ericksson, B. O., Karlberg, P., Nylander, I., Saltin, B., and Thoren, C.**, Girl swimmers, *Acta Paediatr. Scand. Suppl.*, 147, 1963.
55. **Bouchard, C. and Malina, R. M.**, Genetics for the sport scientist, selected methodological considerations, *Exer. Sport Sci. Rev.*, 11, 275, 1983.
56. **Bouchard, C. and Malina, R. M.**, Genetics of physiological fitness and motor performance, *Exer. Sport Sci. Rev.*, 11, 306, 1983.
57. **Komi, P. V. and Karlsson, J.**, Physical performance, skeletal muscle enzyme activities, and fiber types in monozygous and dizygous twins of both sexes, *Acta Physiol. Scand. Suppl.*, 462, 1979.
58. **Klissouras, V. and Weber, G.**, Training, growth and heredity, in *Pediatric Work Physiology,* Bar-Or, O., Ed., Wingate Institute for Physical Education and Sport, Tel-Aviv, 1973, 209.
59. **Klissouras, V., Pirnay, F., and Petit, J.-M.**, Adaptation to maximal effort, genetics and age, *J. Appl. Physiol.*, 35, 288, 1973.
60. **Weber, G., Kartodihardjo, W., and Klissouras, V.**, Growth and physical training with reference to heredity, *J. Appl. Physiol.*, 40, 211, 1976.
61. **Ericksson, B. O., Grimby, G., and Saltin, B.**, Cardiac output and arterial blood gases during exercise in pubertal boys, *J. Appl. Physiol.*, 31, 348, 1971.

INDEX

INDEX

A

Acetylcholine, young ventricular cells and, 52
N-Acetylsalicylic acid and GAG synthesis, 79
ACTH and pulmonary development, 141—142
Action potential changes in developing chick heart, 54—56
Afferent nerve activity in fetal breathing movements, 150
Age-related changes
  in cardiorespiratory capacity, 93
  in coronary artery diameter, 103
  in ventilatory response to exercise, 211—212
Airway dynamics, 175—176
Alcohol; see also Ethanol
  fetal breathing movements and, 156
  as teratogen, 109—110, 117—119
Alveolar epithelium in the fetus, 165—167
Ambient pollution and sudden infant death syndrome (SIDS), 186
Amniocentesis and pulmonary hypoplasia, 141
Amniotic fluid and changes in surfactant system, 167
Anaerobic threshold, 96
Anaerobic work capacity in children, 95—97
Anesthetics and analgesics and fetal breathing, 156
Aortic arch defects, 116—117
Apnea
  of infancy, 182, 184, 186—189
  of prematurity, 185
Arousal failure and sudden infant death syndrome (SIDS), 187—189, 194
(Na, K)-ATPase activity, 52—53
Atrioventricular canal formation, 31, 34
Atrioventricular valves, 38
Automaticity, 54
Autonomic nervous system in surfactant regulation, 169

B

Barbiturates and fetal breathing movements, 156
Behavioral states and fetal breathing movements, 151
Bepridil, 56
β-Adrenergic activity and congenital heart defects, 116
β-Agonists and surfactant stimulation, 169
β-Endorphins and fetal breathing, 155
β-Lipoproteins in fetal plasma, 168
Biosynthesis of surfactant-type phospholipids, 166
Birth and pulmonary liquid dynamics, 139
Blood flow; see also Hemodynamics
  cardiogenesis and, 75
  exercise and, 88
Blood gas levels and fetal breathing movements, 153—154
Blood pressure
  cardiogenesis and, 73—74
  normal at various ages, 95
Blood vessel malformations, 107—132
Blood volume and cardiogenesis, 73
Body temperature and fetal breathing movements, 152—153
Botulism diagnosed as SIDS, 182
Brainstem pathology and sudden infant death syndrome (SIDS), 189
Brainstem reticular formation and arousal, 189, 194
Breathing frequency and exercise, 210, 212—213

C

Caffeine, 116—117, 156
Calcium channel blockers, 56
cAMP changes in embryonic chick heart, 58—59
Canilicular stage of fetal lung development, 132
Capillary supply of human heart, 104
Carbon dioxide levels and fetal breathing movements, 153—154
Cardiac automaticity, 54
Cardiac hypertrophy and endurance training, 87—88
Cardiac innervation, 43, 79
Cardiac jelly, 14—19
  resorption and cell changes, 71
  role in cardiogenesis, 75—76
Cardiac looping, 7, 71
  morphogenesis after, 26—27
  theory of, 19—20
  three key factors controlling, 77
Cardiac output
  equation, 87
  factors affecting, 87
Cardiac tissue differentiation
  heart looping, 19—20
  heart tube, 9—19
Cardiac vasculature, 39—41
  postnatal development, 103—105
  prenatal development, 101—103
Cardiogenesis
  cellular changes in, 69—71
  early, 3—5
  factors influencing, 69—70
  hemodynamic effects on, 41—43, 69—79
  role of cardiac jelly, 75—76
Cardiogenic plate, 4
Cardiorespiratory function, 92
Cardiovascular function
  bedrest and, 87
  development of, 85—97
Ca^{2+} slow channel regulation, 59—64
Catecholamines
  fetal breathing movements and, 155
  lung liquid regulation and, 136—137
  as teratogens, 110, 115—117
Cell death/cell proliferation and cardiac morphology, 71

Central nervous system maturation and risk of SIDS, 191
Cerebrospinal fluid HCO_3 and fetal breathing, 154
Chest diameter and heart size, 93
Chest wall mechanics, 177
Chick heart, early morphogenesis, 73
Chick heart studies of electrical activity, 51—64
Children
　anaerobic work capacity in, 95—97
　cardiovascular capacity in very young, 92—94
　exercise training and pulmonary function, 214—215
Cholinophosphate cytidyltransferase (CCT) in surfactant synthesis, 168—169
Chromosomal aberrations and congenital heart disease, 113
Cigarette smoking; see Smoking
Circadian rhythms and forebrain maturation, 193—194
Compliance of total respiratory system, 175
Conal and truncal cushions, 37
Congenital diaphragmatic hernia, 133, 140
Congenital heart disease
　chromosomal aberrations associated with, 113
　genetic factors in, 113—114
　multifactorial inheritance in, 114—115
　proposed teratogenic mechanisms, 115—120
Contraceptive hormones as potential teratogens, 110
Conus and truncus septation, 35—36
Coronary arteries, age-related changes in diameter, 103
Coronary artery transposition, 121—122
Coronary bed, 102
Coronary veins, 103
Coronary vessels, late development, 38—41
Crying vital capacity, 176
Cyclic nucleotide levels in developing chick heart, 58—59
Cyclic nucleotide regulation of Ca^{2+} slow channels, 59—64

D

D-600 (Ca^{2+} antagonist), 56
Diabetes mellitus, maternal and lung maturation, 142, 170
Diaphragmatic hernia, 133, 140
Diaphragmatic work of breathing, 176
Diazepam and fetal breathing, 156
Diltiazem, 56
Diurnal rhythms in fetal breathing movements, 152
Double-hearted embryo, 5
Down's syndrome and congenital heart disease, 113
Doxapram and fetal breathing, 157
Drugs and fetal breathing movements, 156—157

E

Edema syndrome, 119
EEG sleep spindle, 193—194
Elastic properties of the lung, 175
Electrical activity and ion channels in heart, 51—64
Electrogenic pump potential and (NaK)-ATPase activity, 52—53
Embryonic endocardium compared with chicken heart, 16—17
Embryonic period of lung development, 131
Endocardial blood flow, 42
Endocardial cushions, 32—34
　formation of, 77—79
　migration of, 78
　teratogenic exposure, 119
Endocardium, early development, 13
Endorphins and fetal breathing movements, 155
Endurance performance, genetic factors in, 89—92
Endurance training
　cardiac dimensions and, 87—88
　cardiovascular function and, 97
　maximal oxygen uptake and, 91
　respiratory response to, 214—215
Epicardial cells compared with myocytes, 39
Epicardium, late development, 38
Epidemiology of sudden infant death syndrome (SIDS), 181
Epidermal growth factor, 142
Epinephrine, heart rate and, 79
Esophageal pressure measurement, 175
Estrogen/progestin contraceptives as potential teratogens, 110
Estrogens in surfactant regulation, 169—170
Ethanol, proposed mechanism of injury, 118; see also Alcohol
^{3}H-Ethylenediamine-TTX, 56
Exercise
　breathing frequency in, 210
　cardiovascular function and, 85—97
　respiratory capacity development and, 209—216
　ventilatory response to, 210—213
Exercise training, pulmonary response to, 214—215

F

Fetal alcohol syndrome, 110
Fetal breathing movements, 134, 149—157; see also Sudden infant death syndrome (SIDS)
Fetal lung growth; see Pulmonary development
Fetus
　alveolar epithelium in, 166—167
　appearance of surfactant in, 167
Fibroblast pneumonocyte factor, 142
Fibronectin fluorescence of mesenchymal cells, 36—37
Fick principle and equation, 86
Flow-molding theories of cardiomorphogenesis, 76
Frank-Starling relationship and cardiogenesis, 73—74
Functional residual capacity, 174—175

G

Gas exchange in neonate, 177
Genetic factors
　in congenital heart disease, 113—115
　in endurance performance, 89—92

in pulmonary function, 215
German measles; see Rubella
Gestational age
  fetal breathing movements and, 151
  fetal movement in humans and, 151
Glucocorticoids
  in pulmonary development, 141—142
  in surfactant regulation, 170
Glucose and fetal breathing movements, 154—156
Glycosaminogycans (GAG), 71, 76—79
cGMP in embryonic chick heart, 59
Great artery transposition, teratogenic mechanisms, 121—122
Growth hormone, pre- and postexercise levels in boys, 96
Growth retardation *in utero*, 157
Guild for Infant Survival, 199

H

Heart; see also Cardiac
  differentiation and morphogenesis, 7—9
  early structure and developmental biology, 3—20
  electrical activity and ion channels, 51—64
  late development, 25—44
    coronary vessels, 38—41
    epicardium, 38
    external shape changes, 25—28
    inner shape changes, 28—38
    innervation, 43, 79
    role of hemodynamics, 41—43
  malformations of, 107—132
  primordia of, 4
  septation of, 34—37
  transversely fractured, 18
  vasculature of; see Cardiac vasculature
Heart rate
  cardiogenesis and, 72—73
  epinephrine and, 79
  normal at various ages, 95
Heart tube, structure of, 9—19
Heart valves, 37—38
Hemodynamic alterations and malformation, 119
Hemodynamics, role in cardiac development, 41—43, 69—79
Home monitoring in infants at risk for SIDS, 196—199
Hormones
  fetal breathing movements and, 154—156
  as potential teratogens, 110
  in surfactant regulation, 169—170
Hyaluronate, 78
Hydrogen ion and fetal breathing movements, 154
Hydrogen peroxide and cGMP levels, 59
Hypercapnia and fetal breathing movements, 153—154
Hypoglycemia and fetal breathing movements, 155
Hypothermic conditions as teratogens, 112
Hypoxia; see also Apnea
  fetal breathing movements and, 153
  in sudden infant death syndrome (SIDS), 186—189

  as teratogen, 111—112, 119—120

I

Innervation, cardiac, 43, 79
Inotropic agents in Ca^{2+} slow channel regulation, 59—64
Insulin and pulmonary development, 142
Interatrial septum, 34—35
Interventricular septum, 35
Intrathoracic space and fetal lung development, 133
Intrauterine growth retardation, 157
Ionizing radiation; see Radiation as potential teratogen
Ischemia and Ca^{2+} regulation, 62—63
Isoproterenol, 58

K

Keeshond dogs and congenital heart disease, 114
K^+ permeability, 51—52

L

Labor and fetal breathing movements, 152
Lithium as potential teratogen, 110—111
Looping; see Cardiac looping
Los Angeles County SIDS study, 186—187
Lung(s)
  development of; see Pulmonary development
  elastic properties of, 175
  expansion and lung liquid regulation, 137
  liquid, 133—139
  mechanics of, 173—177
Lymphatic capillaries of heart, 41

M

Male/female differences in ventilatory function, 212
Malformations of heart and blood vessels, 107—132
  environmental influences, 107—112
  genetic etiology of congenital heart disease, 113—114
  maternal conditions, 112—113
  multifactorial inheritance, 114—115
  proposed mechanisms
    in congenital heart disease, 115—120
    in great-artery transposition, 122—123
    in ventricular septal defect, 121—122
Maternal conditions, teratogenic effects, 112
Meperidine and fetal breathing, 156
Mesudipine, 56
Metabolites and fetal breathing movements, 154—156
Methadone and fetal breathing movements, 155
Methylxanthines, 116—117, 154, 156—157
Micropuncture techniques of embryo examination, 74—75
Mitotic myocardial cell, 14
Mn^{2+}-insensitive slow Na^+ channels, 54—58
Morphogenesis

early cardiac, 5—7
late cardiac, 25—38
Multifactorial inheritance in congenital heart disease, 114—115
Muscles, oxygenation of, 88—89
Myocardial capillaries, 41
Myocardium
　early development, 9—13
　trabeculation of, 30—31
Myocytes compared with epicardial cells, 39
Myoepicardial mantle, 78
Myofibrillogenesis, 12

N

Naloxone and fetal breathing movements, 154
Na^+ slow channels, 56—58
National Foundation for SIDS, 199
Neurotransmitters in sudden infant death syndrome (SIDS), 194—195
Nicotine and fetal breathing movements, 156
Nifedipine, 56
Nitroprusside and cGMP levels, 59
Nonpulmonary tissues and lung development, 135
NREM sleep
　diaphragmatic work of breathing in, 176
　fetal breathing movements and, 151
　functional residual capacity and, 174

O

Oligohydramnios and pulmonary hypoplasia, 140—141
Olympic athletes and training response, 86, 88
Opioids and fetal breathing movements, 155
Organ-cultured chick heart studies, 56—58
Overshoot theory of sudden infant death syndrome (SIDS), 192
Oxygenation and fetal breathing movements, 153
Oxygen consumption, maximal in adult, 86—89

P

Paradoxical sleep; see REM sleep
Parallel flow theory, 76
Peripheral blood flow, 88
Peripheral oxygen use, 88—89
Pethidine, 155
Phagocytosed myocardial cell fragment, 15
Phospholipid synthesis in surfactant system development, 166—167
Phosphorylation in Ca^{2+} slow channel regulation, 59—64
Phrenic nerve and lung development, 134—135
Pilocarpine, 157
Pituitary hormones and pulmonary development, 142
Postnatal development of coronary vasculature, 103—105
Potassium levels and hypoxic teratogenicity, 120
Precardiac mesoderm, differentiative events, 4—19
Premature infants and SIDS risk, 183

Prolactin
　lung liquid regulation and, 137
　pulmonary development and, 142
Prostaglandins, fetal breathing movements and, 155—156
Pseudoglandular stage of fetal lung development, 131—132
Pubertal growth spurt
　optimizing training effects, 94—95
　respiratory function and, 92
Pulmonary development
　embryonic/fetal period, 131—132
　endocrine control, 141—142
　physical factors in, 132—135
Pulmonary diffusion/perfusion and exercise, 213—214
Pulmonary exercise studies, 209—216
Pulmonary hypolasia, 139—140
Pulmonary liquid; see Lung(s)
Pulmonary surfactant
　biosynthesis, 166
　control of turnover, 168—170
　in mature lung, 165—166

Q

Quiet sleep and sudden infant death syndrome (SIDS), 188—189

R

Rabbit myocyte during late development, 28—29
Racial/ethnic differences in ventilatory function, 212
Radiation as potential teratogen, 109
REM sleep
　chest wall mechanics and, 177
　fetal breathing movements and, 151
　functional residual capacity and, 174
Respiratory movements; see Fetal breathing movements
Resting potential changes in embryonic development, 51—52
Risk identification for sudden infant death syndrome (SIDS), 181—183
Rubella
　gestational age dependency, 108
　in heart and blood vessel malformations, 107—108

S

Saccular stage of fetal lung development, 132
Saskatoon studies of cardiorespiratory function, 92
Sciatic nerve stimulation and fetal breathing movements, 154
Semilunar valves of heart, 37—38
Septation of heart, 34—37
Septum secundum, 35
Serotonin turnover and fetal breathing, 157
Sex differences in ventilatory function, 212
Sex hormones as potential teratogens, 110
SIDS; see Sudden infant death syndrome

Single-gene disorders and congenital heart disease, 114
Sinusoid-venous system, 103
Sleep spindle, 193—194
Slow channel regulation in embryonic chick heart, 59—64
Smoking
  fetal breathing movements and, 156
  sudden infant death syndrome (SIDS) and, 186
Socioeconomic factors in sudden infant death syndrome (SIDS), 186
Spiral flow theory, 76
Stature and ventilatory function, 212
Stroke volume
  cardiogenesis and, 73
  factors affecting, 87
Subepicardial space, 40
Sudden infant death syndrome (SIDS); see also Fetal breathing movements
  apparent life-threatening event and, 182, 196—198
  epidemiology, 181
  etiologic hypotheses, 187—197
  identification of risk infants, 181—183
  incidence and definition, 181
  management, 197—199
  theophylline and, 154, 156
Surfactant system development, 165—170
Swedish swimmer studies of cardiorespiratory function, 92
Sympathetic nervous system in surfactant regulation, 169

T

Teratogenic mechanisms
  in congenital heart disease, 115—120
  in heart and blood vessel malformations, 107—115
Terminal sac stage of fetal lung development, 132
Testosterone, pre- and postexercise levels in boys, 96
Tetrodotoxin (TTX)-insensitive slow Na^+ channels, 54—58
Thalidomide in heart and blood vessel malformations, 108—109

Thebesian veins, 103
Theophylline, 116—117, 154, 156
Thyroid hormones
  pulmonary development and, 142
  in surfactant regulation, 170
Tidal volume and exercise, 211—213
Total peripheral resistance, 83, 94
Trabeculation of myocardium, 30—31
Trainability, 91
Training response; see also Endurance training
  oxygen consumption and, 86—87
  prepubertal growth spurt and, 92, 94—95
Transcutaneous gases and autonomic variables in apneic and control infants, 198
Transposition of great arteries, 121—122
Truncal and conal cushions, 37
Tubular heart; see Heart tube, structure of
Twin studies
  of cardiovascular endurance, 89—91
  of maximal oxygen uptake, 89—91
  of respiratory function, 215

U

Ultrasound in detection of fetal breathing movements, 149
Uterine contractions and fetal breathing, 152

V

Vasopressin and lung liquid regulation, 137
Ventilation
  as limiting factor in maximal function, 215
  in neonates, 177
Ventilation inflection point, 213
Ventilatory response to exercise, 210—213
Ventricular hypertrophy, 88
Ventricular puncture, 74—75
Ventricular septal defect, 115—116, 120—121
Verapamil, 56
Voltage clamp studies of chick heart, 56